EXTREME HORIZONS

the climbing & adventure essays

First published in 2023 by Monograph Media

www.monographmedia.com

David Pickford has asserted his rights under the Copyright, Designs and Patents Act 1988 to be identified as author of this work.

All rights reserved. No part of this work covered by the copyright hereon may be reproduced or used in any form or by any means – graphic, electronic, or mechanised, including photocopying, recording, taping or via digital or analogue information storage and retrieval – without the written permission of the author and publisher. Every effort has been made to obtain the necessary permissions with reference to copyright material, both illustrative and quoted. We apologise for any omissions in this respect and will be pleased to make the appropriate acknowledgements in any future edition.

All photographs by the author, unless credited. All rights reserved.

Design by Joe Walczak @joe.m.walczak

Cover: Looking west across the high Andes in a clearing storm from Camp II on Aconcagua, Argentina.

All the activities described in this book are potentially dangerous and carry the real risk of serious injury or death. The author and publisher accept no responsibility for damages arising from participation in any of the activities documented herein.

First edition

Foreword

by Helen Mort

I often fear that writing about climbing or exploration has the same pitfalls as writing about love or dreams: we face the challenge of making an intensely, obsessively personal experience feel both true to its origin but also accessible at the same time. Just as nobody will ever be as interested as you are in your recurring nightmare about the house made entirely out of bicycle wheels, or the face you glimpsed across a crowded room, nobody but the climber can be as intensely invested in a blow-by-blow account of a desperate tango with a slab, or indeed the singular appeal of the slab itself. But these particularities are what we live for.

"I am always obsessed by one place or another", says M. John Harrison in his delightfully slippery anti-memoir *Wish I Was Here*. Trying to describe a specific part of Wales between the A496 road and the Rhinogs, he says: "There's a real sense, in this landscape, of haunting, but rarely by anything specific. Yes the trace of use, but the moment you try to imagine by whom, or for what, or begin to believe you might 'bring it to life', it slips quietly back into the twilight downslope, the wind-contorted tree. Every site is very calm, despite the things it must have seen. Even to say that is to say too much…. Where everything you can say is either an understatement or an overstatement, a liberalisation or a fiction, it's not your place to say anything."

In this sense, writing is an elusive game: first we have to catch something ephemeral, then we have to communicate the indescribable to the reader. Why would we try? Perhaps for the same reason we climb mountains. For the view. For the experience. For the joyful folly of it all. For what we learn about ourselves. And sometimes, just sometimes, the elusive feels possible.

In *Extreme Horizons,* David Pickford manages to catch transformative moments before they slip away, back into the textures of rock, into the snow pack, into the ocean swell, into the pub chatter. Psychologically fascinating, questing and acute, in these essays we feel as if we are truly there. Or better still, we feel as if we are simultaneously there and watching from the sidelines.

Pickford's writing is gorgeously lyrical. "Desire is a strange hotel". The beach "carves out" into sunlight. The horizon is a "tidal zone". I would call it poetic, but to do so diminishes the inherent poetry of the everyday. What he is really doing here is noticing fully, and being alive to detail in the world. In different ways, these pieces of writing are reflections on risk, choice, loss and alchemy. They ask the big, never-exhausted questions.

Why do we climb? What value can adventure bring to us? What does it mean for the world around us? Can it change us or do we change with it?

If I didn't care about climbing and the places it takes us already, I would after reading this book.

Contents

Introduction

Into the Shadow Country	06

Climbs

Sea Change	16	Rapture of the Deep	124
The Pillar in the Dream	26	The Jade-Star Kingdom	128
Fjord Fandango	38	Crusoe & the Witch	134
A Flare In the Dark	46	Wind, Sandstone & Stars	140
The Liminal Game	58	The Magic of Falling Mercury	144
The Heart of Nature	74	Flow in Climbing	148
The Full Welsh	84	High Exposure	152
The Viking Hinterland	90	Totem Masters	160
Border Country	96	The Sporting Contract	166
Behind the Lines	100	The Skilful Climber	170
Bring on the Wall	106	Heaven's Highway	174
The Hidden Edge	112		
The Wind in the Pines	116		

Journeys

Facing East	180
A Short Walk in Tierra del Fuego	208
Shakti & Dust	214
Shadows over Sicily	222
Exploration & the Human Spirit	228

Voyages

Days of the Celtic Sun	236
Vectors in the Stream	248
Rock Steady 'Round Eddy	258
Tidelands	264
West by Northwest	270
Stream of the Blue Men	278

Perspectives

The Painter in the Cave	284
Rituals of Faith	294
The Hunter in the Mind	298
A Pattern of Grace	304
The Path of the Warrior	308
A Letter from the Free World	314
The Pursuit of Sport	318
Lifeboat in the Deep	322
Out of Town	328
The Weight of the Dice	330
Outside the Panopticon	334
The Forbidden Terrain	338
The Lone Imagination	340
The Pursuit of Speed	346
The Dream and the Foil	350
Who's There?	354
Constructive Paranoia	360
The Edge of Things	364

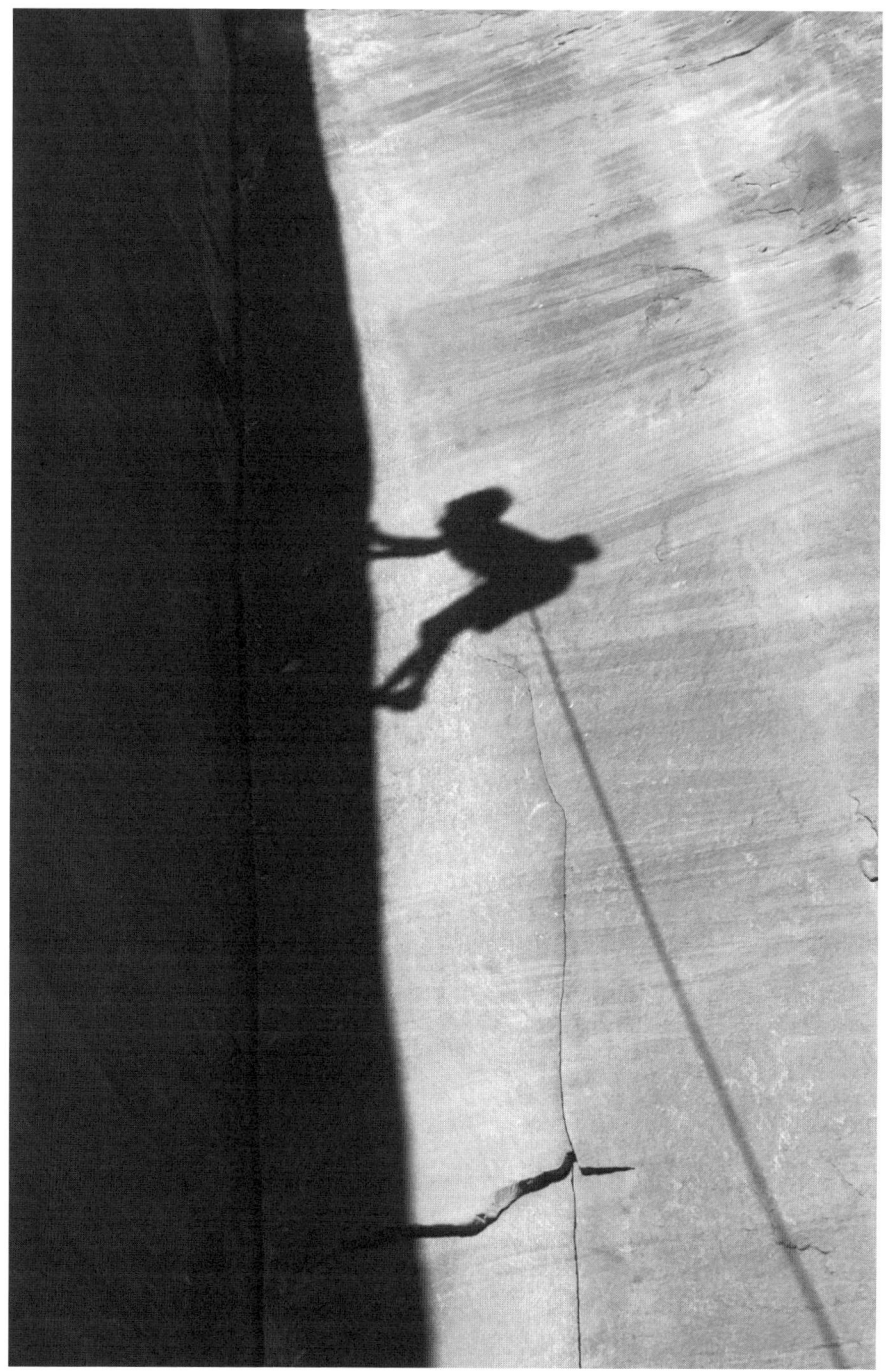

Into the Shadow Country

The dark romance of setting out

Something's happening out there, in the distance up ahead.

A shadow falls across the place where the land meets the sky; the place where things change. Over there, the horizon's tidal zone separates us from what we know and what we can only imagine. As that strange threshold flickers in and out of vision, the boundary is drawn, redefined, smudged out, and made anew.

For those wanderers before us who travelled without maps of any kind, the horizon may have been the edge of the world. Even for us, with an idea of what's on the other side, it remains elusive. Like a tremendous magnet, the horizon beckons and draws us away from home, west or east, north or south. We want to find out what lies on the other side of that most mystical of boundaries; it's a place richer and stranger, in its own way, than outer space itself.

It's late in the evening, perhaps. We're leaving at dawn. A packed rucksack stands in wait at the front door, like a shuttle before takeoff. Compact and powerful, that small device containing our journey's essentials is a portal to a new world. It will remain with us into the outer reaches of our quest. And with it, too, we'll return home. We don't know what's going to happen between now and then, of course. But we're going to find out very soon. In a few hours, the journey begins.

Different skies await us in the unknown country ahead. Seas will break on shores we're yet to find. Forest glades, mountain passes, and city squares appear suddenly, unexpectedly, as we encounter new surroundings. Sometimes we'll lose the trail through forgotten hills. The pathless land will deliver us into a high valley on the edge of the map we don't know about, and have never heard of. What happens there?

The search for a place we don't know is one of the defining concepts of the human desire for travel and exploration. Eventually, of course, our journey will end, and we'll strike back for home. Something will have changed, though, once we arrive there.

Why explore? Why deliberately expose oneself to physical risk? What's the purpose of adventure? These are existential questions. As with many of the most important philosophical queries, they have no finite answers. The essays collected in this book represent two decades of writing around these subjects, and many others, in relation to climbing, mountaineering, travel and exploration.

Whether we're old or young, wealthy or on a shoestring, single or attached, the drive to travel and explore will always spring from somewhat different roots, and the same journey will hold alternative meanings to different people. As I write, the first shoots of spring are beginning to appear, yet the last snow of winter still lies on the bare and frozen ground outside. In the same way, travel and exploration is full of the continual paradoxes that emerge constantly in nature – and in ourselves.

If time is on the side of the younger traveller, for very obvious reasons, it can also be on the side of the older traveller, too. Travelling more slowly, and perhaps more observantly, is just one benefit of a more mature approach to exploring. You see more from a car than you do from a plane, more from a bike than you do from a car, and quite a lot more from being on foot than via any of the above.

The interactions of time and place, people and geography form a resistant yet instructive force for the traveller, like a liquid that changes viscosity as it flows. As St. Exupery put it, "the Earth has more to teach us than all the books in the world, because it is resistant to us."

Whether we travel in relative luxury or on a shoestring budget, we must be absorbed by the places we go and whatever we do there, but also remain conscious of the need to leave them. The Californian hippies of the 1960s called this 'going with the flow' (or something else along those lines, depending on the quantities of psychedelic drugs consumed). Yet this complex process of staying and then moving on would be more accurately described as 'getting in and out of the flow'.

You need to know when to stay, when to move on – and also when to get the hell out. You only know paradise, perhaps, by knowing what it feels like to leave it. The condition of the expert traveller, in this respect, is one of simultaneous presence and absence. A bit like Schroedinger's famous cat, the true nomad must be somewhere and also elsewhere at the same time.

The experience of risk is a central part – sometimes a dominant part – of adventurous journeys and exploratory outdoor pursuits. Many of the essays collected in this book circumnavigate this truly oceanic subject. Is risk taking a good idea? Humans have always, until very recently, needed to take risks in order to survive. Due to a series of complex and interconnected reasons, this is no longer true in many advanced societies. Real risk takers are becoming an endangered species, which is possibly not a very good thing.

Over the course of the early twenty-first century, the trend of so-called 'safetyism' has emerged and gained enormous traction in some Western countries. You could summarise it as the promotion of safety as a kind of core moral value in human life, and it can be observed almost everywhere. It arises in discussions about everything from technology to education, from the benefits of self-driving cars to debates about the care of students by the institutions that enroll them. It can be observed in various physical forms, including the lavish quantities of high-visibility clothing worn by staff at public events, and the proliferation of safety-related messaging.

Depending on your world-view, the cultural elevation of safety as a moral value over recent decades is not necessarily a particularly positive development. Pointing out the existential purpose and value of intelligent risk-taking is more necessary today than ever before. This book, I hope, goes some way towards doing that. I'd like to think you'll find at least one or two good reasons to take calculated risks in the pages ahead.

There are all kinds of physical and psychological benefits of going to places and doing things that are unfamiliar, and perhaps even frightening; going outside our comfort zones, whatever they might be, also extends us in many ways. At the same time, the whole process of exposing ourselves to real risk in wild environments isn't just about testing and proving ourselves physically; it's also very much about pushing the boundaries of what we can achieve emotionally and spiritually. There is a powerful psychogeography at play for anyone undertaking an exploratory quest; new landscapes of the mind are revealed when we go elsewhere. Towards the end of one of the essays collected in this book, I suggest that real adventure isn't just about a route on a map, but also a journey into the heart of your own life. When we experience real risk in the physical world, it develops and educates us in other ways.

Fellowship, courage, and faith are three vital qualities in the adventurous mindset. A sense of fellowship with our companions and a faith in ourselves are necessary to pull off the toughest expeditions, as is the courage to make the imaginative leap to try and do so in the first place. But what if you don't have a companion? Accounts of solitary adventures are also a feature in this book. Solo exploring can be the most challenging and rewarding of expeditionary styles. For it to work, the explorer needs not only self-confidence and self-reliance, but also a degree of enjoyment of being their own company. It's not for everyone, but if you're wired that way it's one of the best things you'll ever do. The six month solo journey around East Asia by two-stroke motorcycle I did back in 2003-2004 (described in the essay 'Facing East') changed my life in all sorts of ways. I'm still reminded by those days and nights on the road about why we should always try and do things that are on the outer edge of our capabilities. That expedition, in one sense, was my version of *walkabout*, the ancient Aboriginal ritual where boys go into the wilderness for a while to make the transition into manhood.

Adventure and exploration also provide unique glimpses into the lives of others. As a traveller in a wild place far from home, we can sometimes enter the private world of those who live very far away both geographically and culturally. A traveller in a far-off land also represents something exotic and often extremely interesting to the person meeting them, so a kind of mutual exchange takes place: we are given a window into someone else's life, and they into ours. This dynamic often makes for a kind of encounter that's simply not possible in your place of origin; an encounter sometimes both confusing and funny, moving and often deeply enlightening.

As so many explorers of the past discovered, the world's wilder regions may be the ultimate proving ground. They're also one of the most uncompromising. I've lost a lot of friends over the years, in accidents that could have easily happened to me. Exploring the limits in adventure sports has consequences, as those I've lost will always remind me. Everyone who takes part in adventurous activities has to accept this fact. It's the dark side of the game.

There's a fine line out there between success and failure. Like the Roman god Janus, his two faces looking in different directions, triumph and disaster are often close associates. As Captain Scott famously acknowledged not long before Antarctica eventually claimed his own life, "Great God! this is an awful place…" Whilst it can be rewarding beyond compare, Scott's remark points out how serious exploring is sometimes not a very nice thing to be doing when you're doing it. At such moments, it falls into the category of type-two fun, unpleasant whilst it's taking place but enjoyable afterwards. When you're on a mountain in a white-out, or at sea in a small craft in a rising gale, you're not having fun at the time: you're in survival mode.

Why is it, though, that a small, largely low-lying island like Britain has produced so many explorers, all the way from David Livingstone, Captain Scott, and Wilfred Thesiger to the present day? Mark Cocker offers a fascinating perspective on this in *Loneliness & Time: British travel writing in the twentieth century:* "in the interior landscape of the traveller, Britain seemed to represent, and to place on his or her experience, some kind of limitation… abroad is always a metaphysical blank sheet on which the traveller could write or re-write the story."

As a boy, the streets of English cities like London and Oxford were familiar ground, and I felt a similar sense of limitation there. As a teenager, long distance cycling, and later an all-consuming obsession with climbing were my own ways out of 1990s England's cloistered, defended terrain. Climbing would later come to shape my life in an extraordinary way, but travelling and exploring are a natural – and often essential – part of climbing and mountaineering. As soon as I was free to travel independently, that's exactly what I started to do. And once I hit the road seriously in my early twenties, I really never looked back.

Many people are naturally disinclined to travel. They perhaps feel, as Philip Larkin once remarked with his trademark irony, "I wouldn't mind seeing China if I could come back the same day." Many people don't like being away from what they're familiar with for various reasons, and often very good reasons: family, friends, work, and all that. I'm the opposite: I enjoy being elsewhere.

The familiar form of the packed rucksack at the door late in the evening never fails to thrill, to inspire, and intrigue. Even when static the rucksack represents a state of flux, of movement, and of discovery. That modest object is nothing less than a time-capsule to other worlds.

Whenever I head out of the door to a place I've never been before, I want to go further, test myself, and perhaps master something new. Most of all, I want to see what I'll find there. Leading an adventurous lifestyle might be my interpretation of what Socrates famously spoke of before he drank the hemlock: "an unexamined life is not worth living".

You don't need to go to the ends of the Earth to experience real adventures, either. During the Covid pandemic of 2020-21, when international travel was put into a medically-induced coma for a while, I was forced (and not for the first time in my life) to find adventure much closer to home. In the spring of 2020, I decided to try to master wing-foiling, a recently developed watersport that involves using a hand-held sail and a board with a hydrofoil attached to it, which is essentially an underwater aeroplane wing, in order to literally fly above the water powered only by the wind. It feels impossible at first, but after a while you learn how to take off, how to control the foil, then how to jibe and turn through the wind, and so on. It's a beautiful and addictive way to travel over a large expanse of water. When I was a small boy, I used to love a book called *The Wreck Of The Zephyr*, about a mythical boat that lay forlorn on top of a cliff yet which could also fly across the waves. With the benefit of modern technology, this strange-sounding concept is actually possible. When out wing-foiling during the Covid lockdowns, not only did I feel I'd found the ideal activity for 'social distancing', but I also felt a powerful sense of what adventure itself might mean. By being creative, by foraging around, and by using new technology, it's possible to do something seemingly extraordinary, like flying across the sea on a five-foot hoverboard powered only by the wind. Small, quiet, personal projects like this might be amongst the most rewarding adventures of all, transcending expectations like the Zephyr itself.

This book is arranged into four sections, and collects some of my non-fiction writing on climbing, adventure travel, and expedition paddling from the past two decades. *Climbs* pulls together some of my writing on rock climbing and mountaineering. *Journeys*, the second part, is focused mainly on adventure travel and exploration. *Voyages* documents various open-ocean standup paddling expeditions. And *Perspectives*, the final section, is a looser amalgamation of editorial pieces and essays that may offer some insights on the exploratory life, the nature of risk, the psychology of adventure, and many other subjects.

This book could be seen as a series of notebooks from places at the edges of the physical and the experiential world: the extreme horizons. Real adventure is, by its very nature, a kind of psychedelic experience. It's a way of seeing things differently, a way of perceiving your own life and the lives of others more deeply. It's also a good way of dusting off the assorted psychological detritus of contemporary Western life.

The desire for exploration is at its best when it's driven by a romantic vision of elsewhere. This book could also be seen as a kind of guide to that psychic space. Here, we'll venture into the liminal zone on the edge of the map where desire finally becomes reality. What happens when we travel there can be challenging beyond measure. And still, as twilight falls, we walk straight in.

Some of these essays might also illuminate why living an adventurous life and taking calculated risks is a worthwhile endeavour. I've always aspired to such a lifestyle, ever since I was a wide-eyed kid packing his rucksack late at night just so he could leave home at the crack of dawn. Perhaps, in the end, this is the only way I know how to live.

The hinterland always shapes our journey, but the country ahead remains more important than the country behind. This is partly because we don't know it yet. Once travelled, though, the dusty road stays in our bones forever. And yet again that shadow country beyond the horizon beckons us into the unknown, to live on the edge of things, and to see what happens.

Bristol, England, 2023

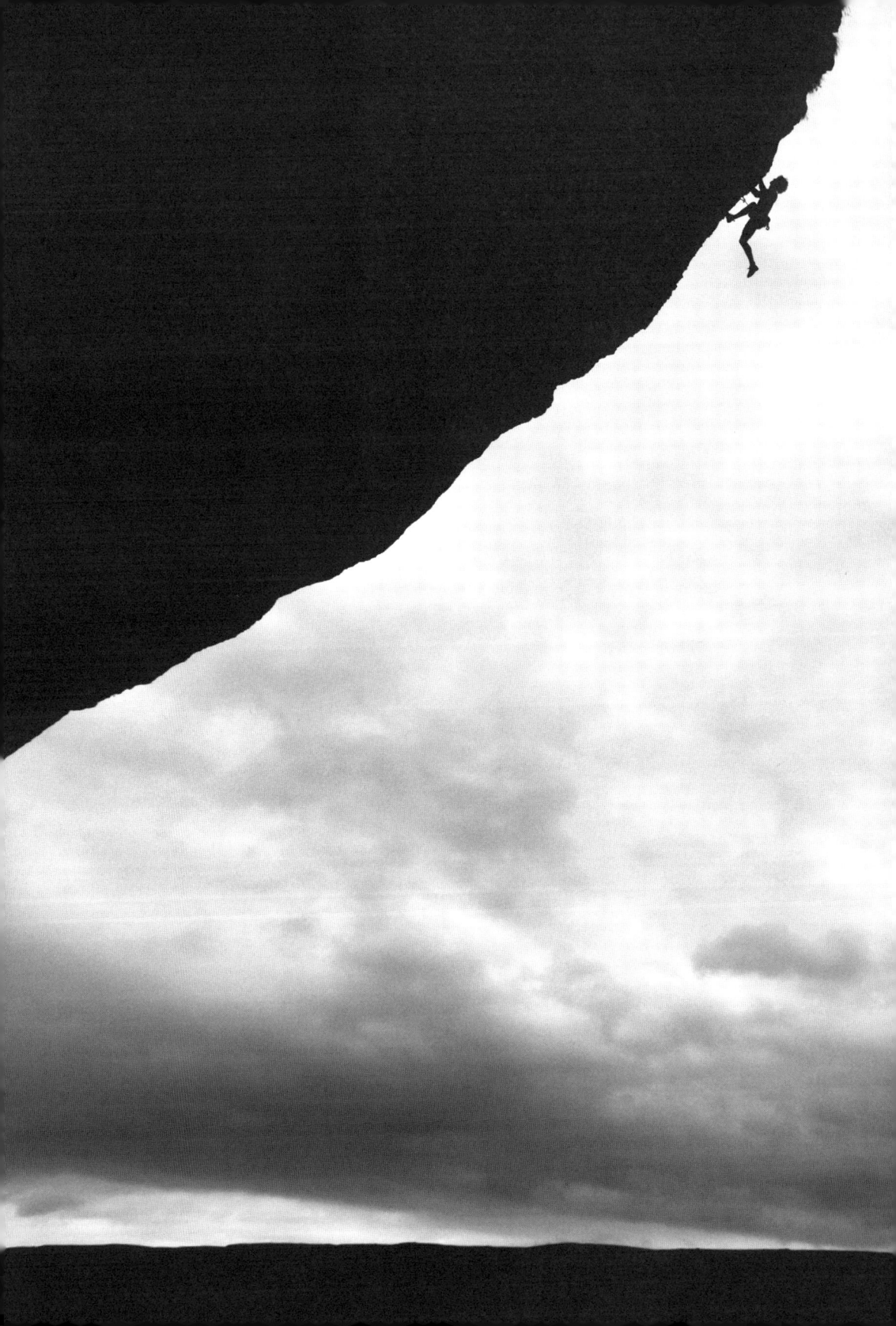

Climbs

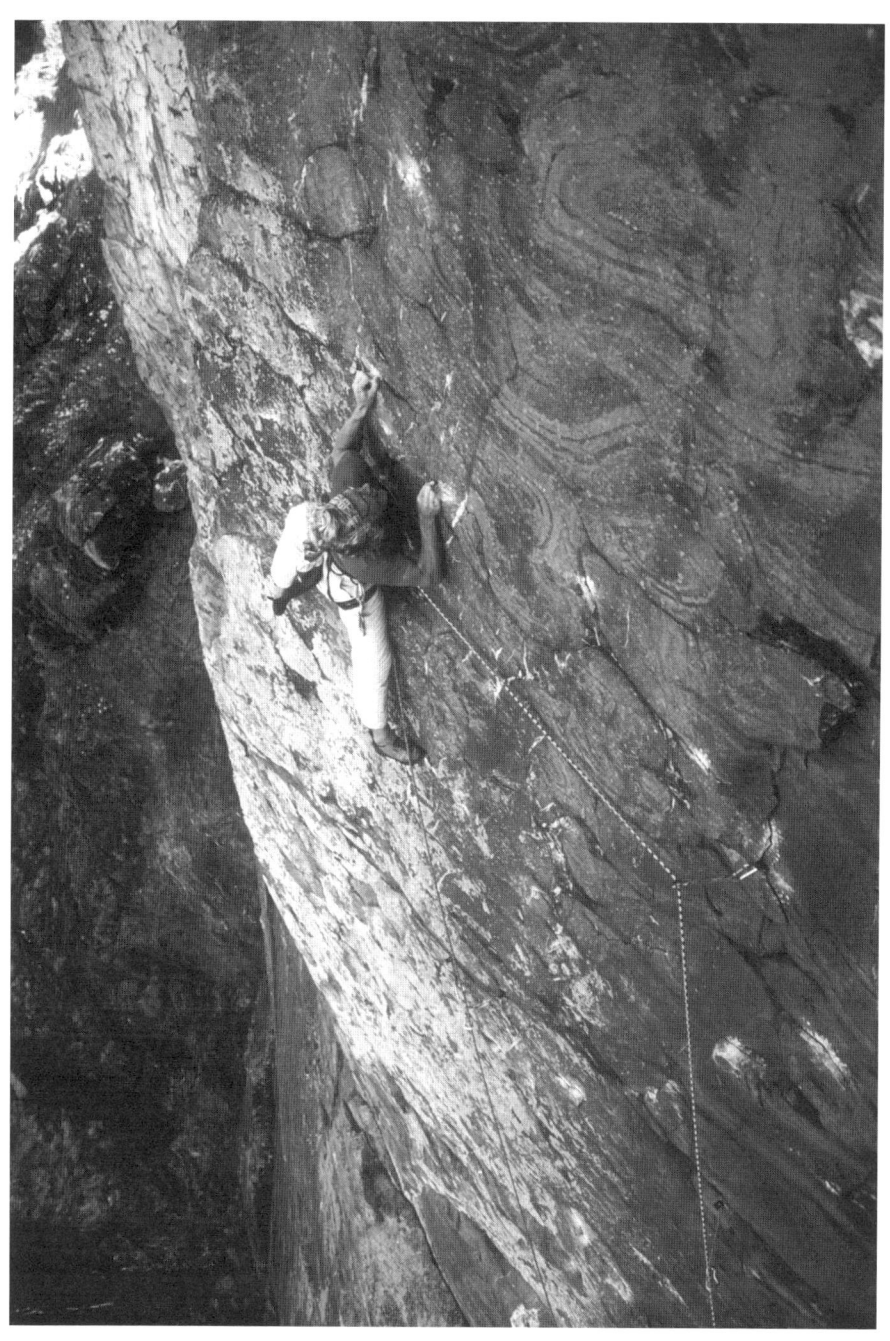

The author on the first ascent of Wall of Spirits (E8 6b) at Pentire, Cornwall, in 2004. Photo: Mike Robertson.

Sea Change

Coming of age through sea cliff climbing

Desire is a strange hotel, a place of echoes and half-opened doors. On a late monsoon morning in Pulau Nias, Indonesia, from an outrigger canoe, I watched the vapour lift and stir over the rainforest. Half visible within it, a dark sheet of volcanic rock rose from the palms and tangled creepers, steepening with height.

At once, the memory of an unclimbed headwall on Cornwall's north coast flashed back at me, like quartz by moonlight. I began to conjure up the line: a spidery crack; an incipient flake, fading out; a traverse left, perhaps, along a smooth shelf.

When I was fourteen, I'd run down that steep seaward slope, my gaze fixed as the Great Wall of Pentire Head reared from the waves. My friend Paul must have noticed how the wall had entranced me, because he handed me the lead on both pitches of *Eroica*, the classic line up the huge flake that splits the heart of the crag. At the top of the final groove, I paused in balance. The frozen magma under my hands was traced with delicate veins, holding the visible, ancient life of the earth. It was as if a fault line had cracked open.

Once you've cast adrift on a sea cliff, there's no quick retreat. That day's enchantment had directed my life as a climber, taking me from Cornwall on to Europe, then Africa, Central Asia and eventually the East. Now, ten years later, as I peered through the mist at the sheet of dark stone emerging from the jungle, the longing to return home resounded hard.

The edge of land: a place where the light changes, where people change. I went to high school in Oxford, about as far inland as you can get in England. The cloistered towers and narrow streets were a world apart from the cliffs and mountains I longed to be close to. And climbing was easier than the awkwardness of teenage dating, anyway. On Friday afternoons, I'd escape the city with a driven urgency, headed for the coasts of western England and Wales. Here, between the mirrored vastness of sky and water, I became part of a visible world that extended beyond reckoning.

During the week, I'd keep a copy of Pat Littlejohn's classic guidebook *South West Climbs* under my school book, ready to read at any chance. Through the 1970s and into the '80s, Littlejohn had the plumb lines on the sea cliffs of western England mostly to himself. With Keith Darbyshire, his closest friend, he solved the massive buttress at the northern end of Carn Gowla in 1973, and christened it *America*. Shortly after this groundbreaking climb, while exploring Nare Head, Darbyshire died in a fall. Littlejohn didn't return to these cliffs for several years.

Stories like these with real heroes and ghosts, made schoolyard gossip evaporate. In maths class, Daniella sat next to me. She breathed a bright-eyed, gorgeous chutzpah. She was way too cool to try and casually impress, so I'd tell her about the Eiger Nordwand instead. *Death Bivouac, Traverse of the Gods, White Spider*: the names sounded better than the signs of Pythagoras' theorem. But I never asked her out; I had to go climbing on the weekend. By the time I was sixteen, I'd sneak off from football at half-time, unnoticed, and cycle head-down for the climbing gym.

One midwinter weekend, sport climbing in the Swanage quarries, I met Mike Robertson. Almost two decades older, he was entirely removed from the weird world of growing up. His ragged hair and stubble gave him the appearance of a dishevelled grizzly bear, and his stories of climbs and travels in Europe, Asia and America held me with a preternatural fascination.

One June morning after my last exams, I jumped on a train to meet Mike for a week's climbing on Lundy. From Ilfracombe, a harbour town on Devon's north coast, the old flat-bottomed boat M. S. Oldenburg sails for a tiny granite island twelve miles offshore at 51°10' 37.8876" north, 4°39' 57.96" west: the coordinates of some of the best sea-cliff climbing in Europe.

I woke to the sound of gulls, and waited on the quay as wooden boxes were loaded in the hold. The west wind blew straight off the Atlantic, full of brine and salt. Quickly, the ocean took hold of my senses, replacing the reams of algebra and lines of Shakespeare that had occupied my past weeks. *Full fathom five thy father lies / And of his bones are coral made; / Those are pearls that were his eyes; / Nothing of him that doth fade, / But doth suffer a sea-change / Into something rich and strange.* Quickly, the Oldenburg cast out beyond the harbour and into the tidal stream.

The men working on the boat had deeply wrinkled faces, drawn by years of running the tides around southwest England. I felt a premonition of my own sea change from schoolkid to climber. By the time we dropped anchor in the bay on the island's southern tip, the tide had begun to turn again, ebbing now. Mike had left a note in the campground: *see you at Landing Craft Bay, we'll be there until late*. Too excited to walk, I ran across the fields past the Old Light and scrambled down the undercliff to where Mike and the others had stashed their gear. As I clipped in to the seventy meter static line to rappel down to the boulders, the exam room that had encased me earlier in the week faded altogether.

A line of low clouds built up in the west. Mike moved rapidly across the strongest line through the center of the granite slab: *Matt Black*. He stopped at a faint weakness and coaxed in a tiny microwire. I was so engrossed in watching his progress that when the first wave splashed my rock shoes I thought it had started to rain. A second later, I registered its source: the tide was roaring in. After Canada's Bay of Fundy, the English and Welsh coasts of the Bristol Channel have the largest tidal range on Earth. This, coupled with the Channel's funneling effect, has caught out countless climbers unaware.

As Mike pulled through the final moves, his last gear fifteen meters below him, I scurried from the water's reach, aiming for a tiny ledge at seven meters. By now, the ropes were a swirling, blue-green mass in the cove, like the tentacles of a gigantic octopus. I rocked up on the tiny ledge just as a bigger wave came in, completely submerging the boulders where I'd been anchored a few moments earlier. With just enough room to swap feet, I wrung the seawater from each of my shoes, balancing precariously on the ledge with the other. Within a couple of minutes I was leaving a trail of wet prints on the smooth rock. The distant cloud bank rolled in over the ocean, thickening the fading light.

"Bet you got wet feet there, heh?" Mike laughed as I scrambled through the bracken to meet him.

"Only a quick rinse. I was trying to watch you up there too, you old rascal!"

I felt I'd just run into the sea's jaws, enthralled by the peril of the breaking wave, then raced back, filled with fear and delight. In that instant before I'd fled, I'd felt myself edging along a polished shelf separating existence and desire, aware of another world that lay beyond them both.

At eighteen, with more energy for climbing than I could usually expend in one day, I was often overtaken by the ocean or by darkness. Late one August evening at Bosigran, Mark Glaister declared he was ready for a cup of tea. Mark, who had introduced me to Cornish granite, was older and far more sensible than I. After we'd packed up our gear I ran back, alone, all the way from the Carn Galver Mines to the top of the Great Zawn and abseiled down a static line to the base of *Desolation Row*, one of the classic extremes of Penwith granite.

The heat of the day had given in to lengthening shadows. I began climbing almost immediately, enthralled by the freedom of moving quickly without ropes or gear. Soon I was high on the slab, balancing on small edges below the crux section, where the crack thins into a spidery seam. The sea glinted darkly from the cavern below. Somewhere overhead, a peregrine shrieked. Lifted by its call, I floated through the crux and suddenly landed back on the col: I was on a roll. An offshore wind blew huge eddies beyond Bosigran Head and north toward Carn Gloose. I raced back over to the Main Cliff and up Joe Brown's 1957 masterpiece, *Bow Wall*. As I swung from the perfect handjam at the end of the diagonal crack, a rush of feathers whistled past my shoulder and fell through the 200 feet of space below. The peregrine had returned. I scampered back down *Doorpost* in the gloom. Soon, I was chalking up below the thin crack of *Suicide Wall*, in the heart of the cliff. It was almost nightfall when I reached the top for the final time.

Huge, parabolic boulders rest on the summit of Bosigran Head like slumbering sea-monsters. Dawn and dusk seem to animate them back to life. I stretched out flat on one of them, feeling the immensity of time in the sharp crystals under my fingers. The air was still. Far out to sea the offshore wind carved whorls in the darkening water. My eyes adjusted, and the sky crackled with stars.

I met Sarah just as we were both on the final straight of postgraduate studies. She, too, was on the brink of a journey. Before she left for Australia and I for Vietnam, we climbed for a few days in West Cornwall. That summer was the longest and hottest in Europe for decades, and the Atlantic was Aegean-blue. I woke at dawn one morning after we'd wild camped on the beach, half-covered in sand and fragments of seaweed. I swam out to the edge of the bay. Held in a lull between the breakers, I watched the granite turn to gold in the breaking light.

Suddenly, a powerful gust of wind blew off the moor over Gurnard's Head. Strong convective winds shark this coast, and in 1964 Peter Lanyon, one of the most talented English painters of his generation, was blown off course. He'd flown Spitfires in WWII, but in the end it was the same place that most awoke his creativity that killed him. Disconsolation moved over me. Out to sea, mist began to move the horizon.

"Let's go climbing," Sarah said, and her bright voice broke the spell.

Later that day, high on Bosigran Ridge, a great jagged line of upturned fangs rising from Porthmonia Cove for 300 metres, I looked back. Sarah appeared and vanished, emerging again as a faint, dark shape on the crest. As I watched her shadow moving against the swirling air, my own was superimposed against the cloud, weird and unreal. It was a rare sea-cliff version of a Brocken Spectre, the natural doppelgänger that haunted early alpinists. Edward Whymper, descending the Matterhorn after the deaths of his four companions, had been spooked by those preternatural silhouettes.

In German folklore, to meet your 'double walker' is a portend of destruction. The English Romantic poet Shelley saw his, adrift in a storm off Italy in the *Don Juan*, and never returned alive. Mountains are often just like the sea, most beautiful when most deadly. Hermann Buhl, high on Chogolisa: the cornice falling like a breaking wave. Or Alison Hargreaves near the summit of K2: the storm encasing the high Karakoram like ocean fog. In the places of our greatest inspiration, death is often a quick companion, a shadow-watcher flashing through the vivid dark.

Summer shifted to autumn, and November came on fast through the English rain. London evaporated as I soon as I'd booked a one-way ticket to Asia.

On the plane, somewhere over Siberia, the grid of Novosibirsk sparked back at me, and I closed my eyes. Hong Kong opened and shut like a clam in the tidal current. Women were hanging out their sheets on the balconies of apartment blocks, like white sails swirling in the sun. When I boarded the plane to Hanoi, a red flag with a five-pointed star in its center shimmered through the static on the departures screen. Philip Larkin's ship came to mind; the one that "went wide and far/ Into an unforgiving sea/ Under a fire-spilling star./ And it was rigged for a long journey."

Four months later, the twin-engined speedboat gurgled out from the pier into the slow current, then accelerated hard as it cleared the sandbar at the mouth of the river. Limestone towers rose along the western horizon. Ko Laoliang, a tiny island in the Surin Archipelago about twenty miles off Thailand's south west coast, was the break from the road I'd been searching for intently, but never quite found. In the few weeks I spent living and climbing on the island, the dusty hills of East Asia I'd travelled on my Vietnamese-registered motorcycle faded into a dream-sequence, locked up with the battered Minsk in the chandler's shed by the pier.

On Laoliang, I'd met up with ex-pat American Mike Weitzman and Trev Massiah, a climbing friend from home. We'd found a hundred-metre wall of unclimbed limestone, honeycombed with caves and draped with gigantic tufas, rising straight from a beach of tide-washed sand: a treasure chest for new routing.

Stretched out from the second set of anchors on our three-pitch route, Mike and I watched Trev on his project: a line he'd bolted up a sheet of blood-orange rock where a series of hollow pockets allows tenuous access to an immaculate headwall. Trev cut loose through the roof at the same moment the morning sun lit up the entire cliff. As he was shaking out before the second crux, Mike called down in jest:

"Hey Trev, you got any bamboo for that last move?"

In the past, local birds-nest collectors used huge bamboo sticks to access these caves ground-up for the raw material once prized by the Chinese for an exotic soup. We found bamboo in numerous caves to which the only access was by difficult rock climbing. The birds-nest collectors are gone now, and the unfathomable secrets of their trade are locked in history.

On a red evening full of dusty rain in southern Sumatra, I followed the road to where an ink-dark lake broods beneath a tall volcano. Most days, I'd be on the bike by dawn and ride on into the dark after the evening mist had swirled low over the rice fields. I rode north, to Aceh, east to the high country of Java, and back to Sumatra and the Indian Ocean. A few weeks later, I waited on the quay at Sibolga for the night boat to Pulau Nias. The next day, in Sorake Bay, I met Mr. Gurung.

As the outrigger canoe broke through the first band of surf, he removed his tattered shirt to wring the water from it. Long ropes of muscle stretched across his thin frame. I blinked: a scar started behind his left shoulder blade and ran unbroken down the front of his chest, across his torso and halfway down his right thigh.

"What is it?" I asked.

"A big shark," he replied. A wide, almost boyish smile spread across his face, deepening the strong wrinkles from sixty-two years spent mostly between the ocean and the sky.

How the hell did he survive that? The shark's bite radius couldn't have been less than a metre-plus. The smile on Mr. Gurung's face had widened further, lighting up his dark eyes with a sudden, wild laughter that said, *Yes, I know.*

I thought I already knew about the sea and its wildness. But his life – and the lives of most coastal Indonesians – had been more profoundly marked by the sea than anything I'd encountered. For me, the coast was the place I'd found inspiration as a climber. Mr. Gurung and his fellow fishermen ventured into more hazardous waters, in search of the oldest source of life itself. His world was far harder and probably more dangerous than that of the globetrotting explorer.

In the outrigger a few days later, Mr. Gurung caught a big kingfish on the trailing tackle. That twisting blade of colour lifted my eyes from the sea, and I looked inland. Vapour wreathed the tallest trees. Then the wall appeared, reminding me of the Pentire line. Mr. Gurung paddled back toward the beach. As the hull of the canoe slewed into a sand bar, I looked up again, and the wall had vanished. That was I wanted to figure out: how to draw life from both land and water, existence and desire.

If I could return to my original point of departure with new eyes, I thought, perhaps I could finally unlock its mystery. But such finality confounded me with strangeness: the unforgiving future of adventuring, of setting out.

Three weeks later, in a small town in Malaysia's central highlands, I logged on to my email for the first time in over a month. A message from Mike flashed up in bold at the top of the page, untitled. It was unlike him not to title a mail. Cold fear swirled through me.

"Damian Cook's drowned off Devil's Cove," the email read. "His funeral will be in two weeks. Sorry to have to write to you about this. Take care out there. Mikey."

Our friend had been caught in a powerful cross-current after pitching off a deep water solo in Mallorca. He was a strong swimmer, but the sea was stronger that day. I paid my bill in silence and walked out. The evening smelled of dust and rain. My voyage out was over.

After Damian's funeral, I returned to Pentire alone. A passionate exponent of bold climbing, Damian would have loved the idea of my route: a faint, inescapable line of weakness through the most imposing piece of rock on the crag. He'd have wanted it to rise straight from the ocean, too.

The morning's rain had cleared and the blade-sharp air washed the city from my senses. I threw a static line out across the overhanging upper wall. A gust of wind trapped it in mid-arc; for a moment the rope was held in equilibrium, almost motionless, but quivering slightly. Three hundred feet below, a massive swell threw spray halfway up the cliff.

I rappelled down to the lip of the headwall. The wind twirled me around like a string puppet before I could catch an edge and place a wire to hold the rope close to the rock. I peered down at the smooth sheet of stone below, gently overhanging for twenty meters. Tight, tiny swirls of quartz distracted my attention from the almost complete lack of gear placements. Sawn-off knifeblades bounced from the minute cracks into which I tried to coax them. One protruded from a horizontal seam by almost a centimetre. I tied it off; it was bodyweight-only gear, for sure.

Far out to sea, shadows of clouds crossed the water. I remembered how Damian had laughed when I'd said a particular crux above deep water was too high, that anything above sixty feet was just like soloing above ground.

"It's not about the height, Dave. It's about relaxing. Just be free up there."

He would have loved this spot. Maybe this was how you got past desire's half-opened door: by finding those places and instances where life and death collide so intensely that they break apart, letting you slip through unnoticed. Like a magician, or a ghost.

Ten metres above the terrace, as I poked the nose of a carabiner into a shallow depression, it disintegrated to reveal an open pocket. A 2.5 cam just sank into it, the teeth biting against the gritty quartz. The wall afforded this one, crucial concession. Then, before I could even chalk up, the first heavy splats of a summer squall hit the rock. Rain sheeted across southwest England through August, and it was several weeks before I could return.

On a cool September morning of perfect conditions, I rapped down for the third time. Mike Robertson was already anchored in the niche, uncoiling the ropes, just as attentive as on our first day climbing together. A fulmar squawked, firing a neat globule of regurgitated fish at him. He ducked from the shoulder and laughed it off. His quiet self-confidence and strength pervaded the air along the narrow terrace.

"Where's the cam, Mikey?" I enquired in a moment of pre big-lead amnesia.

"On that sling you've just clipped to red, you young rascal!"

Suddenly I was moving, as a gust of wind curled across the crag. My fingers hit a crimp. Before I could fully register what I was doing, I irreversibly left the sanctuary of the pocket, casting adrift on the headwall, moving quickly between small edges. My ropes trailed out, jarring brightly against the dark rock. Far below, the hiss and gurgle of the sea disappeared. Even the cawing gulls fell quiet, and I entered a space of vivid silence with no beginning or end in time. The new climb was enacted, and revealed itself for the first time. It hooked me with the powerful lure of pioneering.

Then, without warning, I was in balance once more, this time crouched on the arête that defines the headwall's limit. Tiny quartz edges bit into my skin as I glanced back down at Mike. Below, the Atlantic Ocean streamed with strands of kelp, like the tentacles of a giant squid. Mike's quiet voice folded into the crash of the next wave, and the world I'd left behind returned, suddenly made new.

They say the wave looked small at first, just a strange, wide line beyond the broad circumference of the reef. When it hit the west coast of Nias, it was ten metres high. The Boxing Day tsunami on December 26th, 2004, killed 140 Niasans and circa 225,000 people on the coasts of Sumatra, Thailand, Sri Lanka and India in the world's deadliest recorded natural disaster, caused by the second largest earthquake ever measured on a seismograph.

The Aegean was still that winter evening on Kalymnos, Greece, when I heard the news of the tsunami. I looked out at the darkened beach. Lanterns and shadows appeared as the local men set out to night-fish in small boats.

I suddenly understood what Mr. Gurung had been trying to tell me with his big smile, the morning when I asked him about the shark: *Yes, I know. I know what it means to be taken by the sea. To be consumed by her. And to escape from her, to find a way back to shore. To be a voyager between worlds.*

I've really no idea what happened to Mr. Gurung when the tsunami hit, but I hope he was out fishing. If he was, I'm certain he's still alive.

Alpinist magazine, 2008

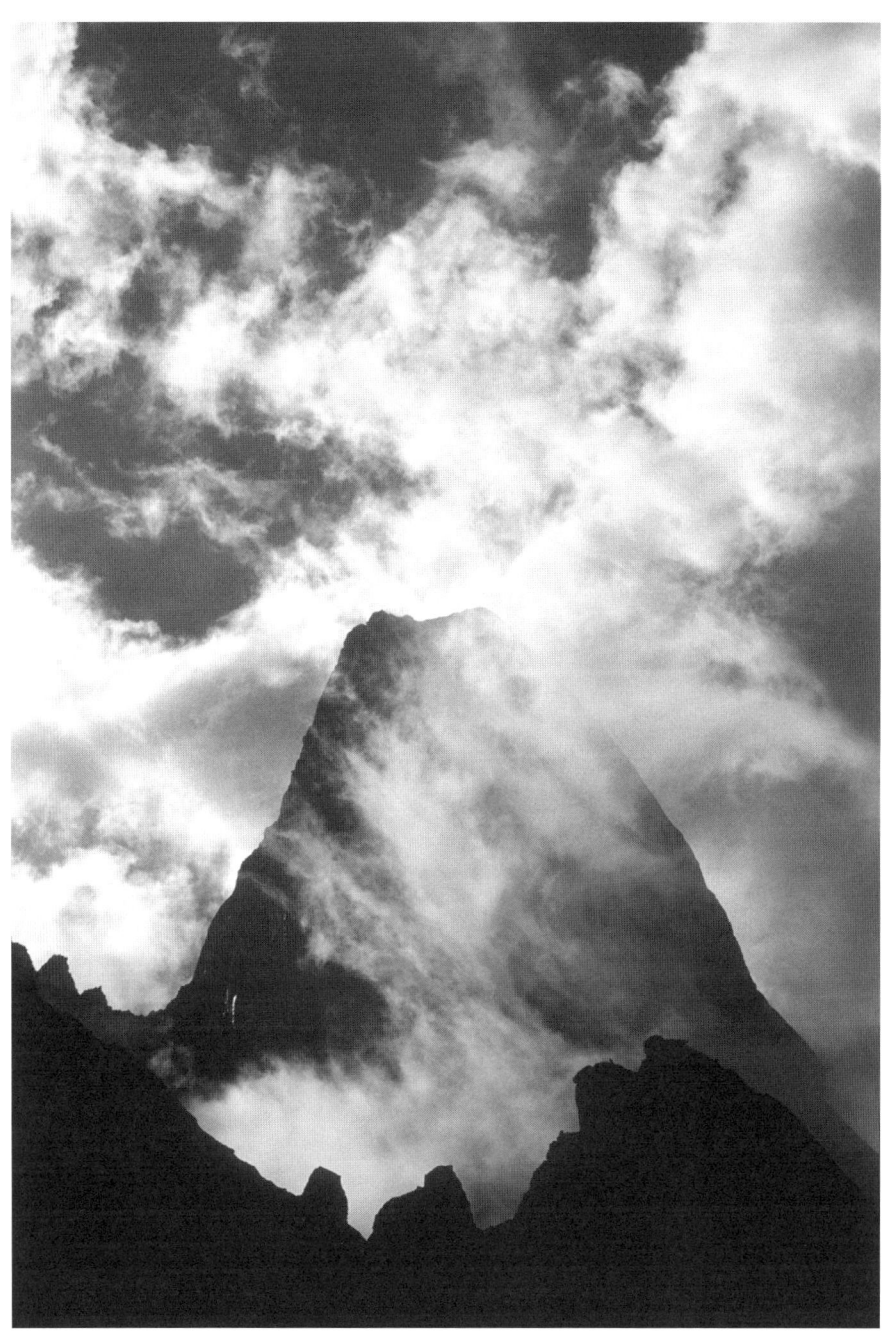
The Russian Tower emerges from a clearing storm, Ak Su valley, Pamir Alay, Kyrgyzstan.

The Pillar in the Dream

Pioneering in Kyrgyzstan's Wild West

At five past four in the morning, the ancient Aeroflot Tupolov clunks into the runway at Manas airport. There's an audible intake of breath from the passengers as overhead lockers spring open, spewing their contents across the cabin floor. The plane is filled with a cacophony of creaking hinges, interspersed with other, less identifiable noises, culminating in a tremendous growl from the bowels of the fuselage. The aircrew look on, unflinching. Clearly this is a perfectly normal landing, Russian-style.

A great cheer rises from the back and curls like a gust of wind through the plane. The trio of Poles in the row behind me are in high spirits. They've been drinking vodka with kamikaze determination throughout the flight and now propose a new round of toasts. After a while, their words slur into cheers, and I smile at the post-Soviet solidarities. Mikhail Gorbachev would have been proud of our rowdy welcome to Kyrgyzstan.

As pre-dawn arrivals go, this is unconventional. We shamble out into the hush of the arrivals hall chattering loudly, like the after-party crowd leaving a nightclub at closing time, and whisk through immigration with the vodka fumes from the Poles building behind us. The revelry continues as my expedition teammates, Sam and Mark, head for an empty bar with die-hard determination.

Having woken the bartender, who had been snoring on his chair with an empty vodka bottle in one hand, they return with several beers. Too tired to stomach Russian export lager, I crash out under the table to the sound of their chinking glasses. The hum of a distant vacuum in the terminal building lulls me into sleep.

A chill wind engulfs the mountains about us. The sky overhead turns dark. The sound of rushing cloud fills my ears. Am I falling?

I awake to the sound of a gigantic industrial vacuum cleaner sweeping the floor beside my table. Peering up, I am momentarily unsure whether my dream has ended, as I see that the huge man controlling it dwarfs the enormous contraption itself.

Five hours later we shamble out into the heat of the day. High summer on the steppes of Central Asia is defined by a fervid haze that begins around 10.30am and lasts well into the evening. The heavy air of Bishkek, capital of the Republic of Kyrgyzstan, hangs just a few centimetres above the asphalt. Finding the Twin Otter at last in the far corner of the runway, I seek shade under the left wing and notice a reassuring bit of Cyrillic translated into English plastered on the fuselage: Chop Here With Crash Axe. Reaching into my pack for a camera, I become aware that a military policeman with the physique of a medium-sized adult gorilla is walking in my direction. As the camera appears, Rambo raises his Kalashnikov towards me. Momentarily forgetting the Russian for 'sorry, old chap', I stash the Canon out of sight. Rambo howls with laughter, revealing a well-polished set of gold teeth. Gold teeth are a hallmark of wealth and status in Kyrgyzstan, a sort of Central Asian equivalent of a Mercedes SUV with tinted windows.

We are bound for Osh, the main town on the Kyrgyz side of the Fergana Corridor, a long strip of fertile land that stretches west towards Samarkand and on to the deserts of Uzbekistan and Turkmenistan. Looking out of the window of the little plane as it begins its descent, the Pamir Alay rises sharply to the south. An immense wall of jagged spires and glinting white summits lifts out of the plains, defining the point where the steppe of Central Asia gives way to the westernmost end of the Greater Himalaya. Two days later my friend Ian Parnell arrives in Osh, having flown in via London on his way back from Everest. The team is thus complete: Sam, Mark, Niall, Donie, Ian and myself – or the 'Kyrgyz Six' as we would come to call ourselves. We're to leave for Ozgouruch next morning.

I wake at dawn to the drift of the call to prayer from the mosque across the street.

We breakfast on black Kyrgyz tea and eggs; the tea is always strong in this country and the men seem to drink it constantly. One local proverb translates roughly as 'If he does not drink tea, a Kyrgyz man will die'.

In the eerie half-light we hear the approach of The Beast. Several minutes elapse between the first sound of the approaching contraption and its tremendous arrival. With a puff of foul black smoke – an exotic mixture of diesel and other, more obscure petrochemicals – a vehicle straight from a Mad Max movie growls into Ury's yard. It is a fearsome sight. Best described as a kind of mechanical rhinoceros with added sound effects, the remarkable Uuaz is one of the engineering hallmarks of the former USSR where it provided a universal form of transport from Ukraine to Ulaanbaatar. With a flash of gold teeth in the dawn light, our driver appears behind The Beast's enormous wheel. He speaks almost no English, but has a gentle, determined manner that instils confidence. Quickly he becomes known as 'Shortcut' due to the constant detours he takes. And more to the point, it sounds good. If anyone can get us across 400 miles of potholed desert roads in this unlikely contraption, Shortcut is our man.

The Beast's starter-motor fires the engine up and we hit the road west towards Ozgouruch (also known as Voruch), the tiny village at the road head to Karavshin.

Less than an hour out of Osh, The Beast grinds to a belching, juddering halt.

"Problem, Shortcut?" I enquire curiously.

"No problem. I fix. Ten minute."

Shortcut looks unperturbed. After 10 minutes of fiddling with the engine, he establishes the fault – a broken distributor rotor – and attempts to fix it with part of a baked bean can (yes, really) fashioned with a penknife. Firing up the engine again, The Beast revolts at his repair and spits out the entire assembly in an explosion of bolts and fragments of tin.

Our prospects of reaching Ozgouruch today are not looking good. Shortcut merely shrugs, wanders across the dusty highway and hitches a ride with a truck back to Osh. A sand-coloured sun rises over the desert to the east as the dust-cloud of the truck disappears. We play football in the road as the chill of dawn fades with the creeping heat of the morning. After a while, another dust-cloud appears in the distance and battered Lada rolls to a halt. Shortcut! Our unflappable driver swings out of the Lada wielding a new distributor cap and bolts it on, cigarette dangling precariously out of his mouth over the oil-caked engine. He turns the crank and The Beast belches into life with a renewed vigour. In the burning midday heat we lurch down an endless road through a desert full of gigantic potholes. At an unmarked crossroads two bored Kyrgyz soldiers emerge from a posting-hut bristling with various bits of ex-Soviet weaponry; the rocket launchers and artillery look forlorn against the barren landscape. The soldiers ask for our passports with a languid curiosity. Clearly mystified by our presence, they wave us on, making a vague gesture of their AK47s towards the mountains to the south. It is almost dark as we reach Isfara. Children are playing in the dusty street, wide eyes flashing with the light from roadside fires. At the edge of the small town we turn south again, then west towards the castellated dusk sky. From here, the road curls tortuously up a steep valley towards Ozgoroush.

Just before midnight, as we cross a bridge of recycled wooden railway sleepers, I hear the Kara Su river roaring through the darkness below. A blast of icy air hits me as I open the Beast's rear window, craning my head. Far above, the ridge-line is in strong silhouette against a night sky dense with stars. Further to the south, the snowy peaks of the Kara Su valley dim the star-sheen slightly. The journey from Osh has taken 19 hours.

Nothing you might have heard about Karavshin, a remote village deep in the mountains of south western Kyrgyzstan, will prepare you for the spectacle of arrival. With tired legs after two long days of hard walking, we emerge from the arid lower canyon onto high pasture. A short way ahead of our horsemen, we stop beyond some mud-brick houses at a small clearing among stunted pines where the glacial Ak Su river levels on to a gravel flood plain. It is almost dusk, and a cold wind is blowing off the high mountains to the south, from Tajikistan. Through the trees, three immense shapes begin to define the western rim of the valley, while to the east loom another two monolithic summits. We set up base camp as night gathers among the boulders under Central Pyramid, its monumental outline cut sharp against the night sky of the Pamir Alay.

After a few days of unstable weather, in which we climb some short multipitch routes, the sky clears one evening after a thunderstorm, promising a change. Sam and I have set our hopes on a slender, unclimbed pillar rising out of the back of the couloir between Central Pyramid and the Russian Tower. We fall into a restless sleep with fingers crossed, daunted and intrigued by the forthcoming challenge.

I wake in the icy chill before dawn, peering out of the tent to see the early light catching the three summits to the west. The air is cold and still, the morning sky an inky, polarised blue. A shiver runs through me: *today's the day*.

"Hey Sam, you awake?"

A furtive reply comes from the depths of his sleeping bag.

"Looks like it's settled."

A local hunter who stayed with us last night has rekindled the fire and is squatting in front of it, warming his hands, breaking sticks for kindling. We gulp down tea with a few pieces of indigestible Kyrgyz flatbread before beginning the steep ascent across the scree-cones under the enormous bulk of Central Pyramid. An hour later, we're gearing up at the head of the couloir. Sam sets off up a slabby rake from which the great pillar rises. He climbs quickly across initial slabs of lichenous granite. As he arranges a belay, I get a chance to take in where I am.

The sun is still hidden behind the icy bulk of the Russian Tower's north wall. The silence of the couloir is shattered intermittently as huge chunks of ice and rock peel away from the wall and plunge more than a thousand feet to the screes. From the safe distance of our position on the pillar I watch the tumbling debris with an alarmed fascination. Occasional chill gusts eddy around the couloir; a pair of lammergeiers circle on an early thermal coming off the moraine under Wolf Peak.

"Okay, when you're ready." Sam's voice brings me back to the rock and the challenge in wait above us. I follow the pitch slowly, kicking my toes against the granite crystals, trying to get the blood back in my feet. By the time I arrive at the belay the sun has just swung over the Russian Tower and we shed thermals, enjoying the warmth. Above our marginal stance soars a perfect open groove, disappearing just before the arête that bounds the right hand edge of the pillar. We swing leads and I climb on into the groove. All too soon, after 60 metres of immaculate climbing on equally perfect granite, Sam warns that I've run out of rope. Above, the groove vanishes in about 10 metres and is capped by a series of daunting overlaps. Sam leads on through, eager to get on with the climbing. Moving out right from the top of the groove, his progress slows, eventually coming to a halt at a final overlap.

"How does it look?" I enquire.

"There's just a sea of granite up here, I don't think it'll go this way."

There is an audible silence; we're both weighing up exactly how disappointing a retreat would be from this point.

"How about reversing back to the groove and breaking out left around the arête?" I suggest from the security of the belay.

Sam arrives back at the top of the groove, the slack already looping out on his right rope, the way ahead uncertain. After a while, he chalks up, cleans his boots and makes a delicate, precise move out left on to the arête. I can hear him breathing with those brief, deep exhalations that say far more than any words can about the gravity of a climb. After another short pause, he says with a quiet assurance:

"Okay, watch me here. I'm going for it."

He then disappears around the arête and out of my field of vision.

The ropes inch out. Vagrant gusts come and go, blowing chill air from the shadows of the icy couloir to the right. When the air is still I can feel the warmth of a thermal rising from the open slabs way beneath us. Occasionally there is a hollow boom from the depths of the couloir as another massive chunk of ice and granite explodes into the scree. Then suddenly there is a different sound from high to my left – a brief shout of adrenaline release, and relief.

"Nice one", I reply, hoping he's cracked the invisible puzzle around the arête.

"It's not over yet, watch me." His urgent words are carried out on the wind.

I detect the urgency of his movements as the ropes pull out quickly, pause for a second, inch back, pull out again, inch back, stop. After an indeterminate moment, they haul out firmly for another metre. As I hear Sam's holler of success, I feel my own nerves settle. Instinctively knowing the meaning of his exclamation, I shout an enthusiastic reply into the silent mountain air.

And all too soon I'm above the groove, back-stepping through those wretched overlaps. I remove the final, small cam Sam has placed high on the right rope with trepidation, watching it loop, slack and useless across the wall to the left. I guess it must be at least 20 metres to the side runner above Sam's belay. Switching off, I reverse the intricate moves back to the groove, soon peering around the arête with some relief.

A faint line of chalk weaves across a featureless slab to Sam's belay in a tiny niche 15m to the left. He beams across at me, laughing:

"That was the hardest bit of onsight climbing I've ever done."

I congratulate him on what was clearly an astonishing lead. Simultaneously, with rather less enthusiasm, I'm calculating the length of a potential pendulum fall from here. I'm precariously balanced on minute footholds. Although not perhaps as dangerous as on the lead, a fall from here would take me on an extensive and accelerating sideways tour of the abyss. Never a fan of bridge swinging, I try to zone out from my position on the arête.

Eventually, I get the better of the weight of my camera and the monster pendulum yawning back at me, and make a series of intricate and balancy moves left to a rest on a tiny edge. Part of it crumbles away and I stab my foot back at it, holding on friction. Sam laughs. I look up left, smiling back with incredulity at the seriousness of the pitch.

"Shit, a fall from here would be a monster."

Slightly safer now, having closed the angle to the last gear considerably, I enjoy a brilliant final sequence to join Sam on the belay. With a sudden flashback to last night's conversation in base camp about James Bond, I compliment him, tongue-in-cheek:

"Nice lead, 007."

As our laughter subsides, I realise we've found a perfect name for the route. It has to be *From Russia With Love,* even though the nearest Daniella Bianchi look-alike would surely be drinking vodka with local mafiosos in downtown Bishkek.

After a while, I set off up the daunting crack-line in the impending wall above us. I'm reminded of those classic extreme climbs on Gogarth's Main Cliff as I move strenuously through wild terrain, studiously avoiding a giant, booming flake apparently welded to the retaining wall by lichen alone. Pulling through the last of a series of small but awkward overhangs, I find a secure belay, content in the knowledge that we have finally unlocked the puzzle of this climb, one of the most perfect, elegant lines either of us have ever seen.

Sam soon appears around the roof, his grin even wider now, and races on up the easier angled continuation of the crack. After another massive pitch, I belay on the apex of the pillar. We revel in our new perspective on the surrounding world of towering granite walls and snow-capped peaks, sharp in the late afternoon sunlight at 4500 metres. Eventually, we find the top of a reasonable-looking abseil line down an enormous corner system we'd spotted from below. Half way down, the evening shadows race up from the couloir to overtake us.

Only a few hundred feet from the scree and we look back up at the line. The last sunlight falls across the pillar, just discernible among the deep shadows of the couloir, a slender orange brush-stroke against the vast, darkening wall. Pulling the ropes on the final abseil, we watch and remain silent. That last moment of light captures everything we might have said then, about why we had chosen to make the long journey to Ak Su, or why we ever go climbing.

At dawn, from somewhere down the valley, comes the crack of a Kalashnikov, probably fired by those half-trained, half-mad Kyrgyz soldiers from Batken. The soldiers arrive and we find we are leaving Ak Su under a very Russian kind of unofficial arrest. We have no permit, they tell us. Where is our permit? Yes, we reply. We tried many times to get one, but you cannot get one in Batken.

Thirty-six hours later I'm in a small cell in the army barracks in Batken with two Kyrgyz officials cross-examining me in a mixture of Russian and Kyrgyz. The fat one on the right calls himself 'The General'. The skinny one – who looks and smells like he's had a few vodkas already – calls himself 'Mister Zulu'. One or two English phrases are thrown in for good measure.

"You zwatch footzbal?"

"Arzenal?"

"Manchzesta United!"

"You have many western girl?"

It is like being a contestant in a Russian version of *I'm A Celebrity* with a reference to Tarantino's *Reservoir Dogs*. The General and Mister Zulu are playing good-cop-bad-cop, but cannot decide which of them is which. Like a couple of teenage boys desperate to impress, they carry on until the absurdity of the situation slowly dawns on them. Mister Zulu slinks off for his vodka and the General opens the door purposefully, now assuming an air of military seriousness:

"You may tell zur frientz zat maybe you stay in zis platz for some days."

I break the news to the others, my throat parched from the dry air and the General's cigarette smoke. In need of refreshment, I go over the road to where a small boy has arrived on an enormous bicycle laden with watermelons. Biting into a slice, I walk back to pass the melon round to the others. They are convulsed with laughter.

"What is it, Sam?" I ask. Turning back towards the barracks, I see Mister Zulu greedily knocking back another shot by the main gateway.

"Now they… they want a couple of hundred som for more vodka…!"

Hours later, we are rattling back across the featureless desert on the Uzbek border when Shortcut pronounces he is lost.

"How can we possibly be lost?" Grimer enquires, to a murmur of agreement. It is nearly midnight and we are all knackered. Shortcut has stopped his interminable chain smoking, so even he must be in need of a break. We shamble out of The Beast and crash out instantly on the sand.

Twenty-six hours after leaving Batken, we arrive back in Osh at mid-afternoon and bid goodbye to Shortcut, tipping him handsomely for getting us to and from

Ozgourosh in such style. He departs with his signature flash of gold teeth. That evening, I read the news on the Internet. A couple of weeks have past since the 7th July Islamist suicide bombings in London. In no mood for sleep after everyone else has retired, I stay up reading. After a while, I hear a woman's voice beyond the curtain separating the courtyard from the house. It is Ury's wife, beginning her late reading of the Koran. Her voice is clear and sharp, lifting and falling through the air like windblown snow. It sounds like the voice of a woman far younger. Fig leaves rustle in the empty courtyard and the vines rattles from time to time overhead. The night air smells of dust and wood-smoke. Another gust of wind blows through the fig tree, colder now. To me, this Kyrgyz woman's hauntingly beautiful reading of the Koran now seems as an elegy for all those who died in the London attacks, and the countless others who've perished in our modern, assymetric wars of religion.

Two days later, we're back in Bishkek. After the wildness of the Pamir and the wild-west characters of Karavshin there is a certain refreshing austerity to the glum Soviet blocks and monolithic heaps of concrete masquerading as public art that line the streets of the capital. Two gorgeous Kyrgyz girls in fluffy boots and matching hot-pants sashay across the boulevard from where we're drinking coffee, snatching cheeky glances at every opportunity. Even the beer is cold. A big city is a very odd place after several weeks in the high mountains of Asia.

It is 4.15am and the hatchet-faced Russian clerk at the Aeroflot check-in desk wants money. She is trying to weigh our baggage and all manoeuvers to try and divert her prove ineffective. Even the two bottles of vodka I produce as a goodwill gesture fail to register. The excess baggage bill is, for our modest resources, pretty huge. Ian is convulsed with laughter as we wait to board our plane to Moscow, stripped of our last reserves of cash but happy to have made our great escape from Kyrgyzstan, the Batken Barracks, the vodka fumes of Mister Zulu, and the ire of the General.

Just before we taxi away to take off, at the back of the plane the door swings open and in strides a somewhat menacing-looking uniformed official.

"Are zere six Engliz Alpinistz on zeez airkraft?" he asks repeatedly as he strides down the cabin.

"Shit" Mark breathes through clenched teeth.

Sam's eyes roll with anxiety and sleep deprivation. Donie and Grimer left on an earlier flight, so at least there aren't six of us. Mark, Ian and Sam have their hand baggage checked for climbing gear, which, fortunately, they are not carrying.

"Alpinists?" Mark asks with a confused expression.

"No, no, we're just tourists", Ian chips in immediately.

"We have been on a wonderful holiday in your excellent country", adds Sam.

Lacking the linguistic ability to contest our explanation, the official nods slowly. After a while, he turns around with the painfully disappointed expression of a man who has just lost a game of poker, and walks back towards the door of the plane. The grins widen and as the door finally closes and we burst into paroxysmal laughter.

"The Kyrgyz Six" Mark proclaims triumphantly.

"Long live the Kyrgyz Six!"

The voices of my friends fade as the engines roar into life, and I'm pulled back into my seat as the decrepit Tupolov lifts into the summer night. I watch the lights of Bishkek fade below until they vanish altogether. Leaning back against the window, I am overcome by waves of sleep.

A chill wind engulfs the mountains about us. The sky overhead turns dark.
The sound of rushing cloud fills my ears. Am I falling?

The Alpine Journal, 2008 (Volume 113)

Summary: An account of the first ascent of *From Russia With Love* by Sam Whittaker and David Pickford (500m, E7 6b / 5.12 X), Ak Su Valley, Kyrgyzstan, in July 2005.

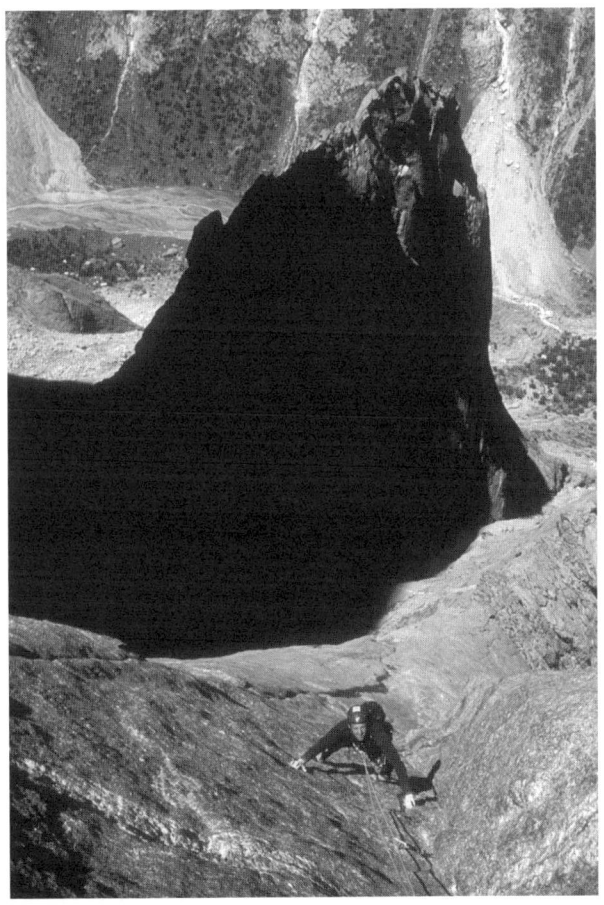

Sam Whittaker on pitch 3 of From Russia With Love (E7 6b, 500m), Kyrgyzstan, on the first ascent in 2005.

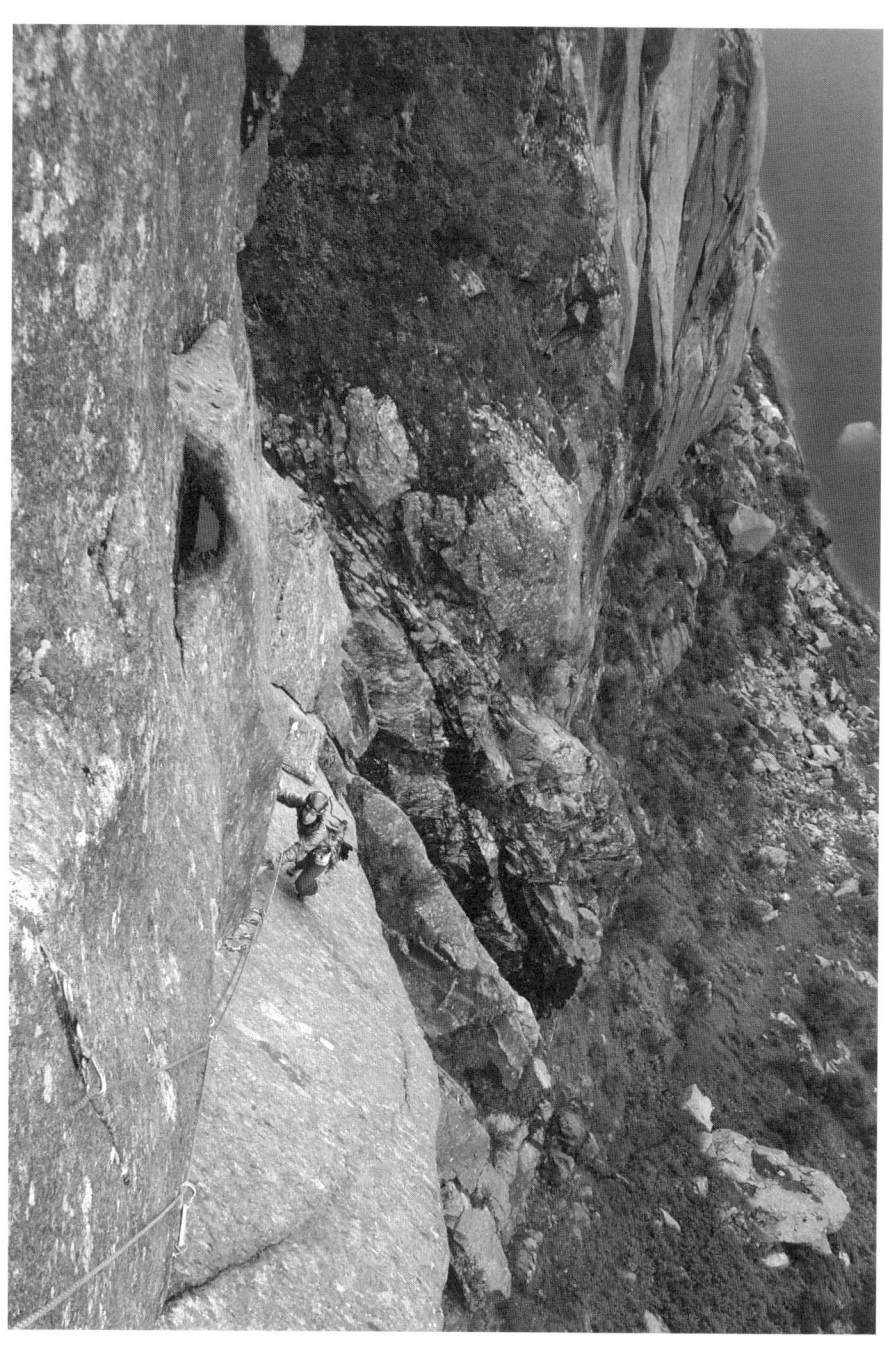

Malin Holmberg on pitch 3 of The Lady of the Lake (N9) on Djupfjord Wall during the first ascent in 2011.

Fjord Fandango

First ascents in the Lofoten Islands

I wake to the Sunday morning light at a quarter to five. Last night's traffic is silenced and gone. The midsummer air smells of the streets and the harbour and of the dirty wild flowers that grow along the edges of the tram lines. I open my eyes. Birdsong drifts through the window of our tiny flat in downtown Gothenburg, Sweden. In a blaze of red hair, Malin's already up, and the scent of fresh coffee drifts through the room. Here, I rise to a different life.

The solstice just past, we're drawn north to Norway's Lofoten Islands. Uddevalla. Trollhätten. Fredrickstad. Oslo. The names on the signboards make hidden signals as we drive beneath them, gold and blue, like the interplay of sun and shadow on Bohuslän granite. Before my first journeys to Sweden in 2010, I hadn't understood the real power of light and dark.

One of Europe's longest roads, the E45 runs for almost a thousand miles north from the fertile land of Sweden's southwest coast to the barren tundra of Arctic Norway. By evening, we're leaving behind Storuman, its windows bolted and shuttered as if the town were already preparing for the long, isolated months of winter, shutting out the featureless gloom of the surrounding forest, the drone of its mosquitoes, the rank stench of its endless bogs, and the impossible vastness of the world beyond.

Midnight. The sun rolls along the horizon when we reach the edge of northern Scandinavia, where the forest gives way to tundra. Huddles of reindeer have long replaced passing cars. Small groups break as we pass; the animals try to shake off mosquitoes with our slipstream.

At 3 a.m., I open my eyes just as Malin drives over the crest of the hills south of Kiruna, and the morning sun hits the central extraction tower of the world's largest iron mine. The Arctic summer light sparks off the quartz shards of humped slag heaps that loom above the town like a monstrous UFO. A mirror of mine debris covers the mountain, and as we turn across the sun, the reflection fills the windshield, blinding us for a moment, then melting across the dashboard into pools of quicksilver light. It is a moment of such extreme, sudden beauty we both remain very quiet.

We turn again and descend toward the town, passing sidings of rusting trucks and the snowmobile salvage yard locked in the bright, sleepless silence of the early morning. Winter grips Kiruna for most of the year. Summer is a stranger here, a traveller from the south country, like me. Only a native can understand the true nature of this isolation at the top of the European continent, or what it's like to spend a winter in one of its hyper-insulated, narrow-windowed, Soviet-style apartment blocks. Malin grew up here. Her dad worked in the mine for forty years, hacking iron from the guts of the surrounding mountains. As we turn into the driveway of her family home, a wry, curious smile spreads over her face: "I know this place" she says.

After a few hours' sleep, we drive west beside the railway that freights out thousands of tons of iron ore every week. The tundra shines in the morning sun. Stendhal once wrote that beauty is the promise of happiness. The first glimpse of Norway's northwest coast seems like a physical expression of this idea. To the west, south and north, a concatenation of mountains and islands stretches into the far distance. Sharp outlines cut the sky and sketch patterns in the blue ocean. We find a stopping place to sleep late in the evening by the shore. Beyond the boulder-studded dunes, the wild grass blows in the salt wind, and the sound of the sea picks up between the gusts. Gulls fill the air with wheeling cries. Hearing their calls under the midnight sun, we feel free.

For the first few days in Lofoten, we climb mainly classic routes. A white-tailed eagle soars behind us on the final pitches of an elegant thousand-foot line of interconnected corners and arêtes. The sun swings around the corner of the wall as we prepare to descend a shady dihedral, washing the slabs with liquid amber hundreds of metres below. At the edge of the harbour, tiny wooden houses balance on delicate stilts. Out to sea, the flocks of gulls make high arcs over bright red fishing boats, scattering as they descend. Crenellated lines of mountains stripe the space between the horizon and the sky into a hundred layers. Everywhere we look, and everywhere we are, light and life surround us.

As we drive around the coast road to Henningsvaer, a north-facing wall shadows the calm waters of Djupfjord. A thin seam splits the center of its upper headwall. "Look at that!" I exclaim to Malin as I pull over.

"Imagine if we could climb that wall directly, finishing up there." I point at the seam.

"I wonder if it's climbable? Looks amazing!" she replies. That signature bright and curious smile spreads across her face.

We run along the shore of the fjord with binoculars to try to get the best view. Tracing the line of weakness from the base, I link features through the strange perspective of powerful magnification: a deep chimney, a bottomless flake, a flying ramp, a crack-corrugated wall, then a rising traverse to the base of that quixotic seam. Even through the binoculars, I can't tell how thin or deep it is. It could be straightforward or impossible, or anything in between. We just don't know.

After a rest day, we pick a route through the boulders along the shore, up around a rocky bluff, through dewy moss and cloudberries, to find the fixed rope that leads down the other side. Djupfjord Wall towers above us in all its shady glory. I contort up the dripping, water-worn chimney to a belay niche. Then Malin sets off up a booming flake. As she scales the huge granite tooth, the sun hits the water of the fjord below. I follow her lead with cold feet, wondering what lies ahead.

On the third pitch, I shuffle up the ramp that lifts into the heart of the wall like a stairway suspended in space. While I fiddle with mircowires and small cams, I try to forget about my frozen feet skidding on fresh, thick lichen. Far from my cluster of tiny placements, just at the point when I don't want to run it out any more, I spy a widening in the seam and a better cam placement a few metres above. I sprint for it and breathe deeply as my fingers sink into the first perfect jam on the entire pitch. The ramp narrows to the left and vanishes into an ocean of blank stone. My only hope is a ledge five metres higher. Searching for clues, I glimpse a depression to the right, and I span out from undercuts to another perfect finger jam: a gift of nature. Without the finger jam, there was no way I could have reached the ledge without direct aid.

At the belay, we try to warm our feet. Creased with tiny cracks, the wall now sweeps toward an atrium where the angle will ease and the final seam will appear. Malin's fingers slip in precarious fingertip jams, and her shoes skitter on small crystals, but soon she's moving up toward the golden light that floods across the upper cliff. I check the time: 5 p.m. The wall should be in the sun until just before midnight, when it dips for a mere two hours below the ridge to the north of the fjord before rising again.

As I climb toward her, warmer now at last, clouds blow in and blot out the sun. Huddled in her tiny belay niche, she's shivering with cold. I give her my balaclava and wind jacket.

We can see the headwall seam, now, seventy metres above us. From here, it looks perfect. Drainage water runs down the lower cracks from the summit slopes. I meander up a twisting path toward the headwall as the mist thickens below. After a while, I pause to take my bearings: the only way to gain the seam is a wild traverse across the apex of the capping slab. Again, I search for clues, and again the wall gives just enough: a rail of tiny, rippled edges leads me rightward, and suddenly I'm atop the sloping ledge beneath the incredible seam.

The cloud has come down around us, and the wind picks up to a near-gale. With chattering teeth, Malin places a micro-cam in the seam. The first drops of rain fall. I suggest we take the easier finish up the wide crack to the right and return for the final pitch the next day. She flashes a bright smile between shivers, and agrees.

By the time we've reached the base of the wall, the weather has cleared, and the wind-rippled fjord has settled to a translucent calm, reflecting the serrated shadow of Budalstinden. I show Malin the time: 1:30 a.m. Her sharp green eyes soften with tiredness. We look back up at the line we've just climbed and smile. It's all there in front of us: 220 perfect metres. The midnight sun is hidden below the bulk of Vestvågøya that blocks the northern horizon. Silence fills the fjord in the brief Arctic summer twilight, broken only by the crunch of small stones under our feet and the call of a lone night bird that echoes among the rocks and across the surface of the water.

Two days later, the morning air is so clear that simply drawing breath feels like inhaling pure oxygen. A short storm blew out by midnight. Bright lines of surface runoff cascade from the highest slabs. The sun is already drying off the upper section of Djupfjord Wall. I know I'll never be able to free the tiny seam with my big fingers.

Malin, though, has a good chance with her thin hands and crack-climbing mastery. We rappel in to the base of the seam, rigging a rope on the way down. Malin cleans her shoes, and climbs the crack on top-rope a couple of times. I know she can lead it. We both take a deep breath and clasp our hands into fists before touching them together, our habit before a hard lead.

"Okay, you're on. Go for it," I say. "Full attack."

"Yeah. I'm going." she replies. And she goes. Fighting from the very first move, she hesitates before pulling hard on a flared fingertip jam. Her foot skids on some lichen, and her hand slips slightly, and for a moment, I think she's going to fall. Instead, she holds the slip and steps up strongly. Two more laybacks allow her to swap sides, facing left, then right, conforming to the seam's slight curvature. Many more desperate moves lead to a slight hollow. Resting on the first real foothold since the belay, she places the first gear for many metres. We both relax for the first time since yesterday. She pauses for a few minutes, breathing hard, and launches out again. On the final hard section, her feet dance across the slab while she makes long, fast reaches between poor fingerlocks. After a while, a scream of joy breaks through the sound of the wind ruffling my hood. She's made it.

An hour later, we're at the base of the wall again, and I'm cleaning lichen and moss from the very toe of Djupfjord Wall. A steep finger crack blanks out by a pair of old aid bolts. A sudden wind picks up off the fjord, displacing the humid, leafy air, and I know I'm in luck. Long spans lead between shallow, positive locks to a resting place. From here, the aid line goes right. I move up and left on undercuts, and the crux sequence stares sternly down at me. I span out and stab my left forefinger into a razor-sharp mono that winks from the blank wall like the eye of a Cyclops. Then I run my feet high on grainy stone and launch out. The fingertips of my right hand just catch the first of a rail of slopers in a flared horizontal break. Feet skittering across the holdless wall, I make a series of wild slaps until I can throw my left heel up, place a cam, and – finally – relax.

The upper crack is easier, although I must clean it as I go. My hands slip from flared jams in the lichenous crack. Soil and decomposing crud pour down my T-shirt. I grope over the top and land on a flat terrace. I can hardly believe it: in one afternoon, we've climbed two of the best granite pitches I've seen anywhere in the world.

Long swatches of evening mist swirl over the fjord. Malin leads a slim pillar split by a thin crack. The sun's just dipped below the opposite ridge. A melancholy chill washes over me: from now on, the dusk will come earlier every day. Sometime in late November, the sun will disappear completely from the north Norwegian sky. But we've had one of the best climbing days I can ever remember, so I should be grateful.

Midnight. To the north, the sun rolls unseen beneath cloud-capped mountains. There's a hint of magic about this north Norwegian summer evening under the indigo sky. For a moment, Malin and I are held still in the breathless rush of the turning world.

After a quick, warm, forgetful kiss at the last belay, we're descending through the vertical beech forest that fringes Djupfjord Buttress. Deadwood snaps around me and fresh branches recoil everywhere, as if the forest were closing up behind us. When we contour the steep slope near the toe of the wall, Malin lets out a cry of joy, and I run over to find her kneeling on a ledge covered in bright orange cloudberries: a final gift from the fjord in the fading light.

As we devour them, the sugar rush from the sweet berries makes us giddy and euphoric. Malin's eyes are bright with laughter and fire. Robert Lowell memorably observed that "happiness is something with a girl in summer". I'd take that cloudberry ledge anytime over the moral high ground, or some lofty ideal of a better life.

I know that soon we'll be driving south into the night, leaving all this behind, going back to wherever it was we once called home. But home in the climbing life is elsewhere. It's not a place you stay, but a place you find. And you must leave it when summer ends and pass it on to someone else, so that they too might discover joy and freedom there another time. Every generation has a duty to pass on the exploratory torch to the next one.

The next morning, we drink coffee in Henningsvaer and listen to the news on the radio. It registers like the sound of a massive rockfall: Anders Behring Breivik has just shot dead sixty-nine Norwegian teenagers in a terror attack on Utøya Island; the most deadly massacre by a single individual in Western Europe since the Second World War. The spell of the magic islands breaks. Against this mindless destruction – an extraordinarily unusual event in peaceful, civilised Scandinavia – neither climbing nor happiness seem to matter much.

In search of silence, we drive north to Eggum where we'd slept three weeks before. Lost in the space between memory and forgetting, we climb a few short pitches beneath leaden skies. Invisible under banks of mist, huge breakers surge on the grey shore. I can't stop thinking about all those kids, their young lives so brutally snatched away. Then I think of my dear friend Woody, killed in a freak accident on a Welsh sea cliff just four months before. And of my grandmother back in England, very close now to the end. Waves of absence crash over me, heavy and cold as the Arctic Ocean just beyond, bearing the open water between the living and the dead.

We pack up and leave the cliff in silence. As we drive south, it starts to rain.

We arrive back at Djupfjord in deepening twilight. The weather's cleared, and we walk along the beach that separates the fjord and the ocean. Malin stops to find bright stones among the seaweed and driftwood. A few men and women are fishing on the seaward side of the breakwater, casting out with long rods in the hope of catching a skulking cod as it silently enters the fjord on the midnight tide.

I walk to the end of the beach, strip off and dive into the cold sea. That sudden rush of saltwater across my body washes away the presence of the dead. I swim as far as I dare into the bay before my head burns and my movement starts to slow.

Djupfjord Wall looms on the skyline to the east, shrouded now in darkness and mystery. I think of all the other, potentially even harder climbs it still contains, and of how lucky we were to be the first to free climb into the heart of this sublime, mysterious cliff. The light is fading fast now, and I turn for shore. Lenticular clouds gather on the surrounding summits like flying saucers, casting violet shadows on the dark water of the fjord.

Malin's standing on the beach as I return, alone and beautiful, looking out to sea. As I swim toward her, I'm reminded of that far greater gift than climbing: simply being here at all.

Alpinist magazine, 2013

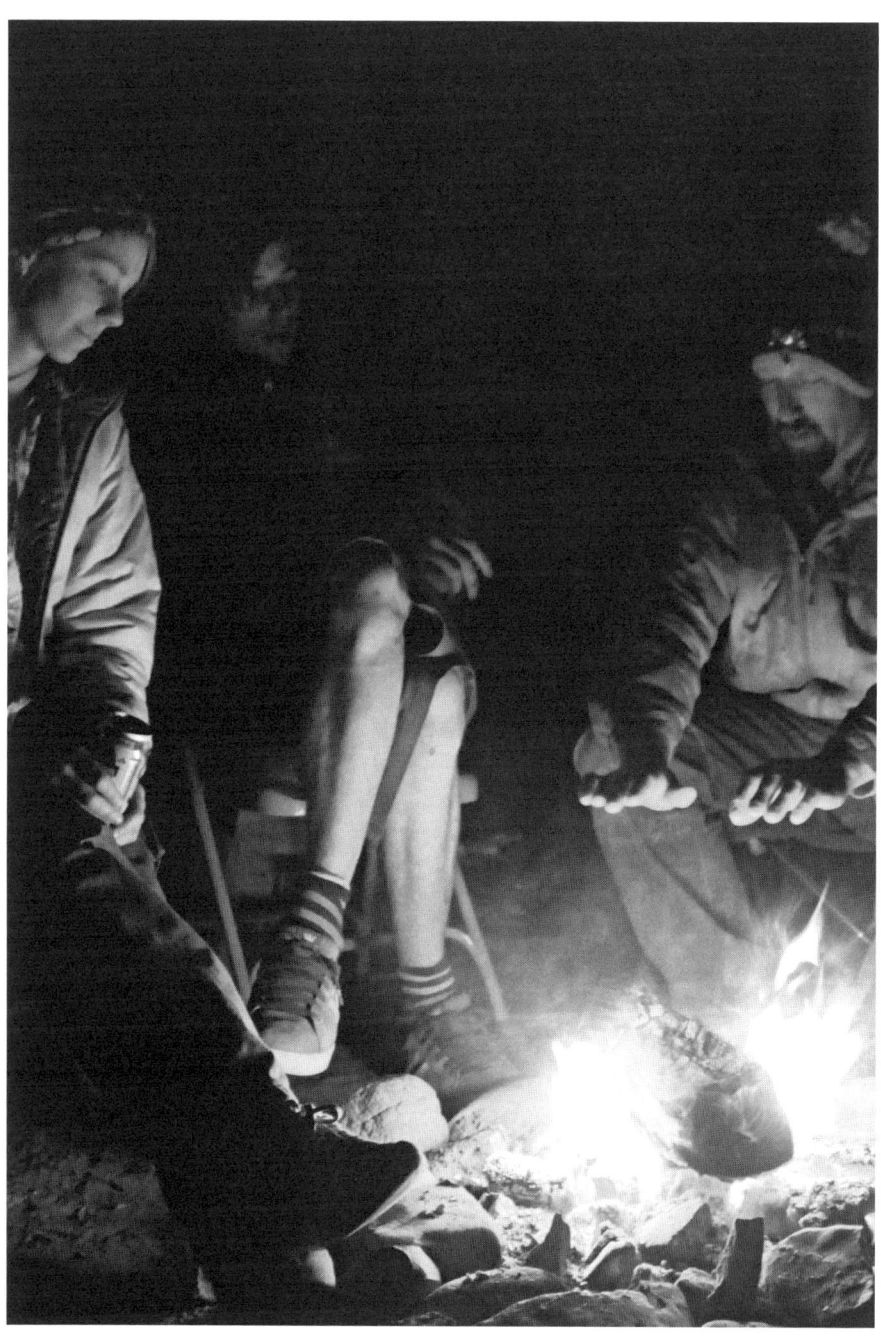
A flare in the dark: campfire stories on a cold spring evening in Indian Creek, Utah.

A Flare in the Dark

On the value of climbing

Somewhere high above, hexagons of sunlight refract and merge. I glance upwards. A thin vein of white cloud streaks the sky above the Russian Tower. I breathe in deeply and exhale. The air is cold up here.

What's happening? As my feet begin cramping, my left boot slips slightly on the smear, and an electric current jolts my limbs back into movement. I've been standing here, near the top of this unclimbed, two-hundred-foot sweep of ash-coloured granite for twenty seconds at least. Too long, probably.

Below, the slab barrels downwards into darkness and distance. I cannot see the last gear in the break forty feet above the deck. The gear doesn't matter, anyway; I'm a hundred and twenty feet up and very much alone. If I fall, I'm going all the way. Crash-landing down there, among the glacial boulders of the Ak Su valley, two days' walk from the nearest inhabited place, would be an extremely bad idea.

"What are you doing here?" asks a quiet voice in my head. It's a refrain that's returned again and again through all my years of climbing, like the disembodied wisdom of a protective godfather. After another five seconds of pondering, I realise that I'm far too high up this thing to even contemplate reversing the sustained, delicate climbing I've done to get here. The only option is to make the moves.

White light blurs the edges of my vision, and I commit.

Thirty seconds later, I'm pulling over to a ledge of sweet-scented alpine grass and the sanctuary of a huge belay boulder. After a while, my heartbeat slows. Listening to the churn of the glacial river far below, I began to take in all of what had just happened on the climb, as I had so many times before and as I would again.

Along with its sometimes life-preserving attributes, the query "What are you doing here?" contains specific, important questions about the essence of climbing. What value – if any – does climbing have? Why do we put ourselves in scary or downright dangerous situations on the rocks and in the mountains? What, really, is the point of it all?

In his book *Alone On The Wall*, published in 2015, the American climber Alex Honnold – and one of the boldest actors in the sport's history – argues that "[Along] with the critique that climbing is selfish is the claim that climbing is useless. But I think that perfecting your skills on rock, ice or snow ends up improving you in other ways. I fully believe what I've learned from climbing translates into other aspects of life." I'd like to explore Honnold's point in greater depth: to ask what we might gain from climbing rather than why we do it, and to ask how what we gain from it might be beneficial at a psychological and moral level as well as a physical one.

Over the years, plenty of perspectives have been offered on the reasons *why* people go climbing. Of these, George Mallory's quip that he wanted to climb Everest "because it's there" remains not only the most famous but also probably the least substantial of the various explanations for why we climb. We might climb a mountain because it's there, but we might use a spoon to eat our cereal because there happens to be a spoon somewhere in the kitchen, and we go to the pub to meet our friends because, well, there's a pub there. Just because something is there doesn't make it interesting.

Of course, a whole host of other, more sophisticated explanations for why we climb have also been set forth. One of the more interesting and psychologically charged propositions was once suggested to me by David Thomas, one of Britain's foremost solo climbers in the 1990s. He raised the idea in conversation that we might become interested in climbing as youngsters "because [our] parents don't show us the world". For some, I think, this provocative and intriguing suggestion might be true. Climbing may be an appealing and also a powerful tool for adventurous or rebellious adolescents to assert their genetic independence from their parents.

For others, though, the opposite could be the case. It has become fairly clear to me, after many years of reading around this jellyfish-like subject, that the only reliable link between the explanations for why we go climbing is the fact they are all quite different, and in many cases contradictory. The British climber, physicist and writer Phil Bartlett captures this situation in his excellent, far-reaching book *The Undiscovered Country*: "One of the reasons for thinking mountaineering a noble pursuit is that it defies our attempts to categorise it and explain it away."

In the spirit of this perspective, and in contrast to the conflicting assertions about individual motivation, the more substantial questions that surround the value of climbing – and their answers – might be found through three key lines of enquiry.

First, what's the value of climbing to those who choose to climb? Second, what value does climbing have, if any, to the rest of humanity; to the vast majority of people who will never climb? And third, what value could climbing have to people in the future and the societies they live in?

About twenty years ago, a friend of mine told me something I'll never forget. He was, incidentally, well known in his youth for his audacious leads of adventurous British trad routes. We were working on a rope access job together when he threw this startling line at me: "You know what… I think climbing saved my life, really".

He'd grown up in a broken home in an area of a major British city known for its problems of social exclusion, drugs and crime; a close relative was serving an extended prison term for attempted murder. Given these facts, when he quit school in his mid-teens the possibility that he'd fall into a life of crime and become socially excluded was statistically quite high.

But then he discovered climbing, which changed everything. It gave him, he told me, "focus, drive, motivation; a sense of purpose in my life, a sense of community with the guys I was climbing with". Today, he doesn't climb as much as he once did. He now runs a successful business. His powerful story shows, I think, how climbing can be a flare in the dark, a kind of guiding beacon that might radiate light into the shadows of existence.

Focus. Drive. Purpose. Community. The association of these qualities with climbing will not be surprising to anyone who climbs regularly. Yet it is not an exaggeration to say that most climbers – myself included – take the covert benefits of climbing entirely for granted. Yet the fact we don't necessarily consciously appreciate them every time we climb does not detract from their importance.

Plato's *Apology*, one of the great philosophical texts of the ancient world, remembers the speech Socrates gave at his trial, in which he chose suicide by drinking hemlock rather than exile from Athens or a commitment to silence. A few moments before his death, Socrates purportedly said "An unexamined life is not worth living".

An activity such as climbing, like anything else that requires the use of skill in a potentially dangerous environment, demands that its practitioners lead an examined life in various ways. It demands that we are physically fit and strong, well organised, and self-critical. It forces us make both considered, rational plans as well as fast, spontaneous decisions, exercising both the so-called System 1 and System 2 aspects of cognitive thought that the Nobel Prize winning psychologist and economist Daniel Kahneman identified in his book *Thinking, Fast and Slow.*

Perhaps most importantly of all, climbing requires that we develop high levels of environmental intelligence, which I would define as the quality of being able to operate efficiently in a hostile environment while leaving no trace of your presence. This is, of course, an incredibly useful, sometimes life-preserving skill, as any member of a Special Forces unit might tell you in more detail.

On a more existential level, leading what Socrates called "an examined life" surely also means that we examine what it is that we truly value, and therefore what we want to do in life, and to develop a belief system based around those values. For many climbers, such a belief system often involves a love of the outdoors and of the natural world, and doing something challenging with friends. For others, more interested in physical training, it might involve working towards certain level of athletic excellence over an extended period.

Yet whatever you believe is most important about climbing, the very fact you are going climbing at all means that you're realising your own version of Socrates' great dictum: you're living an examined life by doing something you believe in, and which might be quite difficult, and which has value to you and those around you. The late writer and journalist A.A. Gill brilliantly suggested that the art of successful living was about choosing what not to participate in; what to leave out, in other words.

This is of course easier to appreciate as you get older, but if climbers often choose to prioritise their love of rocks and mountains over social events and shopping trips, it's perhaps because they're unconsciously following Gill's idea.

Sometimes, too, we can only appreciate the true value of any relationship by its absence. My friend Hazel Findlay, the first British woman to climb E9 and to free climb Yosemite's El Capitan, once explained in an interview how she felt on touching rock again after a six month lay-off in 2015 following shoulder surgery:

"[The first route I climbed after the operation] was ridiculously easy. But it was blissful. Even with my weak body, my whole being drank up those moves like a parched woman licking water droplets off a windowpane." This startling metaphor captures much about the intrinsic value of climbing to those doing it regularly.

Climbing can be an athletic discipline, an exploratory quest, a meditation ritual and a series of social interactions. Its highly addictive nature – for younger people in particular – might be explained by the fact that it can be all these things at once. It fulfils the need for the hunt and also for the dance that are central to many tribal cultures. As such, it is a triumph of the human capacity for repurposing ancient rituals into modern mediums.

At a push, the pursuit of climbing as a formal discipline is no more than a couple of hundred years old. Despite the relatively low participation numbers compared to mainstream sports such as running, it boasts a body of literature and film that can rival that of any other sport, much of which is read and watched by people who are not participants but armchair enthusiasts. This is another measure, I think, of its unique value.

To the second of our three areas of enquiry: what is the value of climbing to the world at large? At first glance, this seems a difficult question to answer; but the response is relatively straightforward. At the most fundamental level, an activity like climbing has value to those who do not climb because human societies still demand stories and myths, in the same way that past societies once did.

The twentieth century American mythologist and academic Joseph Campbell suggested "myths are public dreams, (but] dreams are private myths". Since climbing creates a vast corpus of powerful stories, some of these stories enter the cultural imagination of their time, where they can quickly become myths. Campbell's most famous concept is that of what he called the 'monomyth'. It's best known for being hugely influential on the creator of *Star Wars*, George Lucas. The idea essentially regards all mythic, heroic narratives as adaptations of a single great story that lives at the heart of human consciousness. In this single story, Campbell suggested that a common pattern exists, regardless of the origin or time of creation of any individual narratives.

Central to the monomyth concept is Campbell's notion of 'The Hero's Journey' (detailed in his book *The Hero With A Thousand Faces*), in which a heroic figure goes on a quest, experiences various trials and suffering, and returns with gifts or wisdom to set their people free. Most of the greatest and best-known climbing stories are striking examples of Campbell's hero's journey. It might be a reflection of the importance of Campbell's idea that those rare climbing stories that actually reach the mainstream media are often received by the public with enthusiasm, such as Joe Simpson's mountaineering survival epic *Touching the Void*, or the film *Free Solo*, documenting Alex Honnold's extraordinary ropeless ascent of El Capitan.

Another prime example of climbing myth-creation in recent years is the first free ascent of the *Dawn Wall* in Yosemite in January 2015. The real-time reportage of the climb on social media by Tommy Caldwell and Kevin Jorgeson captured the public imagination in an extraordinary way, partly because it felt as though we were watching a myth being made.

Most who followed didn't know anything about big wall free climbing; not least the President of the United States, Barack Obama, who tweeted enthusiastically about the ascent. While Obama was unable to appreciate the climb's technical difficulty, he certainly could appreciate the mono-mythical power of the story: an epic quest to scale a piece of sheer granite half a mile high.

In a lecture at Kendal Mountain Festival in 2015, Tommy Caldwell explained how following their ascent, he was approached by an elderly woman in a wheelchair who suffered from multiple sclerosis. She told him that his climb had given her the inspiration to get out of her wheelchair and walk again. I thought it was one of the more powerful illustrations I'd ever heard of the possible value of climbing as myth and inspiration to the world at large.

The precise construction of Joseph Campbell's 'Hero's Journey' doesn't apply to all climbing stories. In many of the most famous mountaineering stories, the hero dies before the end, and George Mallory's final attempt on Everest in 1924 is perhaps the ultimate example of this. Yet it remains one of climbing's most powerful myths, largely because we still do not know, and will never know, what the outcome of the story was.

However, even if the climbing hero, like Mallory, does not survive, such stories are nonetheless important; and possibly more so today than ever, because we are increasingly isolated from the reality of death. We need myths of people paying the ultimate price in pursuit of a noble cause – as in the Crucifixion story – because they remind us that our mortality cannot be taken for granted. So the hero does not have to survive and return with gifts to set their people free. The story itself, like Mallory and Irvine's tale and so many tragic climbing narratives after it, could be enough.

How, though, can we summarise the value of climbing stories to non-climbers? After a motivational talk I gave at a school a few years ago, in which I spoke about how dealing with risk, and even the possibility of death, in adventure sports can give us confidence to make difficult decisions elsewhere, a teenage girl quizzed me on my photos from the mountains of Madagascar. She said my talk had confirmed her desire to travel to Africa after her A-levels. She wanted to climb Mount Kenya after volunteering in Malawi. I realised that climbing adventures don't just have value to those who engage in them; because they have the power and simplicity of more traditional myths, the stories that climbing generates have a surprising value to many other people, too, in many different ways.

And so to our final line of enquiry: what value will climbing have as it evolves in our future society? In the Golden Age of alpinism in the nineteenth century, Victorian gentlemen of leisure sought untrodden summits in the Alps – a natural extension of the ambitions of imperial nations. A hundred years later, in 1980s Britain, groups of young climbers would live together in shambolic shared houses in Sheffield and Leeds, their lifestyle of full-time climbing supported by the generous unemployment benefit of Thatcher's government.

This shift away from climbing as a pursuit mainly of the aristocracy, as it had been in the era of the Golden Age of alpinism, also marked the beginnings of the modern era of professional climbing and training. Today, at outdoor trade fairs, young sport climbers and alpinists converge, circling the stalls like hungry hawks, hoping to catch an influential marketing manager's eye with a presentation on a travel-scuffed laptop documenting their achievements. Climbing has now truly gone global, like much else in the modern era. But what can we learn from all this?

Throughout its relatively short history, a single feature of the way in which climbing and mountaineering have evolved has been consistent: all climbing cultures have reflected and channelled the moral and cultural values of their time. Climbing is as much a mirror of the society in which it takes place as it is a transgression of it. All the talk of counter-culture in the British and American climbing scenes of the 1970s was a classic example of climbing following rather than rebelling against the

Zeitgeist; the Yosemite dirtbag era was like a microcosm of the Woodstock Festival with ropes – and possibly even greater quantities of marijuana.

It is likely that climbing will evolve in fascinating ways over the coming century. The extraordinary things that are happening in climbing today, from speed-solos of 8000 metre peaks to 9c sport climbs and boulder problems of extreme difficulty, are to some extent a product of the ongoing development of what technology journalist Kevin Kelly has called the 'technium', describing how technologies don't develop in isolation, but rather evolve together, and create self-improving tendencies in a similar manner to the process of natural evolution. The advancement of climbing and mountaineering cannot be separated from the parallel development of the technology, training methods and equipment used to facilitate it.

For example, new paraglider technology is already changing mountaineering by providing pilot-alpinists with lightweight canopies to fly base-to-base for Himalayan objectives in a few days - feats which would previously have taken weeks. Fine-tuned climbing skills and new technology can interact to create possibilities undreamt of in previous generations.

With a long way still to go, one overwhelming question remains: in what ways will climbing be valued by future generations? Will they love the cliffs and mountains in the same way that we do? And what exactly will they do out there?

Part of the answer to this question is obvious. A lot more people will be climbing in future than there are at present. In 2010, the United Nations forecast world population with three possible outcomes by 2100: two of these outcomes suggest that by the end of this century the world population will have reached circa ten billion or sixteen billion respectively; a huge increase from the 7.8 billion people alive today.

Some people in the outdoor community consider the sheer scale of this predicted human demographic expansion a potentially grave threat not just to wild spaces but to the general ethos of the outdoors at large. Part of the point of wilderness is the absence of other people in it. This problem can already be seen on busy weekends in many of Britain's national parks.

At the same time, in a world with ever more people (the majority of whom will live in urban or suburban areas) the public sense of the value of the outdoors - and of adventure sports - may increase rather than decrease. Since a mountain environment is so very different from an urban environment, more people may become interested in mountain sports simply because mountains themselves seem increasingly exotic to the city-dwellers of the late twenty-first century. Whatever the future relationship between climbing and popular culture, the need to protect climbing areas will remain strong.

However, the current signs for this trend are encouraging: today, globally, there are circa 1840 National Parks; a vastly greater number than there were even thirty years ago. And the IUCN, (the International Union for Conservation of Nature) now represents more than 100,000 protected areas worldwide. The chances that the regions where climbing happens will be preserved for future generations to enjoy looks much more likely today than it did in, say, 1980.

Some British climbers talk about the scope for new routes 'running out'. While this is true on the more popular British cliffs - which are among the most intensively developed climbing areas anywhere on Earth - there are still places even in the UK where excellent climbs have yet to be discovered, particularly on many of the more inaccessible sea cliffs in England, Scotland, and Wales.

Globally, the scope for the development of climbing areas is incalculably vast. Even though a lot of the world's highest mountains have been climbed, a huge number of 6000 metre peaks still remain unvisited, some of which will provide outstanding alpine climbs in remote places.

Climbing in Britain, which was once a truly niche activity guarded by a coterie of self-appointed patriarchs in arcane clubs, has become a far more inclusive activity than it was. This hugely positive development signals the way the value of climbing will grow in the future simply because more people from different ethnic and socio-economic backgrounds will have access to climbing. Almost as many women as men now climb in Europe and North America. The explosion of indoor climbing in cities means that more young people from disadvantaged backgrounds will be exposed to it, creating the chance that climbing will transform their lives. This process will have other implications about the way climbing is projected by the mainstream media and perceived by the wider public: could professional competition climbing compete in future with top-level international tennis for TV audiences?

Some newcomers who choose a life in the vertical will probably develop the value of climbing much as their precursors once did in another era, on a mountain far away and long ago. If this projection is even partially true, then it shows that climbing and mountaineering do have an enduring value to human beings, beyond the shifting cultural space of a single generation - or indeed an entire era.

Climbing can be fantastic, and it can also be awful. Quite a few of my friends have died climbing. As Victorian pioneer Edward Whymper famously highlighted, "There have been joys too great to be described in words, and there have been griefs upon which I have not dared to dwell." And despite all this, the timeless power of life in the vertical doesn't stop enchanting, thrilling, and guiding us onwards.

The Covid pandemic that dramatically changed the world in 2020 and 2021 was an opportunity to reflect on many things. Central to these cogitations for me personally was a subtle but distinct shift in my attitude to travel. I've climbed in more than sixty countries worldwide, and feel very lucky to have done so. But since free-moving international travel was effectively put on hold from March 2020 to late 2021, I began to question my previous need to travel as extensively and as frequently as I could just to get my climbing fix - or simply for the sake of travelling itself.

A lifestyle of constant international travel, in any case, is not sustainable from an environmental perspective. It's also not as good a way of travelling than doing just one or two journeys that you really value. Doing fewer individual trips over the whole year, and perhaps longer ones in the same area, might be a better strategy. And an even better one than that might be to dig into the untapped possibilities in your own backyard. Even in a crowded country like the UK, those possibilities are everywhere.

Towards the end of the final Covid lockdown in England in February 2021, partly because the climbing wall was shut and partly just because I wanted to go back to it, I began revisiting a twenty-five metre boulder traverse underneath a Victorian viaduct close to the river that's just a ten minute drive from my home. I used to use it a long time ago for training, but hadn't been there for ages - maybe fifteen years. The traverse was still there, exactly as it used to be, like an old friend. It had the faintest dusting of chalk; the trace of a fellow traveller in the locked-down edgelands. The traverse had exactly the same precise, fingery, sometimes powerful moves; the skittish, nimble footwork. Everything was still there, just as it had been before, in a different phase of my life. And then, towards the end where you reach the flying buttress, the late afternoon light started to glint in white and gold off the mudbanks of the river to the south. I could hear the tide sluicing in, the lonely rumble of a departing train, and the murmur of the wind cutting through the the winter air. My fingers closed on the final crimp.

There was a feeling of being completely alive in this strange, liminal place.

After a while, I started back on the traverse, heading into the shadows, my mind dancing in the light. Despite having climbed all over the world, it was an old

Victorian viaduct near the riverbank close to my home that reminded me that climbing is always beautiful, wherever and however you climb.

Much of what this essay has explored remains open to interpretation, so I'd like to offer a coda to that initial question of 'What are you doing here?' asked a few thousand words ago, standing on that lonely slab of granite in remote Kyrgyzstan. The American academic and intellectual James P. Carse observed that there two kinds of games in the universe: finite games and infinite games. Finite games end in a win for one party. An infinite game, by contrast, is played to continue the game. It doesn't terminate because there is no winner.

An activity such as climbing, which is diverse, self-organised, difficult, and sometimes dangerous, defies attempts to categorise and explain it precisely because it belongs to the universe of infinite games. Along with the greater advances of human progress - things like science, poetry, medicine, art, and mathematics - climbing is a noble pursuit partly because it defies our attempts to define exactly and precisely what it is. Our desire to climb and also the true value we gain from it cannot, perhaps, be explained at all. And that could be the very thing that makes it so powerful.

Climbing is a macro-game without a clear beginning or an obvious end. While climbs themselves begin and end, our desire to do them doesn't start or finish; it is within us. The desire to climb - like the desire to explore the world, the oceans, or outer space - is part of human nature. And it will be part of our nature long into the future, this creative ambition etched on stone and ice, then distilled into fire. Climbing, in this respect, is an expression of intelligence, desire, courage, and fellowship. These are very good things to value in the world today, and in the unknown global landscape to come.

www.ukclimbing.com, 2022

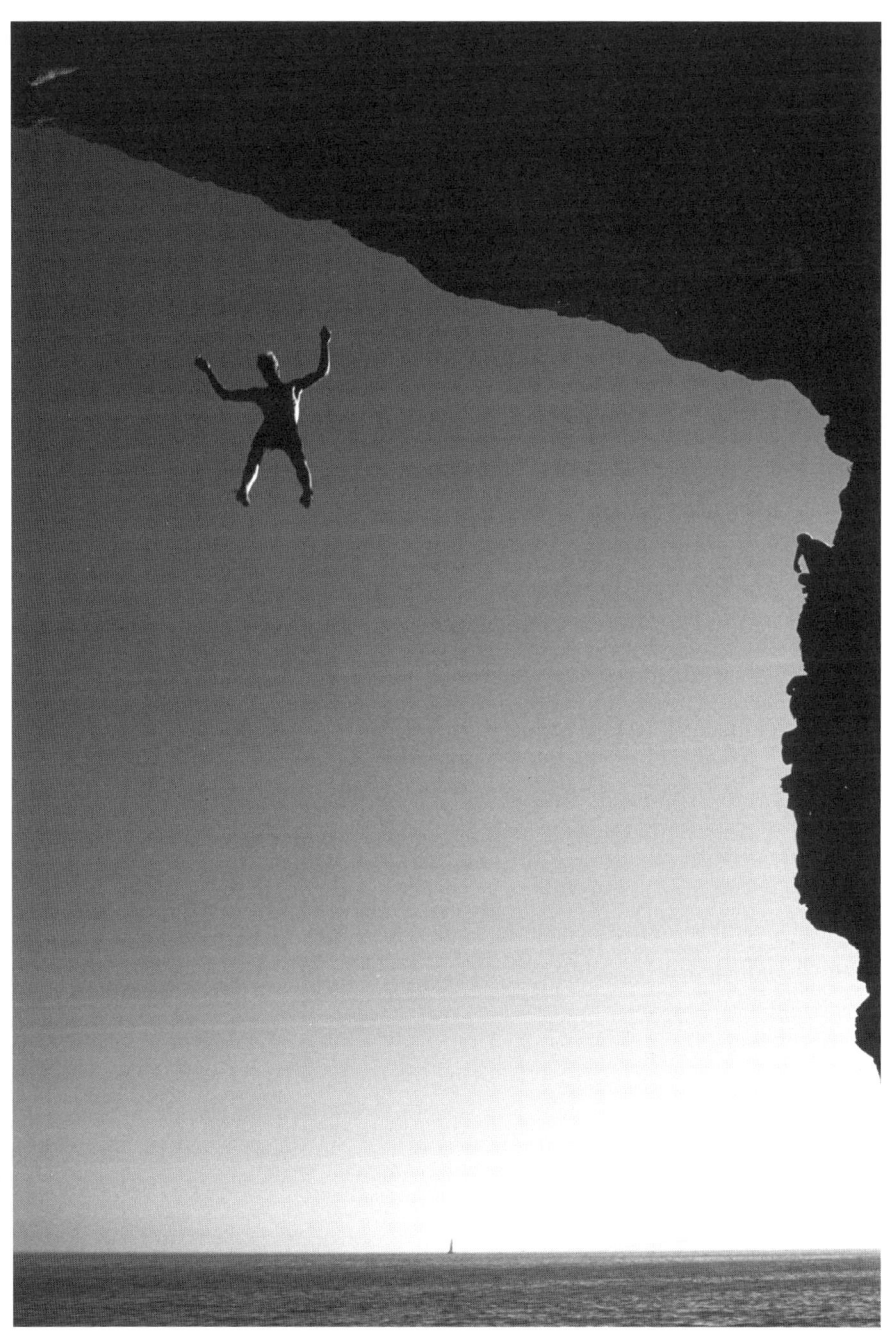

The author takes flight from the crux of Mark of the Beast (7c) Stair Hole, Dorset, in 1998. Photo: Joff Cook.

The Liminal Game

Explorations on the edge of land

It's a blue summer's day, somewhere at the beginning of things.

The beach carves out into the sun. Silence and light merge where the shingle meets the English Channel. An offshore breeze ruffles the wide water of Worbarrow Bay. I run into the sea and swim straight out, with the easy confidence of a nine-year-old boy. The water isn't cold, and refracted sunlight warms my body as I swim. Gulls are flying overhead, diving and turning. The cliffs to the south, over Worbarrow Tout, are gold and topped with tuffs of yellow grass. To the west, the chalk of Flower's Barrow shines in the sun. I can see the land behind and sense the wind moving across the water. Out at sea, away from the noise and haste, I can dream of the world ahead of me. It's beautiful here.

After a while, I look over my right shoulder and notice that the beach doesn't look the same. My idea was to go as far out as I could before heading back to shore. The cross-offshore wind has created a steady current on the surface, which is moving me westwards towards the chalk of Flower's Barrow. I quickly work out that the wind-current is running due west down the beach, and I know that as long as I swim in a diagonal line I will reach the land eventually. I've sailed across it before, and feel I know this place.

I stick to my plan of swimming across the direction of the wind. Sure enough, that distant bank of rust-coloured shingle slowly becomes larger as I close in. My plan is working, even though the wind is stronger here away from the shelter of the cove. I can see a figure on the beach now, his arms waving. As I reach the shore, I roll through the gently breaking surf and feel the slow pressure of the shingle under my toes. My dad gives me a towel and a telling-off: don't swim out like that again in an offshore wind.

But I don't get why he's scolding me. The situation was under control, and I never felt in any danger. Children sometimes know more about the facts on the ground than their parents, as any kid who gets up to mischief knows very well. I understood what was happening in the sea, and felt that the sea understood me, too. That carefree summer's day was the beginning, in a way, of a lifetime's explorations on the edge of land.

That swim more than thirty years ago in Worbarrow Bay prefigured my life as an explorer of sea cliffs and of the sea itself. It also set out why it's a very good idea to respect it. The sea, just like the mountains, doesn't care about anyone or anything. It's a primal force at the heart of nature.

A useful, potentially life-preserving skill for any sea cliff climber is strong swimming ability. If you're going to climb above the sea, you're likely at certain times to enter it both by necessity and by accident. The practice of climbing above the sea is one of the most powerful experiences we can have in a natural environment; it's also one of the least understood.

Why is the sea itself so fascinating? In *The Sea Inside*, Philip Hoare explains why coastal areas exert a powerful existential draw: "The sea defines us, connects us, separates us. Most of us experience only its edges, our available wilderness on a crowded island… And although it seems constant, it is never the same. One day the shore will be swept clean, the next covered by weed; the shingle itself rises and falls. Perpetually renewing and destroying, the sea proposes a beginning and an ending, an alternative to our landlocked state, an existence to which we are tethered when we might rather be set free."

Hoare's proposition configures the sea as a space in which we might attain some kind of spiritual awakening. At the same time, the coast becomes its ritual ground, a place simultaneously sacred and profane where life and death might be blended, reordered, and magnified. It's the ultimate liminal zone; a boundary between land and water, and between life and death.

One of the best ways of entering this zone is through sea cliff climbing, and through one style of sea cliff climbing in particular. Deep water soloing is a very special dimension of rock climbing. The safety net of the water beneath allows us to climb alone and unroped in the most unencumbered style possible. Of course, you can do it only if the sea and rock conditions allow it. To be successful at deep water soloing, just as with mountaineering, you have to be in the right place at the right time. Achieving this can sometimes be extremely complicated, which contributes to the beauty of the pursuit. Climbing is a physical manifestation of Wallace Stevens' concept of "the fascination of what's difficult".

The great thing about deep water soloing is that if you screw things up high above the sea, you'll likely get a chance to fight another day – if you're doing everything right. Some of the most powerful memories in climbing are not the things you did first try, easily within your limit. Rather, they're the climbs that you either just managed to do by the skin of your teeth, or the routes that spat you off in spectacular style and you had to return to realise another day. This is never more true than in deep water soloing, which is to climbing what single-handed sailing is to seafaring: it's one of the most demanding and also one of the best things you can do in the vertical world.

'DWS' really kicked off on the south coast of England in the late 1990s. Back then, I was spending most of my school holidays at my dad's place on the Dorset coast, where I could climb as often as I was able to. In August 1998, age seventeen, I was staring wide-eyed at the 45 degree overhanging wall above Stair Hole's East Cave at Lulworth, Dorset, with a mixture of trepidation and excitement. Pete Oxley's mega-classic *Mark of the Beast* takes this awesome wall head on.

It's a perfect line up a unique piece of rock, and one of the best deep water solos in the UK. I'd climbed quite a few routes of the grade before, but sport climbing is different: up there I'd be on my own. I set off before I could have second thoughts, climbing quickly through the first section; big pulls on big holds through super-steep terrain. I shook out for a few minutes at the half height rest, where you can relax before the upper crux. Then I went for it. The upper section is defined by a series of increasingly powerful moves, the very last of which is the crux. There were a few shouts of encouragement from the ledges below as I climbed higher. I thought the top was in reach – it looked so close. But then, the famous move bared its dragon's teeth: from a couple of shallow pockets, I spied what looked like a hold out left, made a wild lunge for it, and hit something that wasn't a hold at all. Boom! Splashdown.

Two seconds later, I find myself swimming in the English Channel, my friends cheering the fact that I got to the last move on my first try. It was a revelation, and the beginning of a lifetime's fascination with this style of climbing. I realised I could climb a route that was close to my limit without a rope, free as a gull flying across the waves. I was hooked.

A little later that summer, I went back to the East Cave and sent *Mark of the Beast* one evening, as a north-easterly wind blew out through the cave under me, sketching strange shapes across the wide, blue water. Soon, I'd be moving on to the harder line to the left, *Adrenochrome*, one of the finest single pitch climbs in Britain. The fascination of what's difficult never stops.

The revelation of deep water soloing transformed my experience of climbing. From that summer onwards, I continued to push into the twilight zone of my own physical limits and, simultaneously, the psychological height limit at which I was prepared to fall into the sea. The process deepened my powerful addiction to climbing, and at the same time heightened my passion for the sea instilled at an early age while sailing with my dad.

Over the years that followed, I continued to push the edge of the envelope high above deep water. Some climbs were wild, some were foolish, and others were textbook cases of just getting away with it. In 1999, I made the second ascent of *For Whom The Swell Tolls*, another hard Oxley masterpiece at Connor Cove on the Swanage cliffs. The climbing is sustained and technical; the crux, as with *Mark of the Beast*, is right at the top. It's much higher than the Lulworth routes; the hardest move could be sixty feet above the sea at low tide.

I went to the crag alone one evening in July. I abseiled to brush away the thick coating of dusty lichen that accumulates after the winter rains. I swung out quickly along the hanging rail that approaches the edge of the vast sea-level cavern to the west. Inside the cavern, the swell boomed and surged. I reached the point where the route blasts up towards that dauntingly smooth, grey headwall.

I took stock here for a minute or two before pressing on; an oystercatcher flew low over the sea to my right, its shrill call breaking the silence. I used it as a starting-pistol; it's showtime. The headwall was suddenly directly above me. I pulled strongly into the complex series of moves that form the crux. A sequence of layaways and crimps led to the final move: a big lunge for a jug below the final ledge. I could see it, but it looked miles away. I briefly looked down. The sea glinted far below, like a strange beast sleeping beneath the vastness of the wall, its belly rising and falling with the motion of the swell. It's waiting, it seems. Just waiting. It's got time, after all.

But I don't.

Suddenly, my left hand exploded off the intermediate sloper and caught the jug as my left foot popped off a tiny smear. I'd done it. What a route. I was breathing hard and shaking with adrenaline. Sometimes it feels like it shouldn't be possible to climb solo in this style, but it is possible – and that's the thing that makes deep water soloing so addictive.

Anyone can have too much of a good thing, and there's a caveat in all this too. Even in good conditions, deep water soloing – particularly if you're alone – is a very serious business. If you have a bad fall and lose consciousness in the sea, or you end up struggling to exit the water, you're in trouble. In that situation, without a quick rescue even the strongest swimmer can drown.

In early 2004, just after I spent a few weeks establishing the first routes on the tiny island of Laoliang in Thailand's Andaman Sea, I got an email from a friend to say that Damian Cook, one of the pioneers of deep water soloing in the UK, had drowned in Mallorca. I was truly shocked; we'd shared many climbs together. He'd fallen off a route at Cova del Diablo and had been caught in a rip current. It was the end of winter, and the sea was cold; he couldn't exit at the usual place, tried to swim around the point to another landing place, but the current swept him away before he reached it.

Damian was a strong swimmer, and his death was a sobering reminder of the awesome power of the sea, and the fact that deep water soloing is only as safe as the water that lies underneath. If the tide is falling and it's too shallow, or if the swell is too strong to safely exit the water, then DWS can easily be as dangerous as regular soloing. Possibly, in such circumstances it could be – paradoxically – even more dangerous than soloing above the ground, as you may be tempted to a line that's harder than something you'd ever consider soloing with a hard landing below it.

It's a matter of judgement as to what sea conditions you deem to be safe, but one thing's for sure based on my own experience: it is usually much easier to tell if the water is too shallow than to determine is the swell is too big to safely swim out. The sea is unpredictable. And at the base of a sea cliff, where the swell meets the rocks, it's even more unpredictable than usual. All it takes is a big set of waves, and what looks like a safe exit point from the clifftop can turn into a surging maelstrom. The well-known American professional climber, Michael Reardon – who was an ardent soloist – was swept off the base of Fogher Cliff in County Kerry in the west of Ireland by a rogue wave in summer 2007. His body was never recovered. This tragic incident shows that even for the most experienced sea cliff climber, the sea itself may be much more dangerous than the wall you're climbing.

My friend Grant Farquhar and I once had a lucky escape while attempting to make the first ever continuous crossing of the gigantic, 17 kilometre long *Exmoor Coast Traverse* – Britain's longest climb. We were eleven hours deep into our attempt, about three-quarters of the way along, and both mildly hypothermic. It was an hour after high tide, and I suggested swimming a narrow, flooded chasm to the west of Red Cleave to save time. Grant looked at me quizzically: "Maybe best not to, mate," he said. A minute afterwards, a huge set of waves smashed into the chasm, completely filling it with surging whitewater. If we'd been in the chasm at that time, we could easily have drowned.

I've probably used up more than a cat's nine lives in my climbing career, and don't like to take so many chances these days. Every time you go soloing, even above the sea, the risk of a so-called 'low-probability high-consequence event' taking place rises, in the same way that if you sail back and forth across the Atlantic Ocean continuously, the risk of sailing into a hurricane rises.

After my friend Damian drowned in 2004, I realised that a vital rule of thumb for safe deep water soloing is that you should never set out on a climb if you think the sea below it looks too wild to swim in. It's a good idea to look at the sea under the route you're climbing from the perspective of being a weaker swimmer than you actually are. If you've taken a big splashdown from fifty feet after a desperate battle with a crux sequence, you'll be hyperventilating for sure. So your swimming ability will be considerably worse than if you went for a swim off the rocks.

Despite all its potential seriousness, deep water soloing is also one of the most magical experiences in climbing. When you're climbing well, the feeling of flow can be powerful. This is amplified several times over when deep water soloing. It's as if a switch is flicked, and you're suddenly on autopilot. The edge of the envelope is pushed a little further out. You eyeball that distant hold, tense, commit, and make the next move. Then you carry on, climbing higher above that deep blue water, the golden evening light flashing off the waves. It's always beautiful up there, high above the waves.

And if it goes wrong, the sea invites you; at once mesmerising and strange, familiar and utterly mysterious, it is the ultimate landing zone. The sea, as the writer Philip Hoare has pointed out, is where we came from. By falling into it, then, we might be returning home. Embraced by the waves and the refracted light, we are uplifted and set free. Entering the sea, in more than one sense, is like falling in love.

Deep water soloing evolved hard and fast during the 'noughties. Glamorous images of honed guys and girls on the famous DWS cliffs of the UK, Mallorca and Thailand adorned the glossy pages of climbing magazines. People began exploring well off the beaten track; exotic places like Halong Bay in Vietnam and Palawan in the Philippines now harbour some of the most exciting deep water soloing in the world, not least because of the paradisal setting, perfect rock and amenable sea temperatures. Other, even more far-flung destinations will be discovered in time.

One of the most appealing things about DWS is that it can take place on cliffs with absolutely no access other than by sea. In fact, some of the best venues must be accessed this way. In the most extreme cases, such as some of the routes in Halong Bay, the only way to descend the climbs is simply to jump off at a particular point, for example a large stalactite protruding from the wall at 15 metres. There's a unique beauty to this; it's almost a perfect geometric reversal of mountaineering, where a pointy summit marks the top of the climb.

Back home, as the years went by, at some point every summer I returned to soloing, lured back by that magnetic thrill of the dance in the liminal zone between land and water. I pushed the boat out – way too far out sometimes – but always managed to get away with it. Once, halfway along the mega 800 foot traverse of *Zodiac* at St. Govans' Head in Pembrokeshire, I looked down at the thirty-foot waves booming into the boulder-filled cave 100 feet below. Clearly, this was not deep water soloing just because the sea was below me: it was conventional free soloing, where the consequences of a fall are inevitably fatal. (*Zodiac* fell down in a huge rockfall a few years later, but the point remains.)

Other deep water solos on the sea cliffs of Pembrokeshire were equally memorable. One September day in 2001, I raced over to the west face of Mowing

Word, alone. I soloed ten routes, culminating in the mega E5 that takes on the challenge of the steepest, smoothest part of the cliff head on: the appropriately-named *All at Sea*. The tide was falling fast, but I was climbing strongly and confidently. Once I reached the headwall, I glanced down. Big waves were slamming into the base of the cliff, sending spray flying. But what was that yellow colour in the spray? Sand! The tide had fallen so far that the waves were churning up the sandy beach that's only exposed at low spring tide at the base.

I refocused, concentrating hard on the short blank wall that forms the crux. A few sharp pulls on a series of crimps led to a final stretch for a good incut, and the apex of the headwall was in sight. Up here, well over a hundred feet high, I was completely free as I moved alone in a silent, solitary dance under the late summer sun. Deep water soloing becomes conventional free soloing pretty easily on a cliff as big – and as tidal – as the west face of Mowing Word. As long as you can work out where the shadow-line falls between the two, you'll probably be alright.

The art of climbing exploration is, at least partly, about getting off the beaten track; it enables cliffs that would otherwise never be climbed to be explored. Motivated by getting away from the crowd, I've discovered plentiful new DWS routes – and sometimes whole new crags. Over the years I've found great lines all over; that perfect fifty-foot arête in Oman's Musandam Peninsula, rising up from the crystal-clear water like the protruding bow of an Arabian dhow; that crazy, sea-scalloped wall of ultra-compact overhanging limestone west of Loutro on the south coast of Crete, the strange, conch-shell holds forming a natural rising rightwards traverse; the beautiful, gently overhanging pink wall at Clarence Cove in Bermuda that climbed like a waking dream and became, inevitably, *Fifty Shades of Pink*; or the utterly bizarre and brilliant route I found near the North Light on Lundy Island, where I traversed about thirty metres along the retaining wall inside a massive cavern-arch to emerge into the sunlight at a single, protruding hold on the arête on the far side.

The climbing was sustained and superb, and this single hold was literally the end of the road; a lone feature on a canvas of blank, almost jet-black granite. The only options were to reverse the line or jump in: I chose the latter. *The Ungraspable Phantom* was born as I took a refreshing plunge into the Celtic Sea. It remains one of my most thrilling and unlikely deep water solos: a line without an obvious beginning, traversing the interior of a giant cavern, and ending in a no-mans-land of blank granite like a spacecraft cut adrift.

During my voyage around southwest England by standup paddleboard, completed over several summers between 2017 and 2021, I found dozens of new crags along Cornwall's Atlantic coast that would be ideal for deep water soloing, particularly with an approach boat. Many are totally invisible from the clifftop, and only one or two of these crags currently have any documented routes. It simply isn't true that the UK has been 'climbed out'; it just requires imagination and some serious hunting around in the liminal zone between land and sea. There are more new routes to left to climb on Britain's sea cliffs than on our inland crags by several orders of magnitude.

Being creative reaps rewards for the exploratory climber. Back in 2006, a team of us drove over to North Pembrokeshire on a rumour early one morning. A certain Mr. Robertson, one of the key DWS pioneers on the south coast, had found a perfectly-formed barrel of high quality sandstone above a small inlet near St. Davids. It was christened Barrel Zawn, and we climbed around a dozen new routes that day, from some short and sweet bouldery climbs to a long traverse line I climbed into the heart of the cave that featured a sketchy exit, made more memorable by the half-dead pigeon (probably the victim of a peregrine attack) flapping in its death-throws on the sloping ledge after the crux. I pulled through the last moves, and the pigeon squawked one final time before plunging into the sea. *Terrapigeon* thus came into the world, the route's name immortalising the poor bird's fate.

Deep water soloing creates potential for climbing where otherwise there would be none. This particular cliff is a textbook case. It would be useless for trad climbing, as the base is almost always a sea-filled trench and there's not much natural gear. There's certainly no bouldering there, and people don't put bolts in British sea cliffs as a matter of course. Only with the mindset of the deep water soloist can a cliff such as this come into its own.

In 2017, I found a small zawn whilst paddling on the Pembrokeshire coast, close to Raming Hole, that had no recorded routes at all. At high tide, I noticed it had at least three metres of water underneath it; just enough. I did four new solos here of varying difficulty, all of them poised on the seaward edge of the zawn. The lefthand one was particularly good, but I didn't record any of them; I felt they were somehow more special left that way. The cliff is so tricky to find that few people would venture there, in any case. Such is the way of these kinds of cliffs: they are elusive prizes that reward the dedicated and the crafty.

Deep water soloing stands in complete contrast to the classic mountaineering aesthetic: a summit-conquering quest. In opposition to this acquisitive mindset, the soloist might discover perfect limestone overhanging a secret sea cave whilst swimming beneath it, climb a new route completely onsight, and leave the place unmarked and just as mysterious as when they arrived. This is a more appealing aesthetic state, of course, than that of a climber who brandishes a flag on the top of a mountain and takes a selfie.

I had a similar experience in Menorca in 2017, when I found a gorgeous wall of ochre-coloured rock near Cala Morell. I traversed into it and climbed it onsight. It wasn't particularly hard, but the wall had this special feeling of solitude, forming the edge of a small zawn on the outermost point of a tiny island in the middle of the Mediterranean, completely impervious to the yachts and glamorous people in the cove just around the corner.

When attempting harder and higher lines, having company and solid safety protocols are vital. Gavin Symonds, one of my oldest friends and a regular climbing and watersports companion, is a key pioneer of hard deep water soloing in the UK. In 2008, I hung on a rope in Hollow Caves Bay in South Pembrokeshire, photographing him on *San Simeon*, one of the hardest climbs in the region.

Although the route is entirely above deep water, the crux comes at a cool nineteen metres above the high tide mark. That's well over sixty feet. Gavin is a master at the art of the body-angle correction technique needed to enter the water safely in a long fall. I watched him, once again, set up for the big crux lunge up and right to a shallow pocket. As usual for this cliff, conditions were sub-optimal. His right hand flicked out, hovered in a moment of hesitation as his left foot slipped, and he was off. Pirouetting perfectly in the air, he flapped his arms rapidly to attain a vertical 'pole' posture, and locked both hands by his sides as he entered the water like a human bullet.

The height limit I'm comfortable with whilst pushing hard on a deep water solo is definitely lower than the crux of *San Simeon*, and I'm quite happy with that. Some people don't like to swim more than 50 metres to exit the water; others such as myself prefer to solo no higher than about fifteen metres above the water.

In the same way a proficient deep water soloist knows how to fall into the sea safely from height, a skilled racing driver knows how to balance a car in a high speed corner. The two activities are closely linked through a complex interplay between speed, skill, and judgement. There are a number of interesting links between climbing and motorsport; both disciplines involve using high level technique, intense concentration, intelligence and perfect timing in a fast-moving, high-risk environment.

British climber Johnny Dawes has said that a fast car moving through a high-speed corner is "an unresolved work of art in a physical form". The same could be said, I think, of a soloist setting up for a crux move high above the sea. In each act, risk, speed, and motion all combine to create something unique, difficult, and beautiful. My interest in fast cars and high performance driving is linked, I'd guess, to my natural attraction to deep water soloing. The American writer and motorcycle enthusiast Matthew Crawford has explained in his book *Why We Drive* that "to drive is to exercise one's skill at being free". Exactly the same thing is also true of deep water soloing. People with a high-risk, high-reward psyche and a desire to master difficult tasks in challenging environments are likely to be drawn to high-octane adventure sports.

There's a lot that deep water soloists can learn from sailors, too, about operating within a reasonable margin of safety. One of the greatest mistakes in the history of seafaring was the route taken on the Titanic's maiden voyage in April 1912. At the time, the vessel was the world's largest ship, and (unwisely) considered unsinkable. The Titanic's captain, E. J. Smith, reportedly made this remarkable statement some time before the maiden voyage: "I have seen but one vessel in distress in all my years at sea. I never saw a wreck and never have been wrecked nor was I ever in any predicament that threatened to end in disaster of any sort."

Captain Smith's overconfidence ultimately either led to – or contributed to – the Titanic's famous demise. Sailing much too far north in the North Atlantic, she hit a massive iceberg at her cruising speed of twenty-one knots, and sank in less than three hours. 1500 people lost their lives. Captain Smith's overconfidence was possibly a major factor in the most spectacular disaster in maritime history.

If you think you might be smarter than the sea – like Captain Smith of R.S. Titanic – the sea is probably going to take a shark-sized bite out of your arse, and possibly help itself to your head as well.

The American sociologist Diane Vaughan wrote about risk management in *The Challenger Launch Decision,* a forensic study of the 1986 Challenger Shuttle disaster.

Vaughan identifies a process she calls "the normalisation of deviance" as crucial to its evolution. Vaughan shows how the design team conducted analysis to find the performance limits of the solid rocket boosters, but evidence initially interpreted as a "deviation from expected performance" was later reinterpreted as "within the bounds of acceptable risk". This decision had terrible consequences.

The normalisation of deviance could also apply to climbing ropeless above a double-overhead swell, or sailing a very large ocean liner through iceberg-infested waters – which we then get away with. Then, believing it's safe to make the same safety shortcut a second time, we do the same thing again. And again. Repeat this process indefinitely, and something will eventually go wrong; the ship sails into the iceberg and sinks, or the climber falls off – and drowns.

Climbing competence is the concept that you and your partner will be able to extricate yourself from any given situation. If you can't do the route, you have to abseil off it, and so on. In DWS, this is complicated in that you might have to exit the water in a difficult environment whilst fatigued. It's always worth thinking about before you set off into the unknown. Deep water soloing is not necessarily as 'safe' as it's sometimes assumed to be. If the water isn't deep enough – a common problem in tidal waters such as the British coast – then you're in trouble. And if the water is deep enough but the sea is too rough, you're also in trouble.

Mountains and deserts can be very wild, often inhospitable, and sometimes dangerous places. But they don't have the same dynamic quality that the sea has, which makes it so alluring, so magnificent, and also potentially deadly. Tracy Edwards, skipper of the first all-women team to compete in the Whitbread Round-the-World yacht race, memorably remarked that "the ocean's always trying to kill you… and it doesn't let up." If you share this perspective, rather than Captain E. J. Smith's, then you can become proficient at the great game of exploring on the edge of land.

The best deep water solo in Britain, and possibly one of the best in the world, is *Rainbow Bridge* at Berry Head in Devon. It's a traverse roughly 850 feet long across the most perfect, compact, technicolored limestone imaginable. There is barely a single loose hold along the entire thing. It features a tough crux section about a third of the way along, but most of the climbing is glorious jug-pulling above deep water. Even at low tide, it's still safe to climb it.

I first climbed *Rainbow Bridge* in 1998, when I was seventeen. It was a perfect late summer morning and the whole wall was bathed in blue and green light reflecting off the sea. After I left my friends on the clifftop, there was nobody else around; I was alone on the best piece of rock in southwest England. Gulls and a few oystercatchers floated on the water. The climbing was even better than I'd been told it was. Once I reached the big ledge after the Terminal Zawn, I tied my chalk bag around my head to keep it dry and swam back to the beginning. That afternoon, Mr. Robertson and I did *Caveman*, the spooky trad climb that takes on the challenge of the huge cave of Old Redoubt head-on. The combination of these two completely different routes made for one of the best days of sea cliff climbing I have ever had.

In the late summer of 2022, I returned to Berry Head on a cool September afternoon as cloud-shadows moved across the water. I realised I hadn't actually climbed the whole of *Rainbow Bridge*, start-to-finish, since I first did it as a teenager. I'd been back to the cliff quite a bit since then, but for some reason I hadn't repeated the entire traverse.

I clambered down the descent line, and set off across that gorgeous pocketed wall before the Pink Grotto, the strange sequence of upside-down solution pockets both familiar and at the same time distant. As I climbed, I thought about Philip Larkin's point that "truly, though our element is time / We are not suited to the long perspectives / Open at each instant of our lives."

The route was still here, exactly as it had been. I, of course, had changed. That's the thing about climbing; the routes stay largely as they are, but we don't. I thought about the blonde-haired boy who'd barrelled along here that late summer morning at the end of the last millennium. Who was he?

I still knew him, of course, but he was far away in time and space, adrift in the twilight realm between memory and forgetting. I quietly admired him, though. I smiled when I thought about his endless enthusiasm for climbing, his courage and resilience. As Geoffrey Hill said of his younger self, "The boy I was / shouts Go!".

The even younger boy who swam far out in Worbarrow Bay in an offshore wind that summer long ago prefigured the exploratory climber I'd later become. I made it back to shore that day because I didn't panic, understood the sea, and trusted myself at the same time.

Reminiscence, of course, wasn't going to help me get across *Rainbow Bridge* without going for an involuntary swim. I knuckled down and made the slight descent to the crux pitch. It's all slopers and awkwardly-placed footholds; it was trickier than I remembered from twenty-four summers ago.

Perhaps that's one of the things about age; we forget how much easier some things were with the lightness of youth, and also how much harder life was back then. Philip Larkin, again, reminds us of "the strength and pain / Of being young; that it can't come again / But is for others undiminished somewhere". In climbing, as with so many things that involve risk and reward, Larkin's gnomic statement is profoundly accurate.

Then, all too soon, there was a brief tussle with the actual crux – a deceptive hand-match on a weird diagonal hold – and suddenly the freedom of the groove that leads up and out to the next section. I'd crossed the mid-point of *Rainbow Bridge*. When I first passed this way I was at the beginning of things; today, I was somewhere in the middle of my life. What had gone before would not come again, and I couldn't know what might happen in the years to come.

If I climbed this route again in another twenty-four years, I suddenly realised, I'd be approaching old age. I'd be exactly the same age as my grandfather was when he died, quite unexpectedly, one late summer afternoon. An afternoon perhaps not so different to this one. So what? I just hope that if I make it that far, I'll still be able to find my way across this most beautiful of climbs one more time.

Continuing on, I took the high variation before the Terminal Zawn, swinging footless and fancy-free along that glorious line of widely-spaced holds, my feet arcing out above the sea. The swell sucked and hissed in the unknown ventricles of the cliff's deepest heart.

We cannot know what is deep within those places, those submerged clefts in the underwater stone. They are the dark doors to a shadow-world beneath the sea that we may never enter, marking the strange boundary between what we can see and what we can only conjure up. The boundary between knowledge and imagination is a watershed we cross again and again in the vertical world, yet we come closest to this threshold, of course, on world class climbs like this one.

The way was now clear to the Terminal Zawn. This final section of traverse is different to the rest of the climb. The movement is fine but the rock quality isn't quite as good. And the very last section – a short down-climb at the apex of the zawn to the big ledge on which the route ends – is actually quite dangerous. It's as if *Rainbow Bridge*, in its final metres, wants to remind us of our own mortality after the surreal, serene, almost otherworldly climb we have just completed. I climbed that last section carefully – very carefully – and took the final step down to the ledge.

I breathed in deeply and exhaled. The climb was done.

Climbing is much more than just a sport. It's a kind of ritual, a spirit-quest, a journey into the heart of your life. It could even be an art form. Deep water soloing is a style of climbing that comes closest to all these things. Like an astronaut experiencing weightlessness for the first time, with the gift of water beneath their feet the soloist enters a new realm in which the conventional laws of physics no longer apply.

Like Peter Pan in Neverland, climbers moving above the deep may transcend place and time, remaining ageless as they strive to master the shadow-world between the land and the sea. The great polarities of human life begin to swirl around us in this most sublime of earthly regions: love and loss, success and failure, innocence and experience. They are written down, these stories – all of them – in vanishing print on those imagined lines in the rock we climb, in the faces of our friends on and off the wall, and in the constant churn and surge of the sea beneath it all.

Here, amid the edgeland between earth and water, we are possessed by the action of wave over stone, of wind against tide, of refracted light flooding the darkness of the chasm we enter. Here, we can find out what it means to be truly alive. And here, too, we can grasp with crystal-clear certainty that one day, some day, we will die.

This is the liminal game.

www.ukclimbing.com & The Blue Cliff, 2023

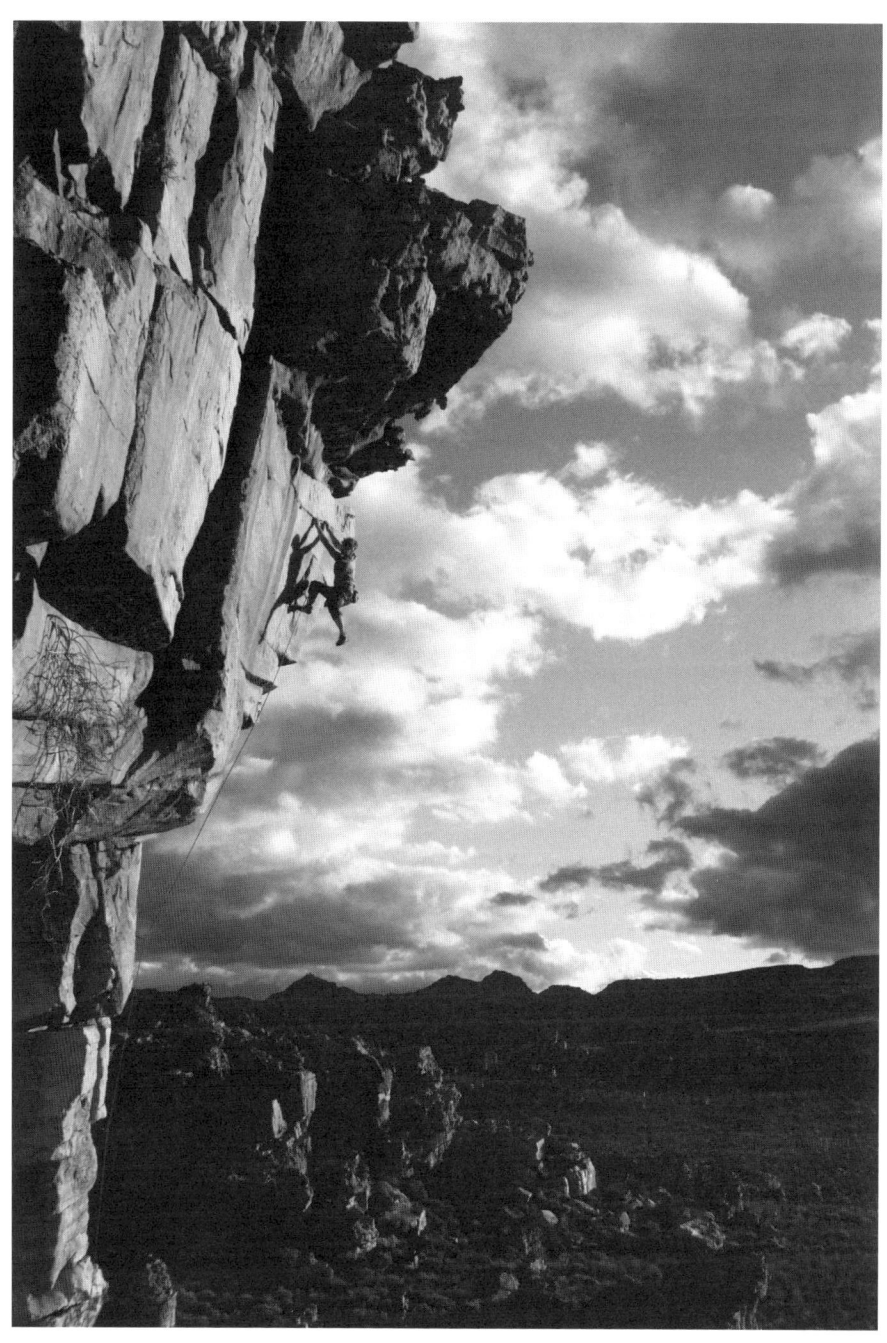

The author on a striking line (7c) at Truitjieskraal, Cederberg mountains, South Africa. Photo: Ramon Marin.

The Heart of Nature

Climbing as a paradigm of the human condition

In the dream, splinters of starlight surge from the summit ridge. After a while, they burn out and are lost in the desert night.

"Andy, you awake?"

"Uh-uh...Yeah."

An hour before dawn, February 2005, I sit upright in my sleeping bag under a gnarled acacia tree, by a smouldering fire at the edge of a goat paddock in Oman's Hajar mountains.

Six thousand feet above, the blue dark of the lukewarm winter night is broken by the outline of a serrated summit ridge like the upturned spine of a stupendous primeval beast. Below the crest, a wall of pale limestone plummets more than 3,000 feet to vast scree cones.

Jebel Misht is the biggest cliff in the Arabian Peninsula: a massive, topographically complex and often loose concatenation of interconnected walls and towers, extending for almost two horizontal miles above the date palms of Wadi Al Ayn. Our intended line runs up a system of jumbled pillars and sinuous grooves on the southeast face, crossing two existing climbs to gain a sheer 200-metre headwall: a remote, smooth, ethereal, almost delicate thing.

After a quick breakfast of coffee and dates, we pass the small prayer room on the south side of our host Mohammed's garden just as the Muezzin blares out on his tinny transistor radio. Beyond the edge of the rickety barbed wire fence, a ravine rises quickly toward the scree mounds of Jebel Misht; in the purplish predawn light, they look like colossal avalanche cones, dark with grit and latent menace.

The shadows of two skin-thin dogs follow us. As the approach steepens, they fall away into the breaking light. A mile and a half away, the sun bursts across the immense north flank of Jebel Akhdar, the highest peak in the Arabian Peninsula. Synclines and anticlines rise from the fading gloom across the corpus of the world in swoops and whorls, the visible memory of the colossal tectonic shift that made these mountains. As the rising light hits the west summit of Akhdar, a pink cloudburst detonates across the pale grey stone, quickly draining down to iridescent red.

Five hundred feet above us, a solitary raven croaks twice.

"He's telling us to get a move on, mate," Andy says. "All this sightseeing isn't going to get us up there."

Andy and I look slowly up the face, taking in all the features. It takes twenty seconds, this way, to trace a line from base to summit. The cliff is truly huge. If you put El Capitan next to it, the world-famous granite monolith would look smaller.

We rack up at the apex of a narrow shoulder. Even though it's only 6:30, we climb in T-shirts. Heat radiates off sun-blasted wadis and fluvial plains. Ahead, the shallow cracks become more and more vertical. It's the kind of rock that makes you check every hold, as if to test the depth of your judgment: the siren lure of exfoliating fissures that lead nowhere; the sudden, visceral recoil from the boom of hollow flakes.

Two pitches up, I breach a roof, finding solid cracks between wobbly blocks, and I make a belay in a shallow alcove. Andy follows quickly, bright-yellow and bright-pink ropes looping down when I can't take in slack fast enough. All at once, the ropes fly out and my belay plate locks hard. A block the size of a microwave oven arcs into space. Andy swings out, a metre from the roof. As the rock hits the talus, a dull explosion echoes around the face.

"Shit" he yells. "Seemed solid...." He's breathing hard, visibly shaken, at the belay. I offer to take the next lead. Tiny dihedrals peter out into nothing. Sinuous cracks weave innumerable blind alleys across a broad pillar where the pale limestone merges with the milky haze of the desert sun. I run out our sixty-metre ropes to a narrow pedestal. Tendrils of high cloud build in the morning sky.

We creep through a vertical maze of narrow pillars balanced precariously one upon another like a spiral staircase of giant Jenga towers. On a long, slim ledge 350 metres above the scree, we drink water and share dates and apricots. A thin film of cirrocumulus softens the rust and tungsten swirls of the stone contours across Wadi Al Ayn. After a while, Andy heads toward what we think might be the deep groove and crack system of Geoff Hornby's 2001 route, *Intifada*. We round a corner, and *Intifada's* impeccable crux dihedrals tower above: an alien refuge of smooth, solid rock amid this vast citadel of choss.

I bridge, arm-bar, and jam through the steep upper section of a split chimney which is actually one of the best bits of limestone corner climbing I've done. Above, we move together across the base of a gargantuan amphitheatre. With our light rack, we might not be able to retreat down this fragile, disconnected wall. There's no one here to rescue us. If either Andy or I had a bad accident, we'd likely die here.

Ten metres to my right, a falling stone makes a weird whistle as it flies past. A lost rider on the mountain's silence. I flinch instinctively, thinking about Alex MacIntyre, one of Britain's best alpinists of the 70s, who was killed by a single falling stone on the South Face of Annapurna in 1982.

Then I check the time: 3:45 p.m. Night falls at 6 p.m. sharp in the Arabian winter. Up here, the temperature will drop like a stone. We are a pair of spiders hanging in the atrium of a gothic cathedral. Just two minute figures poised in a vast cauldron of turreted rock; this high place is truly a crucible of the wild.

Shivering in my windproof, late afternoon clouds swirl around us. On the best rock we've yet encountered on the wall, I run it out between cams in horizontal breaks up a seventy-metre band of compact, weatherworn stone: a mantelpiece of solidity on the crumbling edifice of the mountain. Somewhere behind me, a sharp, solitary croak cracks through the light wind and resounds: the raven has returned.

We reach the summit ridge after eleven hours of continuous climbing. Red light slices through the cloud above Jebel Akdhar, a slow-motion laser shifting the spectra of evening. Eighty miles to the north, the low, elongated caul of Jebel Fahud rises from the sand like an emergent Kraken on the surface of an ancient ocean. As our eyes trace the eastern edge of the Rub' al Khali, we can see the curvature of the Earth. This is the fabled Empty Quarter, the world's largest sandy desert, which stretches some 650,000 square kilometers across the Arabian Peninsula.

Scree-sliding down the long, complex descent, navigating through cliff bands and around maze-like gullies by headlight, we're both lost in our thoughts. The descent is complex, with lots of places to seriously screw up if you go the wrong way. Seventeen hours after we'd set out from the goat paddock outside Mohammed's garden, we reach the dirt road that runs parallel with the edge of Jebel Misht's northern flank.

Despite the astonishing shift that's taken place in Oman over the past century – a people transformed from desert nomads to the relatively affluent citizens of a petro-state – these mountains have remained largely unchanged. Later that night, before I fall asleep, the chromium slice of a crescent moon rises over the jet-black spikes of Misht's summit ridge. A few degrees to the south, Orion is travelling across the lunar terrain of Akhdar's highest slopes: a bright barb hooked somewhere, somehow, on the outer edge of our infinite world.

The new line we climbed that day, *Inshalla Salam*, was certainly not the best big wall I've done. It was discontinuous and loose in places, and it incorporated sections from two existing routes. Even so, it was one of the best days of pure climbing adventure I've ever had.

Afterward, I wrote a couple of short features in the British press about climbing in Oman, but I never wrote anything about that route. All that was recorded was the following one-line entry by Geoff Hornby in the 2007 Alpine Journal: "On Jebel Misht's south east face, David Pickford and Andy Whittaker linked together *Intifada* and *Eastern Promise* and added a 100m finish up the tower to the summit ridge to provide *Inshalla Salam* (1000m, 5.11 R, ED VII+)."

I don't have any photos of the climb, as I had run out of 35mm slide film the day before (back in 2005, many outdoor photographers were still using 35mm film). Yet nobody questioned the veracity of our ascent, mainly because it was a moderately difficult climb on an obscure cliff in Oman, of interest to only a few. But there's another reason nobody asked us to "support" our ascent with "evidence": in 2005, Facebook and Twitter were not yet part of our cultural mainstream. Back in the mid-noughties, as a former editor of *The Economist,* John Micklethwaite, put it in his farewell editorial, "social media had something to do with a very good lunch."

An extraordinary phenomenon, one that's gone largely unquestioned in the vertical world, has taken place over the past decade as a result of the explosion of climbing imagery and videography freely available through social media: the way we document, share and process stories about our activity has shifted. The great majority of people in the developed world now have a smartphone, and thus a camera. If we'd climbed *Inshalla Salam* yesterday, our account might be regarded as spurious. The demand for visual evidence of ascents is not unique, of course, to the digital age: in 1910 Herschel Parker and Belmore Browne famously disproved Frederick Cook's Denali claim by matching a picture from a lower peak with the one he described as his summit photo. Yet in recent years, the sheer saturation of digital media has assumed an unprecedented level of control over the making of climbing history. Often, the documented image now appears to confirm the existential event, far more than the words or the memory of the climber himself.

This conundrum represents a profound conflict between the between the direct experience of a climb and the alternative narratives of its representation, between the actual and the perceived. The notion that an ascent can't truly be captured in words or images is a longstanding theme in climbing literature. Any form of representation in any medium obscures reality to a certain extent. Multiply those forms exponentially, and do you get something closer or farther from the truth? "The digitised discourse is more complete," explains the Canadian climber Michael Down, "but it can also seem more inaccessible, with so much noise, so much chatter, reams of it, layers and layers."

Now and then, a clear and brilliant voice, hitherto unheard, breaks through the din. In April 2014, Jemina Diki Sherpa's blog post, "Three Springs," a response to the deaths of sixteen Nepali expedition workers on Everest, helped tear away foreign fantasies associated with Sherpa climbers, revealing the realities of their lives to readers around the world. At times, writers working outside the fray can thus use digital media to shatter the myths of more dominant groups.

I belong to the last generation to have grown up with books and newspapers as my primary source of written tales. As such, I hugely enjoy the experience of uninterrupted, solitary reading. I still love print; and I love great stories. Against the natural interest in narrative, 'content' has become a somewhat sinister, slippery word in journalism today. It often comes to denote the uneasy but potentially profitable no-mans-land that now exists between real journalism and content marketing. The rise of AI language models only complicates this further, making it even easier to generate readable, bland content about anything, anywhere, extremely quickly.

But what happens when content generation almost becomes the story itself, as is increasingly common in the new media landscape? The account of the first free ascent of the *Dawn Wall* is now inextricably bound up with "the story of the story" of the route – the reporting by the mainstream press, the almost real-time retelling online. Tommy Caldwell and Kevin Jorgeson's regular updates from the wall gave extraordinary momentum to the snowballing of the news. Caldwell's post on January 10, 2015, sums up the remarkable power of social media for the instant transmission of climbing history: "The last few days have been some of the most memorable climbing days of my life. Yesterday I finished the last two 5.13+ pitches.... I kind of lost it when I pulled onto Wino Tower, knowing that this seven-year dream is looking more and more like it could become a reality."

Dawn Wall is a truly great climb: the hardest route on the most famous big wall in the world, a sidewinding dance up the steepest and smoothest part of El Cap. As many people said, the *Dawn Wall* coverage represented one of the rare occasions that mainstream media attention focused on a legitimately important climb, instead of on the latest 8000-metre peak disaster or made-for-media expedition. Corey Rich's dark, spooky shots of Caldwell and Jorgeson climbing some of the hardest pitches by headlight will be remembered as classics of twenty-first century climbing photography.

Yet my suspicion remains that the most meaningful ascents, in a purely human and private sense, are those that are not supported or validated by countless images, helmet-cam videos or professionally shot footage. Instead, they're the climbs that take us into the heart of nature, and ourselves. As a photographer myself, I'm acutely aware of the paradox. The majority of my best climbing experiences were those in which the distraction of a camera was not even present, allowing me to climb to the full in the presence of wild nature, uninterrupted by the burden of documenting what I was doing.

It is true, at least to some extent, that climbing in the digital age has become more meritocratic and democratic, since anyone can today create publicity around their ascents. Yet in the process, the value of the narrative and the importance of the subject can become secondary to its sheer proliferation. Climbs that are easily hyped up by ascensionists or sponsors sometimes gain a level of coverage disproportionate to their actual significance. Groundbreaking routes may be left almost completely unreported, such as Norwegian climber Sindra Saether's astonishing 2010 all-free ascent of *The Arch Wall* on Trollveggen. No one's really heard of this, because Saether doesn't like publicity, and the route was barely reported.

Increasingly, I enjoy the priceless inner solitude that climbing offers me in a world in which digital technology is an almost constant interruptive force. In contrast to the incessant sound and fury of the digital age, climbing provides us with something utterly real and profoundly valuable: those moments of undistracted connection with the physical world that we each encounter during our best days on the rock and in the mountains. Now, as ever, our pursuit can remain as private or as public as we want it to be. You can climb the plumb line on a busy crag on a Sunday afternoon and update Instagram from the parking lot, or you can solo a secret route hours from the road and not tell a soul.

Perhaps because I once edited a climbing and mountaineering magazine, *Climb*, for many years, and I still spend a lot of time listening to climbing stories, I get the strong impression that the more grassroots segment of the climbing community is increasingly interested in people who don't shout about what they do. Out there in cyberspace, there are tens of thousands of two-minute video-blogs of teenagers climbing hard boulder problems to a tech-house soundtrack. The thing is, most of these stories aren't that interesting. But the tale of an unknown girl or guy who quietly solos a remote, interesting line miles from a road, leaving only fleeting trails of chalk or axe and crampon marks – now that's an interesting story. It has all the key elements that make for a great narrative: mystery, uncertainty, enigma. But then it probably won't get told, and maybe it's better that way. Does the telling of an experience dilute how true it is? Does it change what it means?

During a trip to South Africa in 2014, two friends and I climbed at Truitjieskraal in the Cederberg Wilderness Area on sandstone that gleams like the scales of a fossilized blood-orange dinosaur. The crags lie in the rock-strewn, wind-enchanted country between the mountain ranges of the Wolfberg and Tagelberg. Narrow dirt roads cross a landscape of red stone, dry grass and colossal skies. In the late afternoon, rising pillars of cloud built up over the Tafelberg like fire-blackened Doric columns, casting violet shadows over bunchgrass plains.

We didn't see anyone else there: the high Cederberg is a truly wild place. The second morning, fresh leopard tracks and droppings appeared in the sand. The big cat had surely passed this way overnight. That evening, as the light was falling, we came upon a shallow cave adorned with painted human and animal figures. Nobody knows exactly how old they are, but we do know they were made by the Khoisan, the indigenous inhabitants of this part of Africa, quite possibly long before the first white men ever set foot in the Cederberg.

An hour after sunset, a huge full moon rose over Rocklands to the east like an orb of fired glass, globular, opaque and pale-bright. It was one of those visions that, years later, you'll recall in a moment of idleness: a thousand sandstone towers washed in pallid, spectral light. A few strands of high cirrus black against the inky blue air. Orion was sloping off the Tafelberg. The Southern Cross burnt high and bright overhead, a tall rider on the infinite dark.

It grew cold quickly. After a while, my friends walked back down from the huge, flat-topped boulder where we'd been sitting. Just for a while longer, I sat and listened to the night. There was no sound at all except the whistle of the light wind in the dry grass. After a while, I heard what sounded like a footfall. It was very quiet, like the soft thud of a large pebble dropping in a sandpit. There was another, and then another. I slowly turned around.

Two wide, bright green eyes hovered in the long grass, perhaps thirty feet away, or maybe closer.

The leopard was watching me. He'd been watching, I think, for a long time.

No more than three seconds after I rose up from the boulder, I heard him dart away into the shadows. As soon as he vanished, an unfathomable emptiness filled the night air. The east wind rose slightly, rustling the dry grass, and a beautiful, elusive ghost floated across the moonlit Cederberg.

It was time to leave.

On the plane home a week later, I looked through the photographs I'd taken in the Cederberg. There were some brilliant shots, as the place is incredibly photogenic, but none of them captured the essence of the experience. Similarly, Hermann Buhl's grainy photo of his ice axe, shoved into the snow somewhere on Nanga Parbat's trapezoidal summit rocks in August 1953, doesn't tell us much about the landscapes of his mind during that lonely summit push. It's very hard, and maybe almost impossible, to capture the visceral experience of climbing through photography and videography.

If I'd tried to photograph the Earth's curvature along the Rub' al Khali from the top of Jebel Misht in 2005, I don't think it would enhance in any way the memory I have of that huge desert wall; the existence of such an image might in fact change the way I remember the climb itself. And perhaps if I went back and climbed another line on that wall today, taking two hundred pictures of the route on a digital camera as I climbed, I might not recollect much of it at all.

"Whenever we use a tool to exert greater control over the outside world, we change our relationship with that world," writes Nicholas Carr in *The Shallows*; we can lose of some of our original abilities of relating and thinking. In the same way that the traveller who takes an endless stream of selfies can't experience a place at the same level as one who doesn't, there's a possibility that the more photographs we take and videos we shoot of the climbs we do, the less we might remember of those routes.

If this is even partially true, it's important to keep some climbs for memory alone. Secretly and silently, without evidence or epithet, the coiled spring of those compressed hours on Jebel Misht almost two decades ago jump back at me from the past, as if they took place only yesterday. The route we climbed on that huge desert wall has been distilled into the high backcountry of memory; a unique, precious, and singular moment in my life.

The powerful experience of something entirely private, pure and free is one of the reasons I first chose to climb, and that I continue to do so. A sudden gust of wind catching a rope as I fling it from a desert tower. The first sight of an unclimbed wall shining bold and black above a fjord. An asteroid of light on the ice as I strike my axe. The raven circling overhead, somewhere on a massive wall. Moments lost in time that are part of who I am.

I love the athletic quality of climbing, sure. But this aspect alone is an insufficient explanation for why this pursuit is such a powerful presence in my life. Perhaps the real reason is that it is through something like climbing I might come closest to that immense, inexplicable force at the heart of nature; to the sea-green discs of a leopard's eyes burning back at me under a star-emblazoned sky.

Alpinist magazine, 2015

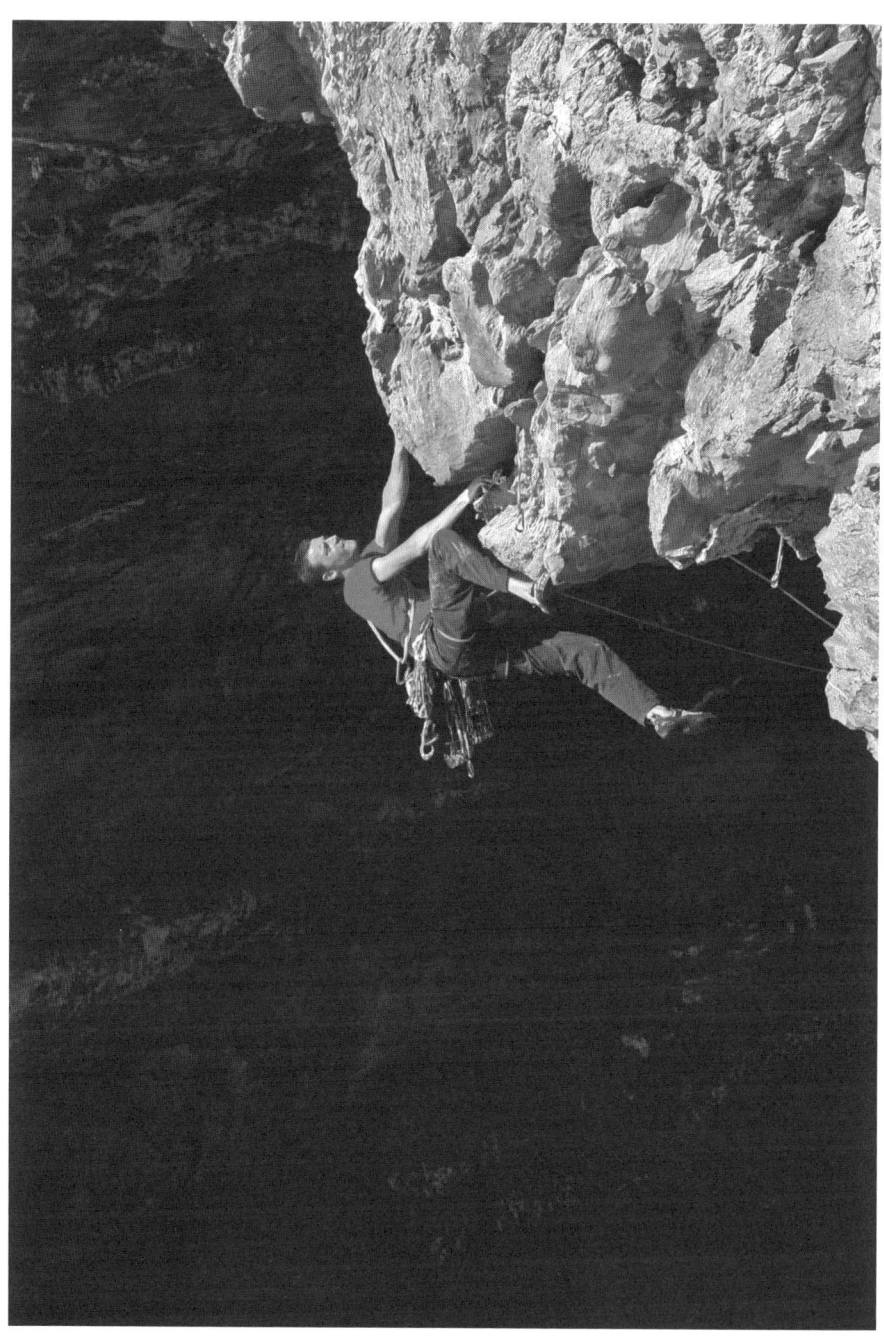

Jack Geldard making the second ascent of Pre-Cambrian Wrestler (E7 6b) at Penlas Rock, Gogarth, in 2007.

The Full Welsh

Reflections on Gogarth

August 2000. I'm a 19 year old rookie, and I've just hit Gogarth for the first time. Peering over the top of the crag, my first sight of Wen Zawn was a bit like my first proper toke on a spliff, only a whole lot better. The headrush from the crashing surf amplified by three hundred feet feet of concave space overwhelmed me. Gulls flew and wheeled and dived. When I was 19, climbing was all I really loved, and all I really loved about climbing was right here in front of me.

As I would come to realise, Gogarth is a place for those who seek to experience everything the vertical world might offer; a place for those who are hungry for it, for those who want the loose rock, the big runouts, and the sketchy topouts.

It's the Full Welsh.

The last day of that brilliant week in North Wales, with local expert Glenda Huxter as our guide, my friend Kevin Avery and I headed down to North Stack Wall. To a kid who'd spent too much time reading 1990s climbing magazines and pouring over the pages of *Extreme Rock*, this crag was the philosopher's stone of British trad climbing. After following Kev and Glenda up *Blue Peter*, I spend an hour swinging around from an abseil rope on the blank headwall of Andy Pollitt's terrifying masterpiece, *The Hollow Man*.

Perturbed by the almost complete absence of any protection worthy of the name on the upper crux sequence, I decide to leave it well alone for the day, and finish up with an onsight ascent of the epoch-making route of the wall, and the route that perhaps best defines the anarchic, maverick, punk-rock spirit of 1970s British climbing: *The Cad*. If a rock climb could wear flares and sell you LSD from a Sunbeam Rapier, it would be this one. First climbed by Ron Fawcett with the controversial use of a bolt to protect the crux, which comes high on the route, the bolt has now rusted away to nothing more than a dark red shadow on the pale quartzite, but the crux is still exactly where it was when Ron first climbed it all those years ago.

The sound of the crashing surf on the boulders towards the seaward edge of the wall anticipates the incoming tide as I sprint up the initial, easy section to the big flake at half-height. I spend a few minutes here, placing as much gear as I can in the flake, glancing nervously at the white face vanishing overhead into the middle distance. In the thundery late summer light, I can see narrow strands of yellow and green lichen dancing against the pale stone, enticing me upwards.

After a while, realising I can't put any more gear in the flake, and that there's no way the crux is going to get easier just by hanging around, I set off into the unknown of *The Cad's* decisive final runout. All I knew is that a fall from the hardest moves, high on the wall, could land the unlucky leader on the boulders far below – at least on rope-stretch – which from ninety feet is not a welcome prospect. Committing to a route like this is a bit like hitting the throttle of a race car on the final straight, holding pole position but not knowing how far behind the other drivers are.

I suddenly arrive at the crux. Below it, there were long moves on small, positive crimps on which I spread my weight like a waterborne insect holding the surface tension. Then, suddenly, nothing. The series of holds that lead here, it seems, simply run out. There's a couple of decent footholds, about half the size of a matchbox lighting-strip each. But for six feet above them, there's nothing but blank, pale grey quartzite. So this is it, I think to myself. This is why I came here. To be right here, right now, high on North Stack Wall, trying to figure out what's going to get me out of here alive.

I shuffle around on the footholds for a while. The half-matchbox-lighting-strip edges seem to grow slightly smaller as I shuffle. Time passes, and I scan the rock above me, looking for clues.

What seems like a long time later, but may have only been five minutes, I realise there's a good diagonal crimp hidden in the lichen just half a metre out of reach, then a good edge another metre above it; a single long move, trusting everything in faith, balance, and friction, is all it'll take to reach it.

I shut my eyes for a couple of seconds, trying to relax on the half-matchbox-lighting-strip edges. When I open them, a gust of wind whips across North Stack Wall, and I step up and reach for the diagonal edge.

Two minutes later I'm pulling over the flat clifftop to the sound of evening gulls and the flash of North Stack Lighthouse sparking up for another night. The offshore wind is blowing big swoops and whorls in the calm water of Gogarth Bay, marking my now invisible passage across this wall in its own indecipherable language. As I shout down to Glenda and Kev that I'm safe, the shadow of Wen Zawn reaches far out into the sea, darkening as it extends into the west.

Now, a very long time later, *The Cad* still makes a strange signal from the hinterland when I think of it. Like a lone flare from a unknown ship, this audacious climb – like Gogarth itself – remains a wild thing beyond time and understanding.

Penlas Rock juts out into the Irish Sea at a strange angle, pointing towards the sky at forty-five degrees then slinking down into the waves. From the top of Yellow Wall, it looks like the dismembered bow of a stricken ship recently run aground. It's a piece of geology so clearly in the process of fragmentation at the hands of the sea that the very idea of climbing on it seems, at least initially, a bit ludicrous.

It was midsummer 2007 when my friend Jack Geldard proposed the idea of going down there to try and repeat George Smith's creation *Pre Cambrian Wrestler*, an inverted masterpiece from the master of upside-down North Wales climbing that had lain dormant since its creation in the late 1990s. Being a fan of super steep routes, and a Gogarth addict to boot, I could hardly refuse Jack's suggestion.

The abseil descent to Penlas confirms the strangeness of this unique crag. From belays atop the rock, we descend a sixty metre, slabby wall of greyish, talcum-powder choss and sea grass: looking around on the way down, there's no evidence of any crag here at all. It's only when we arrive on the sea-washed platform that slopes towards a tell-tale trench at the very base of the rock that the shadow of a vast cave, seeming to penetrate into the bowels of the cliff, begins to take shape.

It's only when you're underneath it that the scale of the feature becomes apparent. The back of the cave, at its deepest point, must be at least thirty metres from the outer lip; the walls above the roof itself overhang between sixty and forty degrees.

Technicolour bands of stone swirl across the ceiling, merging into one another like brush strokes from a deranged painter. *Pre Cambrian Wrestler* takes the only line of weakness through the right hand wall of the cave, beginning up a small hanging corner before breaching the roof via an inverted slot, then traversing out right across the lip and up the hanging wall above. On the recommendation of the guidebook, we've left a hanging rope at the end of the pitch, where the angle changes and the rock deteriorates back into talcum powder.

Jack heads up first, using a mixture of free and aid climbing to figure out the line. The huge rack of gear on his harness quickly diminishes as he reaches the upper wall. Superlatives and expletives drift down in the sea breeze. Jack and I have time for another go each with the beast, working the moves thoroughly: this thing just climbs so well, I think, as I piece together the wild headwall to reach our hanging rope belay. Lowering off, I'm pretty gassed; one glance at Jack confirms he feels the same way. We decide to leave the gear in place and return tomorrow for a proper shot at it. Because of the extreme steepness of this route, which overhangs more than twenty metres by my calculation, removing all the gear for a lead attempt after working it is completely impractical, so we make the choice of the style of our ascent through necessity rather than design.

The next day dawns fine and clear, with a fresh westerly breeze promising perfect conditions in the cave. We arrive around midday and descend the ab rope just as the sun is coming around. Our gear swings in the breeze, the karabiners chinking against the rock, enticing us up. I feel good today, and after a quick warm up traverse along the jugs on the back wall, I tie in and go for it. With some of the gear pre-placed, the crux roof section goes by surprisingly easily; placing the gear on lead would be a different story.

The headwall proves much more pumpy than expected, the typical Gogarth sloping breaks are particularly unhelpful in terms of trying to recover after the hard crux section. I spend a few minutes shaking out in an awkward alcove before the final hard sequence before the end of difficulties. I'm glad I did: I arrive at our hanging rope belay with absolutely nothing left in the tank. The pitch is a real monster, a primeval leviathan of a climb that more than lives up to its name.

Jack makes his own ascent immediately after mine, dispatching the pitch with style and efficiency. After he lowers off, we then spend the rest of the afternoon removing the gear from the extremely steep lower section, a task which is almost as hard as the actual climbing. With the pitch cleaned of protection, all that remains of our epic two-day duel with *Pre Cambrian Wrestler* are the white dabs of our chalk on the crucial holds, the technicolour swirls of the immense cave roof marked by our passage like an Aboriginal dot-painting.

Soon after we depart, of course, these brief traces of our presence will be washed away by the next storm sweeping in off the Atlantic and across the Irish Sea, returning this place to its secrecy and wildness.

The route we've just climbed has waited a decade for its first repeats. And perhaps, I think as we pack the gear and leave, Penlas Rock won't be visited by any climbers again for a few years. It's hard to get to, tough to find in good conditions, and there's no easy way out. Perhaps there's a sense in which the most memorable crags, like this one, are also the most elusive.

The White Cliff, 2018

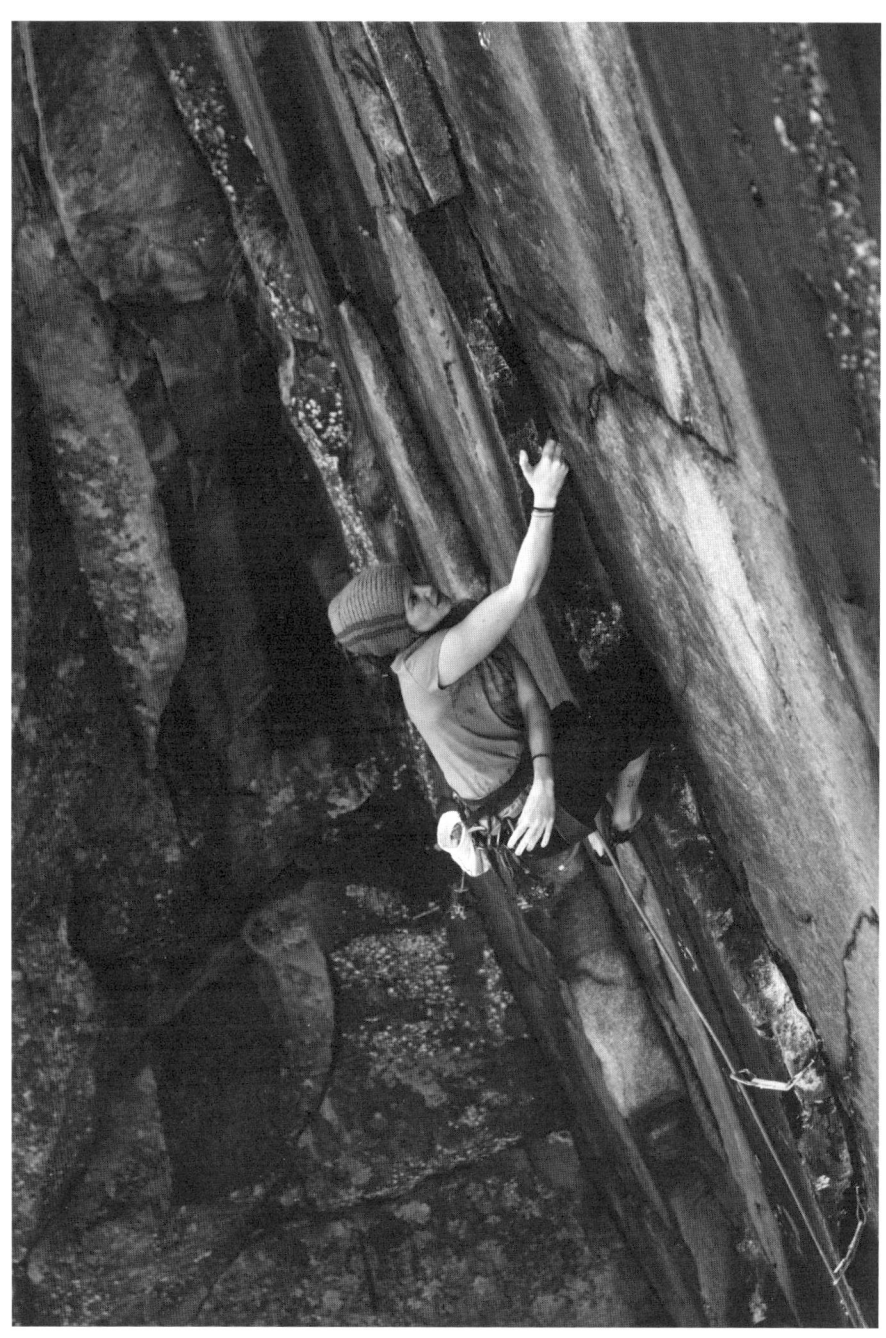

Hilde Bjørgås on Doktor Feg (Swedish 7+ / E5 6a) at Hella, Sweden.

The Viking Hinterland

Romance in the stone on Swedish granite

The forest under Skälefjäll was cool and damp the morning we walked up to the wall one last time. Somewhere up there, among the tall trees shrouded in cloud, was one of the most imposing crags of Bohuslän: a north-facing, forty metre escarpment of steep, smooth, shady granite. Big drops of condensed water clung to the silver birch leaves, playing tricks with the shifting light, and steam rose from some of the tallest pines. Our feet skidded from time to time on the bright lichen blanketing the narrow trail. The trees thickened as the wall began to loom in the middle distance, stilling the air and making our breath condense. Puffs of water vapour hovered in space for a second or two before disappearing into the surrounding gloom.

Suddenly, the wall raises its head above the eager pines that claw and lean against its base. And there it is: *Electric Avenue*. The route follows an impossibly elegant parallel seam that rises the full height of the face, culminating in a bulge at the top before morphing into a cavernous water-runnel. It was there just as it had been before, this most perfect of perfect lines; just as it was before I first came here, or before the first climber ever climbed here, or before anyone ever climbed.

The air was still damp and cloying as I tied on, attempting to remember the mysterious sequence I'd found that unlocked its secrets. I started up the first, easy section with a mixture of trepidation and excitement. At the same time as I reached the break where the wall steepens, marking the beginning of the difficult climbing, an unlikely breeze began to shake the canopy of silver birch and pine behind me. It was a strange serendipity, this sudden wind out of nowhere in the middle of a humid August day. Both times I'd tried to lead the line previously, I'd been slapped off the tiny crux holds by the creep of condensation under my fingertips. But this time – my last attempt before I had to leave to catch my plane home – I knew there could be no such excuses.

As soon as I stepped into the awkward fingertip layback that leads into the crux, my fingers felt cool and dry on the holds for the first time. I chalked up once: this time, once was enough. Instead of battling with the holds as I had before, I began to use them with exactly the right amount of pressure. The gear below disappeared into the enveloping mirage of the trees, and at the limit of reach I caught the small edge at the point where the twin flakes merge into a single, wire-thin seam that cuts through the upper headwall. After another three dynamic moves, I'd caught the shallow one-finger pocket just below the horizontal break, the final rest before the climb's last defence.

Here, guarding the door to the top of the crag, an all-out layback off small vertical edges and poor smears allows a good incut to be grasped. I breathed out hard, tightened up, and launched. Before I could register I'd done it, the fingers of my right hand sank into the hold. Both my feet cut loose, skittering across the granite like skimmed stones, and I palmed up into the huge water-worn atrium at the top in astonishment. I couldn't actually believe I'd done it, right at the eleventh hour.

Some moments in our climbing lives stand tall above the rest. They're not necessarily the biggest grades, or our greatest or boldest accomplishments. Instead they are the experiences that strongly reflect the choices we've made in order to be there to encounter them: the moments that simply tell you who you are, and why you are here, doing what you're doing today. They are the routes that remind us why we still climb, and why we will always climb. For me, *Electric Avenue* was one of those climbs. It is an incomparably brilliant route, and marked the beginning of a personal journey of discovery that would lead me into the heart of one of Europe's greatest climbing areas, a mystical stone kingdom hidden among the forests and islands of Sweden's west coast.

As I sat at the departure gate in Gothenburg's Landvetter airport a few hours after climbing *Electric Avenue,* it seemed as if the connection I'd formed with the cliffs of Bohuslän wouldn't fade with distance or time. I'd also fallen for a Swedish girl who happened be an exceptionally good climber, and Malin Holmberg was the best possible partner for exploring Bohuslän granite. As the plane took off, the pilot banked to the west, and I pressed my face to the window. Sweden's west coast stretched north beyond the horizon's limit: a maze of narrow channels threaded between the islands, reflecting light against the green blur of the surrounding forest. In a few places, white flashes of granite sparked from the trees and along the shoreline. As we climbed, they merged into the approaching blue of the open sea. After a while, the Swedish coast disappeared completely, lost in cloud.

As the summer surged on, granite towers hidden between the trees and hills of forgotten islands and creeks entered my dreams. The urge to return to Bohuslän grew with every passing day.

A few weeks later, on a cool and blustery afternoon near the end of September, I stood nervously below the huge, impending dihedral of *Rätt Lätt* at Häller. First climbed in 1996 by Swedish trad climbing's visionary modern pioneer, Richard Ekehed, it was Sweden's first traditionally protected grade 9. Häller is a cliff that no committed climber can ignore. It has a commanding presence outweighing its moderate size, and its cleanest lines have a monolithic grandeur more akin to an alpine face than to a granite wall barely a rope-length high.

I battled my way to the top of *Rätt Lätt* the afternoon I arrived back in Bohuslän, and the journey I'd started earlier that summer picked up speed. The final crux comes at around thirty metres, at the apex of the steep arête bounding the right wall of the dihedral, and about ten metres out from the last gear at the back of the corner. I forgot my foot sequence on the final, heart-in-fingertip lunge, and very nearly blew the last move, just managing not to clock the full fifteen metres of airtime.

A haze of mustard fields filtered through the oak canopy, and a glance down at the farm below the crag with its old red Volvo parked outside checked me back gently into reality. Bohuslän granite had become as deeply interwoven into my climbing life as anywhere else in the world, even my local cliffs in southwest England.

Later that week, when I reached the top of the classic thin crack of *Catch* on the main wall at Hällinden, the web tightened again. This huge single pitch follows the central crackline that splits a wide expanse of bullet-hard rock, and is as good as any single pitch at that grade I've climbed on Californian granite. Like Bohuslän particularly and Swedish granite generally, this superlative climb remains almost completely unknown outside the small, close-knit Scandinavian climbing community. I continued exploring the area that autumn, encountering new cliffs and even more unlikely climbs. We parked by a huge barn full of rusty machinery on the edge of the woods, walked through a meadow of long grass, and were suddenly in the enormous cave of Granitegrotten. Massive roofs dwarfed the big pines and beeches growing in front of the cliff. Soon I was cutting loose on the wild lip traverse of *Mad Rock* with abandon, throwing heelhooks and searching for kneebars amid the vast ceiling's outlandish structures. Later that day, with weary arms from horizontal sport climbing, we drove up a dirt road that winds up the hill to the hidden enclave of Välseröd. There, we jammed and laybacked splitter cracks evoking the spirit of climbing in the American desert. Later, we ran down to the sea in the gathering dusk, and perhaps Bohuslän's most beautiful cliff, Ulorna. Rising from a tiny, secret shingle beach on the shore of a long saltwater inlet, the small collection of high standard trad routes here represent all that I had come to love about Swedish granite. The more I explored and discovered of this place, the more my enchantment grew.

Climbing stories should reflect the places we climb in as strongly as they evoke our adventures there. Whilst we are all inspired by the thrill of the runout, the focus of the crux, and the finality of the top, every climb is bound up in a complex psychogeography of desire and possession far beyond its mere physical nature. Smaller, less dramatic cliffs often have the strongest effect on our imagination.

There is nowhere I've felt this phenomenon as much as among the crags of Bohuslän. As I write, I can still smell the mossy, mushroomy scent at the base of the cliffs, and see the quicksilver flashes of quartz from the granite as the first sun hits the wall. I can still feel the unmistakable bite of the hard, fine-grained granite as my fingers close on a crimp, and hear the smoky caws of wood pigeons and rooks from the forest as I'm stepping into the first move.

That's the essential magic of the place the Vikings called Ránríki, the sacred land of the goddess of the sea, and Álfheimr, the home of the elf. Climbing has arrived in their realm very recently, yet these presiding spirits of place are old. They've inhabited the woods, inlets and islands of Sweden's west coast – at least metaphorically – since the first Viking explorers imagined them there. And we should count ourselves lucky that we may still explore the green kingdom of their enchantment.

Climb magazine, 2011

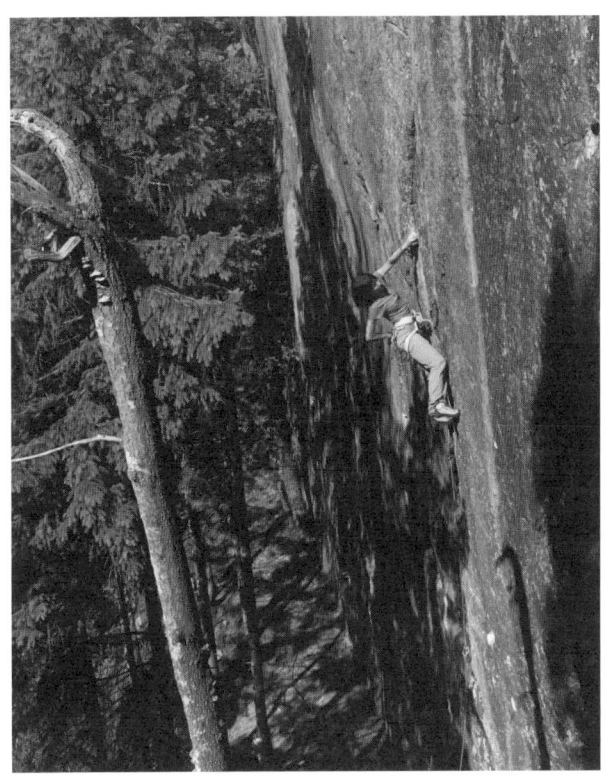

Malin Holmberg making an early repeat and the first female ascent of Event Horizon (9-) at Hallinden.

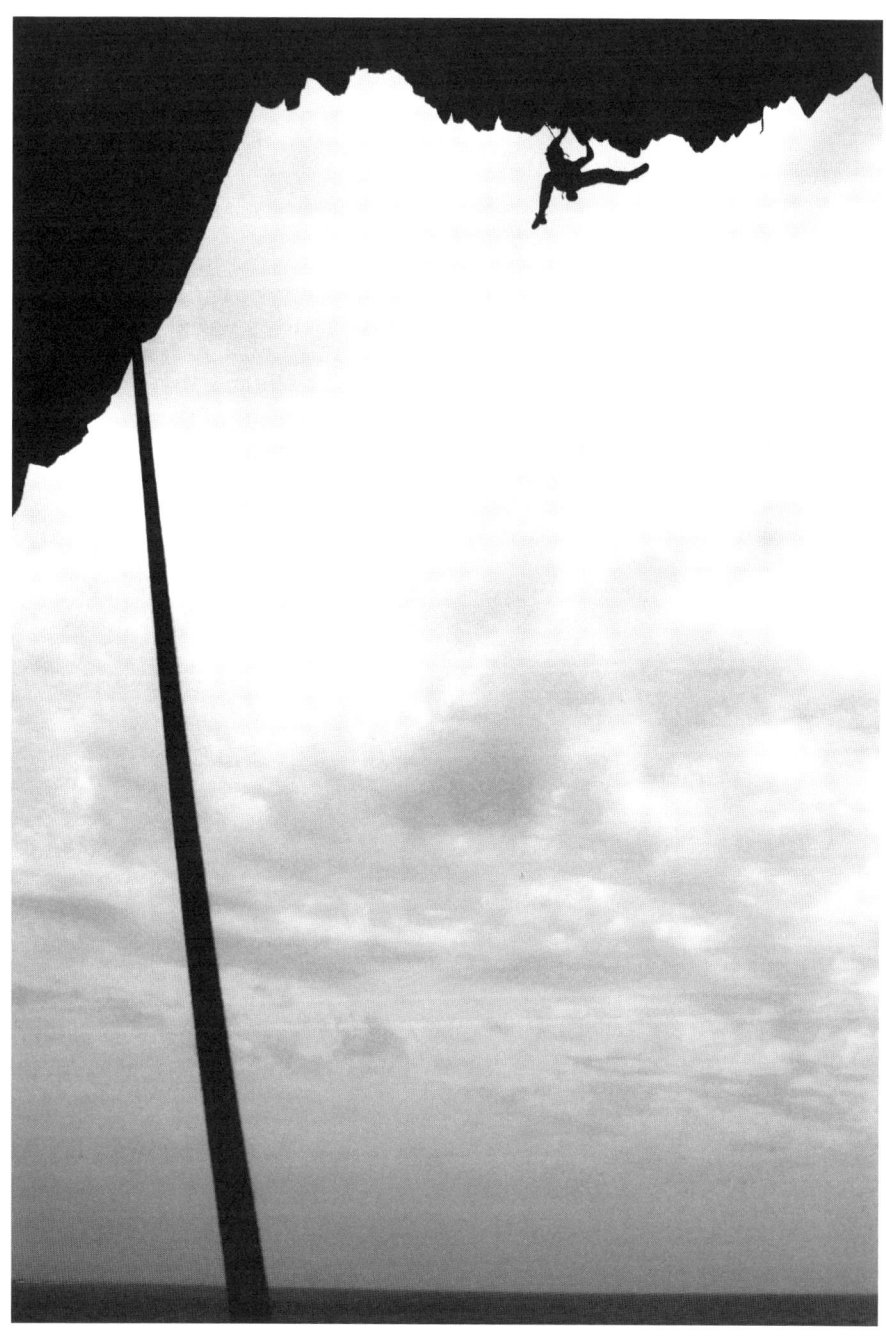
Bob Hickish breaking out of the darkness on the crux of Palace of the Brine (8b), Swanage, England.

Border Country

Chiaroscuro in Boulder Ruckle

zawn

1. (regional Britain) A deep and narrow sea-inlet in the British Isles, cut by erosion into sea-cliffs, and with steep or vertical side-walls.

Could this be the most powerful and mysterious word in English geomorphology? Tightly drawn, thickly webbed, and wrought from strange materials, it's hard to know whether it describes place, creature, or planetary object. Might it describe all three? A zawn is a chamber of metaphysics as much as it is a dark figment of the perpetual battle between the sea and the land. Is there an evil spirit in the heart of nature? If there is, then The Black Zawn – a dark and apparently bottomless cleft in the limestone cliffs near Durlston Head on England's Jurassic Coast – might be as good a place as any to find it.

I first abseiled into the zawn as a rookie fourteen year old, cutting my teeth as a climber on the formidable extremes of the Swanage cliffs. It's a testament to the power of climbing that memory can recall an event of more than half a lifetime ago with clarity. Peering over the edge, the sea gurgled and slapped at the base of the wall, invisible in the depths below. It took a while even to reach the start of the route, as my climbing partner, Darren, and I were both somewhat unfamiliar with the whole free hanging abseil concept. Eventually, though, we'd constructed a belay, and watched with trepidation as our abseil rope arced out into space and out of reach. Like a pair of astronauts cutting the line back to the ship, we'd cast adrift from the human world and entered the time-machine.

After a while, we'd made it across the initial traverse and had established a second hanging belay beneath the acutely overhanging dihedral system that the main pitch tackles directly. Craning my neck, I couldn't see the top of the zawn and could barely see the sky. Down here, only the merest chink of light broke through from the world above.

Totally absorbed by the climbing, a great deal of time passed as I laybacked and jammed my way up the bottomless corner. At the final roof, facing the crux and with the strength in my arms fading fast, I resorted to throwing in jams karate-chop style into the crack, hoping they'd stick. At least there was lots of gear to hold a fall, I thought, despite the unavoidable fact that losing contact with the rock would turn me into a swinging puppet above the void. Somehow, my teenage confidence or beginner's luck prevailed, and I belly flopped on the square-cut ledge at the top and rolled over, lying flat on my back. The sky above was an impossible shade of blue. Out to sea, whitecaps formed under a strong easterly. The air smelt fresh and strange. More than four hours had passed since we'd entered the zawn, and we'd emerged from it all older and wiser somehow – a pair of youthful voyagers between worlds.

In those formative years, the Swanage cliffs were my personal proving ground, a place where I could continually improve as a climber. This stretch of Dorset's Jurassic Coast still remains a kind of landscape of the heart. Just a short drive from my dad's house in a nearby village, this stretch of coast is a uniquely English kind of accessible wilderness. It won't feel very much like one on a summer weekend when the sea glints and the clifftop paths are dry and dusty, but in a rising autumn gale, or late in the evening as the light is falling fast, it enacts a spectacular transition from the benign to the sublime. This coastline is a shapeshifting, surprising space of immense beauty forming a direct boundary between the land and sea. Every time I go back to the Swanage cliffs these days, I feel very lucky to have been able to cut my teeth here as a young climber, and cherish the thought that others will be able to do so, too.

The first time I hit the headwall of what might be the best route in the eastern section of the Boulder Ruckle, *Soul Sacrifice*, I was unprepared for the steepness of the runout up that curving line of perfect solution pockets: I ran out of gas on the first big move and crashed out on a number two cam. A few years later – a little older, not much wiser, but a few degrees stronger – I abseiled in to Boulder Ruckle Far East on the afternoon of a near-total solar eclipse, hoping to encounter the headwall again on different terms. My climbing partner, Richard, and I had just uncoiled the ropes on the big boulder under the wall and were sorting our gear as darkness began to fall. The sky was overcast, so we couldn't see the moon's shadow crossing the sun. Yet the effect on the marine environment of the crag was just as powerful as totality neared: every single airborne bird out to sea returned immediately to the cliffs, believing that night was falling in the middle of the day; a truly extraordinary occurrence. Along the mile-long line of Boulder Ruckle, gulls squawked and settled on the mid-height ledges, then fell eerily silent in the few minutes of near-totality. It was as if the marine world had been teleported into a different dimension of time and space of almost unimaginable rarity.

The light grew quickly as the eclipse passed, as if dawn were breaking in speeded up slow-motion film. Less than ten minutes later, broad daylight had returned, and I set off up the tricky initial wall that guards entry to the bottomless chimney at half height. This time, the route was more benign than it had been on my first foray as a fourteen year old.

Climbing often presents what is, at heart, a deeply subjective reality: the reality of mind. With strength to spare, I arranged some bomber gear at the apex of the chimney and pocket-pulled my way up the immaculate headwall. It's as good a finish as any on a pitch of this grade in the UK.

Reaching the top, the gulls were all back out to sea, wheeling and diving. High over St. Aldhelm's Head, the late summer sun was trying to emerge. Climbing through a solar eclipse is a truly one-off event. If anything, such an experience underscores the sheer good fortune of being able to go climbing at all.

It had been a good day to return to Boulder Ruckle's wild border country of stone, sky and sea; a tidal hinterland to enter in the twilight zone, and from which to emerge anew.

Hard Rock , 4th edition 2020

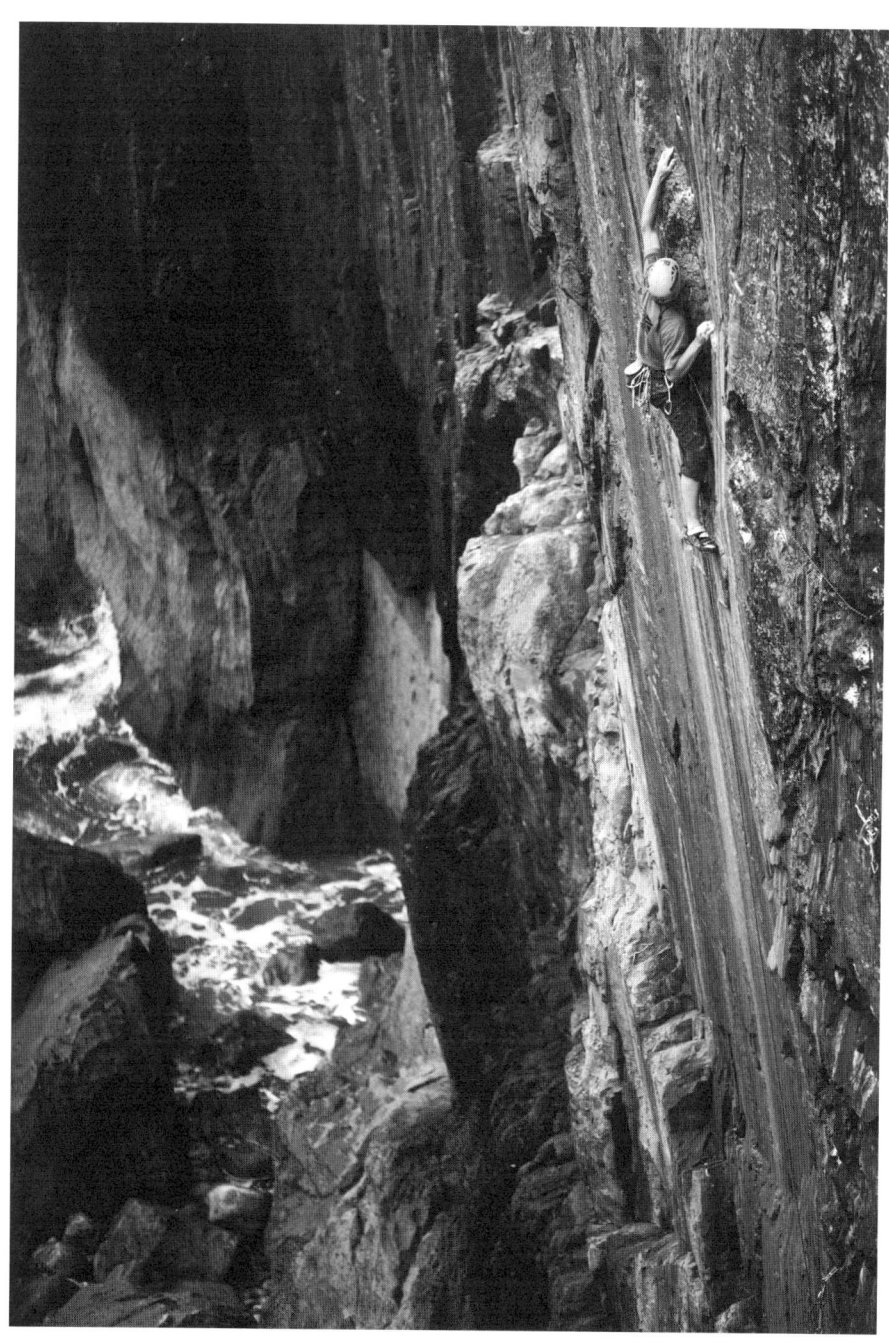

The author on the first ascent of Point Blank (E8 6c) at Stennis Ford, Pembrokeshire, in 2009.

Behind the Lines

The secret architecture of climbing & mountaineering

As the Intercity sped north into the dusky November afternoon, its tail lights faded into an eerie, reddish gleam. The main line out of St. Pancras rumbled with the vibrations of other trains approaching and disappearing. Leaning over the parapet of the railway bridge, my six year-old feet dangling off the ground, I stared at the space where the railway lines met the horizon. Tracing them slowly back to the bridge, I noticed that their apparent straightness was an illusion: the lines formed a very gentle curve, like the earth's surface seen from space. Just then, the bridge shuddered with the vibrations of another northbound train. I watched as kinks appeared at each of the intersections between its carriages, highlighting the hidden, sweeping arc of the tracks. After a while, the train vanished in the same, strange smudge as the previous one. I jumped down from the parapet, shouldered my satchel, and shuffled off into the gloom of the north London evening.

I must have learnt something on that bridge, because I never looked at a line in quite the same way again. In the years since then, lines on cliffs and mountains came to shape my life: lines of stone, lines of language written about those geological and metaphysical lines, and the interplay of shadow-lines through the camera's lens. The idea of lines is central to the adventurous mindset: for the course of the voyage on a nautical chart, or the path of a journey over mapped or even unmapped terrain, lines are important tools by which humans find their way and break new ground. Like sailors and explorers of the past, climbers and mountaineers create, follow, and reinvent lines. They may also break, reconnect, and disrupt them. Lines generate our ambitions, distill our fears, and define our successes and failures. Lines on cliffs, boulders, and mountains are the heartbeat of the climbing life, and the cartography of its infinite map of possibility.

When I first looked at the line in Pembrokeshire, Wales, that became *Point Blank*, there didn't seem to be a way through the apparently featureless limestone of Stennis Ford's central headwall. After an hour of dangling on a rope, contemplating moves, the sorcery happened: a single, unlikely line of weakness appeared in the blankness, a gentle arc swooping through the blue void of the headwall. After making the first ascent the following day, I stopped for a while on the opposite side of the Ford as my friends walked on. Looking at the route more closely, I realised the line was in a crucial sense just like those railway lines I'd stared at as a boy from the parapet of that London bridge. It appeared almost straight at first glance, but it was actually quite the opposite: kinked and twisted by the forces of nature, its true form only became visible through the movement of a climber across it. *Point Blank* is a kind of line of least resistance through a wall of mirrors, half-disguising itself like a figure in a masquerade ball. Dances in the vertical are choreographed by the great game of lines. This is a sketch of its most significant players.

The Broken Line. Whilst many of the finest lines are based on a single, continuous feature, sometimes it's the art of locating the missing piece in the vertical jigsaw puzzle that creates a great route: turning a 'broken line' into a major climb is one of the most satisfying experiences in the process of creating new routes. Perhaps the best example of a perfect 'repair' for a broken line is Nico Favresse and Sean Villaneauva's solution to their stunning 2008 Yosemite free route *The Secret Passage*, which finds a way up the hostile righthand section of El Capitan. The name derives from a short section of the 5.13+ crux pitch which seemed apparently blank until Favresse discovered a tiny seam, just deep enough for his fingertips to hold, which made the entire route possible as a free climb.

In 2004, Mike Robertson and I made the first ascent of *Wall of Spirits*, a bold face climb high on the Great Wall of Pentire in North Cornwall, England. But there was a missing piece in the jigsaw: an independent first pitch that would link into the crux headwall pitch and create a complete route. For two years, the missing piece remained unfixed. I returned in 2006 and climbed an independent first pitch, gaining the ledge directly beneath the crux. With the line complete, *Wall of Spirits* became something more than that isolated pitch on the headwall, and the challenge of the first continuous ascent still remains. Like many of the most appealing broken lines, its history is still incomplete.

The Grand Line. Perhaps the most powerful example of a grand line is *The Nose* of El Capitan in Yosemite: the most famous route on the most recognisable and iconic piece of rock in the world. Grand lines are routes that keep on giving, regardless of popularity, difficulty, or status. First climbed in 1957 after a massive siege, involving a 15 hour lead through the night by Warren Harding up the final headwall, the free ascent eluded the leading rock stars of the 1990s until Lynn Hill's groundbreaking achievement in 1993. The route's grandeur is shown by its immense popularity today with all kinds of climbers, from big wall newbies to the world's best: the line is at once compelling and accommodating, approachable and desperate – if you want to free climb it, anyway. Today, *The Nose* remains the focal point in the culture of big wall speed climbing: it's the racetrack of choice for big walling's Formula One. The much-coveted 'Nose Speed Record' cannot be put down to American exceptionalism alone. The grand line matters, the race seems to say. And it matters a lot.

The Landmark Line. *Geminis* ascends a gorgeous forty metre line of interconnected stalactites, welded calcite patina and fused solution pockets through the steepest section of Gran Boveda, the immense frozen wave of overhanging limestone that dominates the lower part of Rodellar's Mascun Gorge in Aragon, Spain. It's an iconic route, first redpointed by Belgian climbing legend Nico Favresse, and commands centre stage in one of the world's best limestone sport climbing cliffs. In late October 2009, the first big storm of the coming winter had left some sections of Gran Boveda black with seepage. Malin Holmberg, who's holding my rope on my first proper redpoint attempt, vanishes into a tiny speck on the dusty floor of the cave as I approach the final crux; a powerful, technical cross-through sequence on a sloping, letterbox-shaped pocket. The right edge of the pocket is a slimy mess of chalk paste after yesterday's rain. At the instant my left hand hits it, I cartwheel out into space.

There was a monumental quality about this line that stood out above other sport routes I'd climbed. As Alex Honnold put it to me, "that downwards diagonal traverse before the crux feels kinda like being on a big wall." Eight months later, I returned to Mascun to complete the route. There's something about the 8b+ French grade that denotes a 'next level' moment when you break into it; it's American 5.14, after all. Landmark lines like *Geminis* are important partly because they redefine our perceptions of what we think we can achieve.

The Impossible Line. The idea of something being 'impossible' to climb in the first place could be attributed to the special brand of skepticism nurtured by the British climbing community. There's a latent Anglo-Saxon resistance to change that runs deep in British climbing; there's less of it today than in the past, but it still exists. The concept that certain lines will never be climbed has now lost most of its historical currency, largely driven by an awareness of the extraordinary rise in global climbing standards. Even so, hundreds of unclimbed 'projects' scatter the crags and boulders of the world, and some remain perceived as impossible lines. Probably the most famous impossible line in British climbing is the square-cut, virtually holdless arête at Burbage South in the Peak District known locally as *Wizard Ridge*. It was first envisaged as a climbable route by Johnny Dawes over two decades ago, who once gave me a detailed description of a single hold that you would need to use seven times, in different ways, to make the crux moves. Will it ever be climbed? I, for one, wouldn't like to suggest it's impossible.

Mountaineering's answer to the rock climbing concept of the impossible line is the idea of the forbidden line: a route so treacherous that it becomes regarded as completely unjustifiable. The greatest example of such a route is Everest's so-called 'Fantasy Ridge', named by George Mallory due to its otherworldly, unattainable nature. American climber Ed Webster described the ridge in no uncertain terms: "It's a very, very long and narrow ridge, *double-corniced nearly the entire way* [Webster's italics], with a really long way still to go to the summit. It's awkward, dangerous, with no options for retreat when things go wrong".

When the talented team of Dave Watson, George Dijmarescu, and Dawa Nuru Sherpa approached it in 2006 they looked at it for some time before calling the expedition off. When Watson was later asked why they did, he curtly and rather brilliantly replied "we came here to climb, not to die."

Due to the very rapid progress in the sport over the last thirty years, climbers now understand that describing almost anything as 'unclimbable' is unwise. Similarly, mountaineers know that the concept of a forbidden line only really goes as far as the next weather window or the next season, and climate change is a big factor here too. The objective differences between mountaineering and rock climbing can sometimes separate the two activities, but climbing is a dream-making machine whatever you're doing.

The Magic Line. A very few lines somehow capture in a single route everything that's great about climbing. I've been lucky enough to have climbed a few magic lines over the years, but I've spent a lot of time searching them out. Near the top of the third pitch of a new route I climbed in Norway's Lofoten Islands with Malin Holmberg, *The Lady Of The Lake,* there was an impasse before a tiny ledge three metres above me. If I could reach it, I knew the route would link up as a free climb. If I couldn't, it wouldn't. I spanned out from undercuts, and my fingertips sunk into an invisible fingerlock, allowing me to reach the ledge in a couple of moves. It was a sure sign of the line's magical characteristics.

In the same way that the true nature of those railway lines I'd watched as a kid was only revealed by close observation, magic lines often have something of the invisible about them. Magic lines distill the essence of what we seek when we climb. Shaped by nature and crafted by human endeavour, these routes become the catalysts of climbing ambitions that span continents and generations. Of all the myriad structures in the vertical world, magic lines are its strangest. They represent all we dream of, and everything else we have to lose.

Climb magazine, 2014

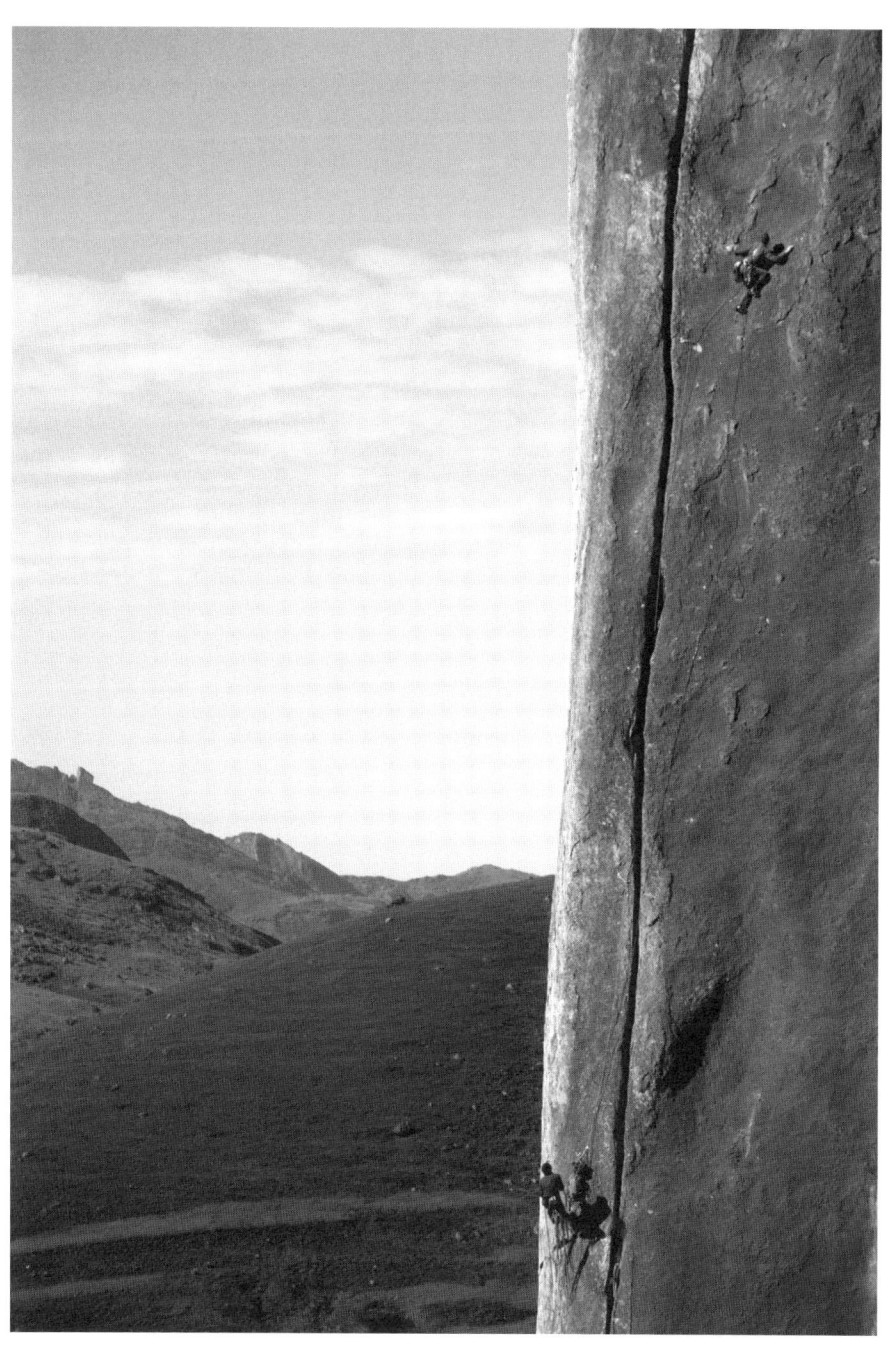

Jack Geldard & James McHaffie on Rain Boto (7b+ max, 380m) on Karimbony, Tsaranoro.

Bring on the Wall

Pioneering in Madagascar's Tsaranoro Massif

We'd been driving for sixteen hours straight when it started to rain. Threatening thunderheads were building in the east and the night air crackled with electricity. We pulled over to tie tarpaulins over the luggage on the roof of our jeep.

"Good idea", our driver murmured in heavily accented French, and a wry smile spread across his face. "To stop bandits!"

We laughed at the idea of a few tarpaulins fending off raiders. We were more concerned about our crucial supplies of climbing rope and equipment getting soaked by the storm than the possibility of kidnap. Satisfied with the job, our driver lit another cigarette and started the engine. The rain drummed on the roof of the jeep like a jackhammer. As we rolled off into the night, a ten second burst of sheet lightning illuminated the wild country to the west. Along the horizon to the south we could see the jagged shadow of a mountain range: the Tsaranoro Massif. I couldn't have expected a more dramatic first sight of the place that has become known to climbers across the world as the Yosemite of Africa.

The climbing world is a bit like a medieval court. Tales of derring-do are relayed from distant lands, then passed around, debated, and cross-examined ahead of the next quest. The idea of this exploratory rock climbing expedition to the mountains of southern Madagascar came about in early 2008, after my friend Jack Geldard and I heard about the potential for establishing new routes there. We completed the team with James McHaffie, one of Britain's top climbers, and another friend, Stephen Horne, to make up a team of four. Preparations lasted several months and we arrived in Madagascar at the beginning of April. We had one month to complete our main objectives: to establish a completely new big wall climb on the Tsaranoro massif and then attempt the uncompleted line dubbed *Tough Enough?* on Karimbony, a 400 metre high granite monolith that dominates the right-hand side of the Tsaranoro Massif. This climb already had an almost mythical status among some of the world's best rock climbers in the late 'noughties.

We arrived at Tsaranoro just before midnight, after a twenty hour journey south from the capital, Antananarivo. The electric storm that had lit the final part of our journey had passed; from my sleeping bag I could see the shadow of the Tsaranoro Massif towering thousands of feet above us. The stars seemed bright enough to burn holes in my mosquito net. It felt incredible to be here at last, in the heart of one of the world's most exciting exploratory rock climbing regions.

After making ascents of some established routes to acclimatise to the subtropical heat, we went in search of an unclimbed cliff where we could create our new route. After a day of trekking around the massif, cutting new trails and scoping the crags with binoculars, we found a perfect natural climbing line on a cliff called Lemur Wall, so named because the endemic Madagascan primates, the lemurs, populate the jungle at the base. It stretched upwards for over two hundred metres, following an impressive yellow streak in the black granite.

During our first day on the route, Jack and I established the first pitch with a view to free climbing it after we had practiced the moves. This initial section of the wall seemed to be the hardest, involving some very difficult climbing on microscopic holds. We tried again the next day but failed due to the mercilessly rough rock and hot conditions. That night at base camp, I suggested we needed an alternative approach to the main challenge of the line in case we couldn't free climb the desperately hard first pitch.

At our advance base camp under the wall, Jack and I looked up at the rock on either side of the very difficult first section. Defeat seemed to stare back at us. The wall to the left seemed even more blank and impossible than the one we'd been trying to climb earlier. Then I had an outrageous idea. To the right, a gigantic vine hung down the side of an overhanging gully. If we climbed it, I worked out that we could probably make a traverse out to a point just above the impossible lower section of the wall. I placed a single bolt for protection on the featureless granite of the gully, made a leap of faith, and grabbed the vine, swinging out on it and then climbing hand over hand, Tarzan-style, to a tiny ledge about 40ft up. Amazingly, despite being weighed down with equipment and carrying a heavy battery drill, it didn't snap, much to Jack's disappointment. Climbers have a natural tendency towards *schadenfreude* in these kind of situations.

I soon realised that this lucky 'Tarzan vine' was our key to free climbing the entire route. We made a belay at the tiny ledge and I set off on a spectacular traverse that swooped out across the wall to regain the yellow streak. It was getting dark by the time we'd finished, so we abseiled off and returned to base camp by head torch.

The following day, Jack led off on what would become the hardest section of the whole route. Despite the threat of a huge fall, he climbed brilliantly and successfully made it through the crucial impasse. The final pitch was a relative breeze, and we knew the line was ours. The first climbers to pioneer a route are obliged to give it a name for the record; we had to find a title that fitted with the febrile theme of the wall. The easier companion route to the left was called *Ebola* after the particularly virulent tropical virus. Because of the compelling streak in the granite that had originally drawn us to the line, we decided on *Yellow Fever*. We had created the hardest climb on Lemur Wall at the time, and the whole experience of devising the route was certainly a memorable challenge.

Flushed with the success of our new route, James and I turned our attention to *Tough Enough?* on the west face of Karimbony. This huge, featureless expanse of rock gleams in the morning sun, as if a giant had taken a cheese slice to a colossal chunk of marzipan.

On first acquaintance, it's hard to imagine how it would be possible to free climb the west face at all. No consistent line leads up it. Spidery flakes and tiny cracks spiral through the bright green lichen, heading nowhere. There is barely a single ledge anywhere on the wall that is big enough to balance on without a handhold. Many of the holds are just minute crystals of granite, sometimes only a few millimetres wide. The route has seen a one-day free ascent, virtually onsight, by Adam Ondra, one of the world's best climbers. But it's still a super tough climb that lives up to its name.

We had a hard but rewarding day on the lower portion of the climb, in which I free climbed the first and second pitches; the second one featured desperately thin climbing on a steep slab at around 8a, and remains to this day one of the hardest things I've personally climbed in this style, which is certainly not my forte – I prefer steeper routes with bigger holds. James then dispatched pitch three with his trademark efficiency.

The next day, we hiked up to the top of Karimbony to approach the upper part of the climb from above. Abseiling over the edge of the west face gives a major overdose of exposure. Sliding down our fixed ropes from the flat summit plateau, we instantly became a pair of tiny puppets on a piece of string, with over a thousand feet of sheer, barren granite between us and the ground. Climbers are used to being in high places, but this certainly required some mental adjustment. If you get vertigo, the upper reaches of this wall are not the place to be. James managed, impressively, to free climb the ninth pitch of *Tough Enough?*, and one of the most difficult section of the entire route, later that afternoon, after a huge fall ended his first attempt high on the headwall.

We both almost fell asleep with exhaustion on the summit that evening – the physical and psychological strain of being on the massive granite wall was beginning to take its toll. We coiled our ropes, stuffed our climbing gear into our packs and stumbled down to the fixed ropes above Lemur Wall in the falling dusk.

The solitary lights of base camp began to float up through the darkness. Stopping for a rest on a huge boulder on the way down, we looked back in silence at the shadowy form of Karimbony. Towards the end of our adventures in the Tsaranoro massif, we all felt lucky to have been part of the conception of climbing in this extraordinary area, which would remain on the radar of the world's most talented climbers for years to come. For my own part, I had been as captivated by the extraordinary landscape of high Madagascar as by the exceptional climbing in the Tsaranoro Massif itself.

Esquire magazine, 2010

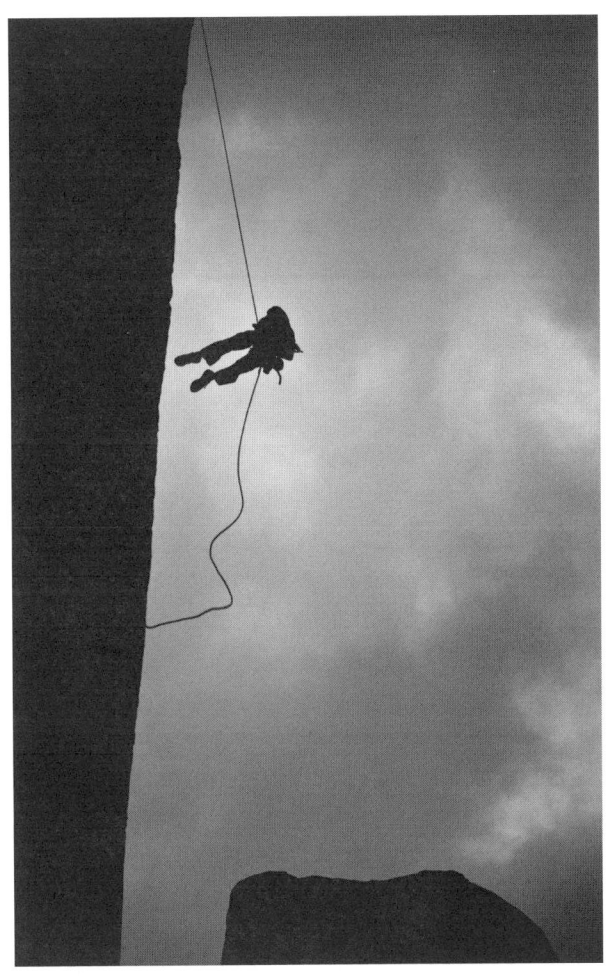

Jack Geldard abseiling off Lemur Wall as dusk falls over the Tsaranoro Massif.

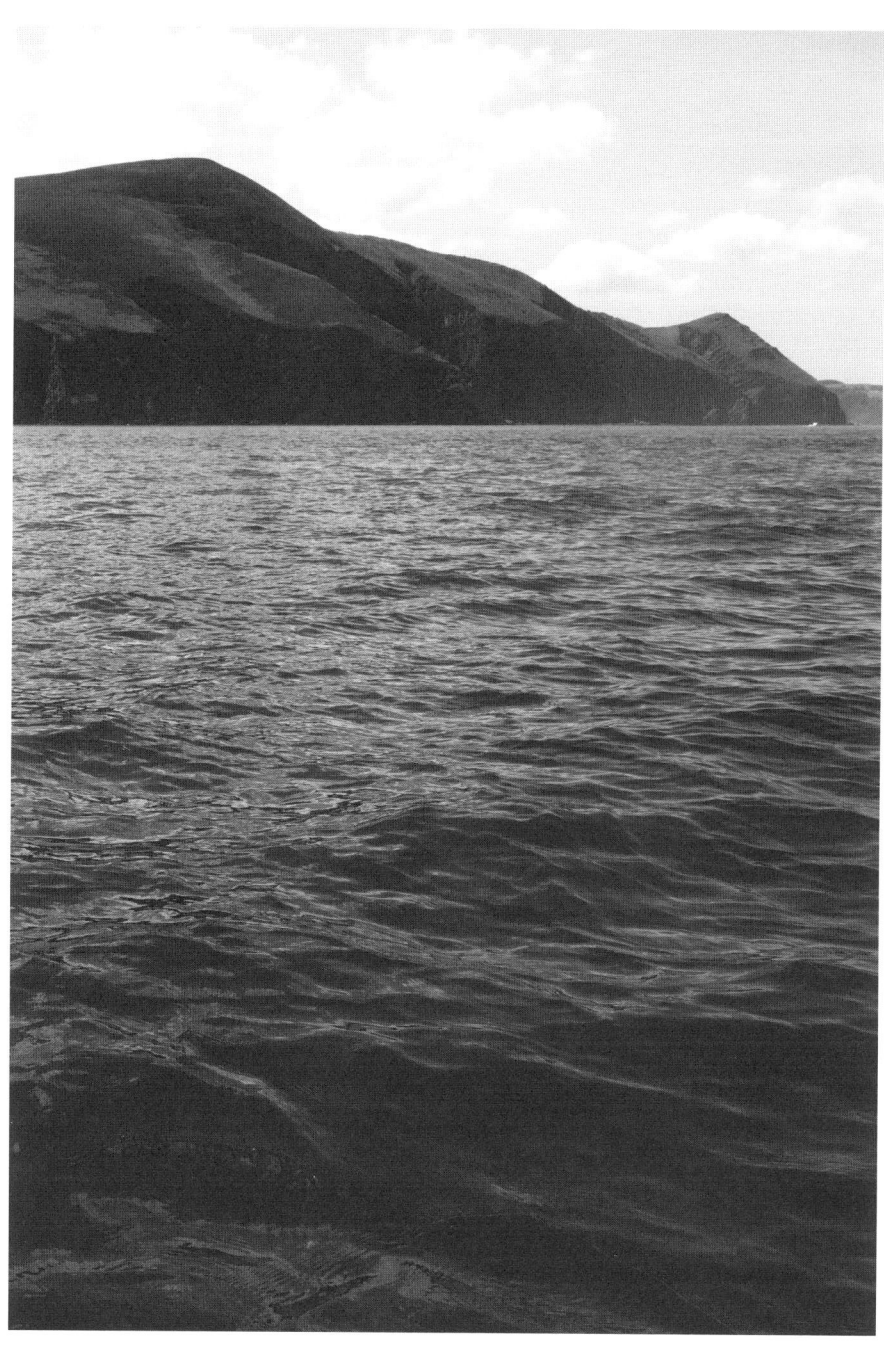
The western Exmoor Coast Traverse from the sea. Great Hangman – the highest cliff in England – is on the left.

The Hidden Edge

Aqua-alpinism on the Exmoor Coast

The heaving swell slaps at my feet, licking along the narrow sandstone shelf like a dragon's tongue before swirling into a through-cave. Reflected light glints on the water on the other side. Swimming across the cave, I point out to the psychiatrist, will cut off a few hundred metres of climbing, saving us precious time.

The sun has started to sink into the sea with the coming of evening. Just then, a massive swell set slams into the cavern, and the constricted entrance takes on the form of a cataract on the upper Brahmaputra. The psychiatrist looks at me quizzically.

"Maybe best not to" he suggests with a shrug. We're both getting cold. I look back at him, shivering. Another huge wave surges into the chasm, and we decide to abandon ship. In case you're wondering, I'm not recounting a flashback from rehab. The psychiatrist is my partner in crime, Dr. Grant Farquhar. He is also largely responsible for the fact that we're traversing towards the base of the highest cliff in England, the Great Hangman, less than three hours before sunset after eleven hours on the go. It's just after high tide, and we both have mild hypothermia. In June 2013, when this mission took place, the waters of the Bristol Channel were still circa 12 degrees Celsius after one of the coldest winters in decades.

We've been on the move since 6 a.m. on the Exmoor Coast Traverse, Britain's longest climb. We're attempting a one day ascent of the most demanding section of the route: Lynmouth to Combe Martin, a distance of some 17,000 metres: around twice the height of Everest from sea level. It's a mammoth undertaking, and has much of the atmosphere and seriousness of a big alpine climb on an unexplored peak. It had never previously been considered feasible in a single, one-day push, and nobody had attempted to do so.

The history of the Exmoor Coast Traverse spans more than a century, and is as complex and convoluted as the route itself. Following the lead of Victorian pioneers James Hannington and Edward Arber, the route was first completed by Clement Archer and Cecil Agar in 1954, but only in different sections climbed on separate days.

A couple of decades later, in 1978, Terry Cheek, Trevor Simpson, Graham Rogers and Robert Simmons made the first continuous ascent over *four and a half days*, using ropes and an expedition-style approach with a support team. The route involves serious and in places quite difficult climbing on wet rock, with only a handful of safe exits in its entire length.

Our approach, by contrast, was fast, light, and totally unsupported. We climbed solo, wearing modified 2mm wetsuits and approach shoes, and carried compact dry bags for our food and water. David Kester Webb and Elizabeth Webb explain the extremely committing nature of this route in their book *The Hidden Edge of Exmoor*:

"[This] is a serious mountaineering venture that is compounded by a tide that can rise vertically over six feet an hour and by cliffs that tower over six hundred feet in places. Out of sight of civilisation, it is an awe-inspiring wilderness, boasting the highest cliff in England, a waterfall as high as Niagara, and a colony of ancient stunted yew trees that may prove to be the largest in Britain."

The seriousness of the undertaking is multiplied by the fact there is no phone reception whatsoever at the cliff base anywhere along the Exmoor coast. In this respect, as in others, the traverse is actually more serious than a big climb in the Alps, where helicopter rescue is just a speed-dial away. If you had a serious accident anywhere along this desolate shore, you could easily die long before help arrived.

We made good progress, covering three quarters of the distance between Lynmouth and Combe Martin in eleven hours. But with mutual respect for the power of the sea, we called the 'Brahmaputra Cave' our full-time whistle, and made our escape up the steep incline of Red Cleave, a fine eleven-hundred foot, fifty-degree bramble-festooned couloir.

Completing the traverse from Lynmouth to Combe Martin in a day, or 'The Exmoor Coast Integral' as Dr. Farquhar and I have called it, would be feasible by a highly proficient team of ideally two or three, and under more favourable conditions than those in which we attempted it. This alpine-style adventure, completed without boat support or support crews, is undoubtedly one of the biggest challenges of physical endurance, logistics, commitment, and route-finding skill in Britain, and presents the possibility of a whole new sub-genre of climbing: you might call it 'aqua-alpinism'. For many British climbers and mountaineers, doing the entire Cuillin Ridge on the Isle of Skye in a single, one-day push is regarded as one of the great challenges in the British mountains. In the same way, doing the Exmoor Coast Traverse in a day is one of the biggest sea cliff challenges on our islands.

In the same way that para-alpinism links climbing with BASE jumping, 'aqua-alpinism' links climbing with coasteering to produce challenges that are beyond the scope of either activity alone. In order to complete the Exmoor Coast Integral, you need to be a strong, experienced climber and an equally strong and experienced coasteerer and open-water swimmer.

When the late Swiss alpinist Erhard Loretan pioneered his 'night naked' approach for climbing hard routes with maximum efficiency in the Himalaya, it's unlikely he imagined his theory might be used on a sea level traverse of an obscure section of the southwest coast of England. At some point, someone will finish the project that Dr. Farquhar and I attempted in 2013, and complete the non-stop Exmoor Coast Integral. This epic, multi-dimensional journey along England's wildest stretch of coastline still remains for the taking. When it is finally completed in a single, aqua-alpine push, I think it'll be one of the more impressive feats of adventure climbing achieved anywhere in the British Isles.

The idea of aqua-alpinism might sound weird, but it's actually a unique approach to having massive adventures on sea cliffs. The beauty of climbing in general – and alpinism, perhaps, in particular – is its stubborn refusal to conform to a single definition of what it might be.

The Blue Cliff, 2023

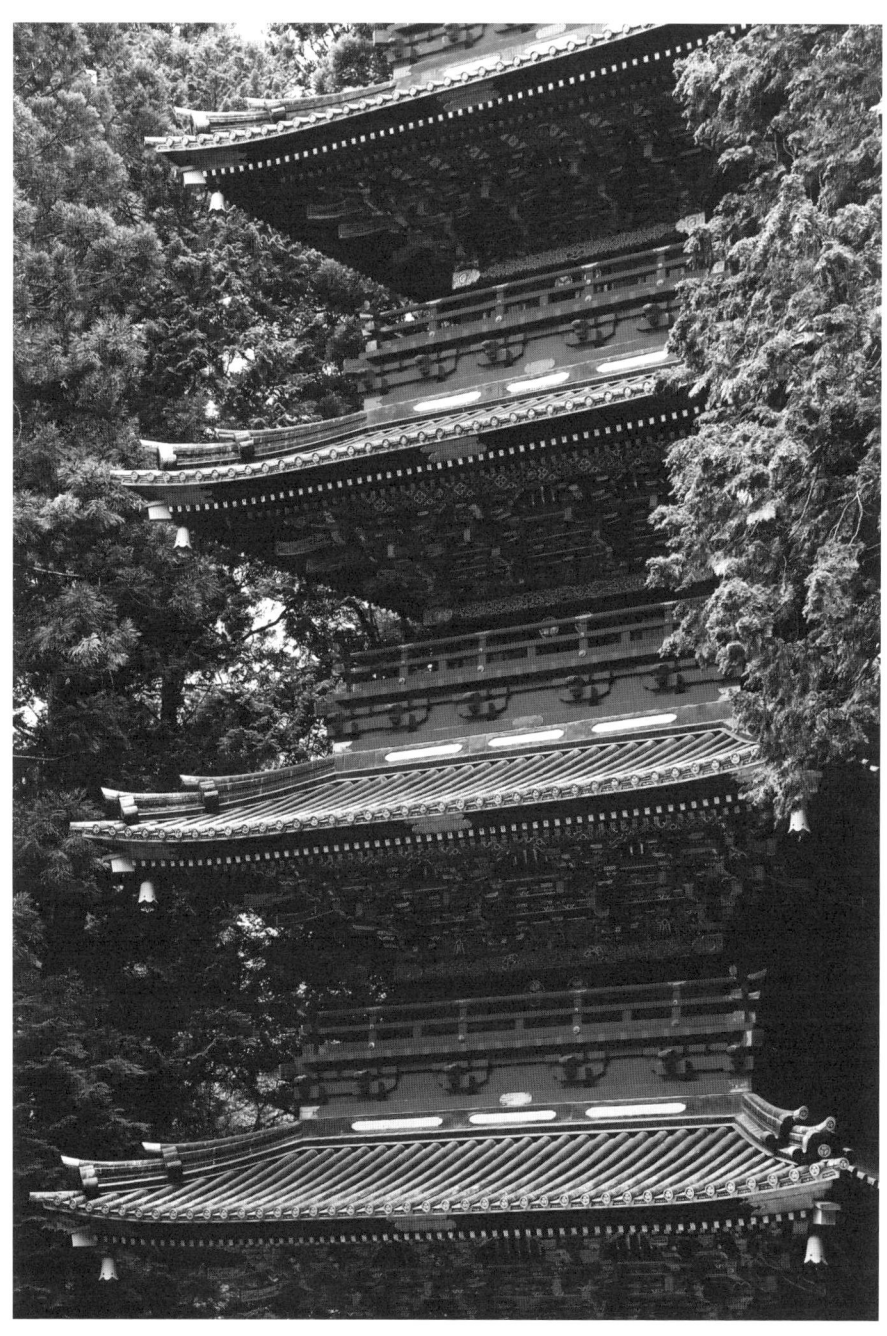

The pines move in the autumn breeze around the five-storied pagoda of Tōshō-gū, Nikko, Honshu.

The Wind in the Pines

The beauty & mystery of climbing in Japan

It's early autumn, and the beginning of night. The time of shadows.

From the subway window, the Tokyo darkness flashed past in a series of shifting lights. A switching billboard; the curtains moving in an apartment window; a fast train from another direction; the neon glow of a small arcade.

After a while, we arrived in the quiet suburb of Zoshiki that would be our base camp for the next three weeks. As a big Japanophile, I'd wanted to visit the country for a long time. My girlfriend at the time, Kelly, had lived in Tokyo for eight years, and it was through her fluent Japanese and local contacts that I had the good fortune to have an insider's view of The Land of The Rising Sun.

The next afternoon, in a corner of the Tokyo National Museum, I stumbled upon Yokoyama Taikan's monochrome silk canvas, *The Moon At Izura*: a copse of black pines clings to a rocky spur leaning down towards the moonlit ocean. I'd seen it before, but only a replica. The original is different, as is often the case. Look carefully at the painting, and the land itself seems to merge with the sea. To me, this obscure image is not only one of the finest examples of traditional Japanese landscape painting; it's also a perfect expression of the psychic space of sea cliff climbing, which would be our first point of contact with Japanese rock at Jogasaki.

We left Tokyo a few days later, catching an early train to Izu Kogen, a small town on the east coast of the verdant and mountainous Izu Peninsula that extends south from the Yokohama suburbs for over a hundred miles. The first photographs of Japanese climbing I saw were Uli Wiesmeier's shots of a young Stefan Glowacz on the volcanic cliffs of the Jogasaki Coast, in Wiesmeier's classic photographic book *Rocks Around The World*. Just two hours by train from south Tokyo, this place was the obvious first base on our whistle-stop tour of climbing in Japan.

The train is on time, as all trains in Japan almost always are. The Japanese railway authorities apologise on the loudspeaker if the trains are *early*, which rarely happens, but can you imagine such a thing in the West? We drop off a few things with Takamune, in whose house we are staying, and make the steep descent down to the coast passing Izu Kogen's nationally famous 'Teddy Bear Museum'. This is the sort of establishment that it would be impossible to find outside Japan, a country in which anything that might be reasonably considered *kawaii* (cute) is given a degree of reverence bordering on the unhinged. In Japan, cats have their very own hospitality venues: cat cafés. Felines are peak *kawaii* in Japan, and a whole ecosystem of cultural cuteness follows them around.

At Jogasaki, you can smell the Pacific before you can see it; the lush subtropical forest that overhangs the crags is dense and blocks out the light. Suddenly, just before the Kadowaki suspension bridge, a slice of blue appears between the trees, and the unmistakable sound of water on shingle confirms you've arrived on the coast. Finding the hidden descent route to the bay in which most of the climbing is located is another matter; we missed it first time around and ended up in an overgrown graveyard on the outskirts of Yawatano fishing harbour with a very strong resemblance to the set of *Indiana Jones and The Temple of Doom*.

After a while, a steep scramble and short abseil lands us on the secluded boulder beach surrounded by steep basalt cliffs up to a hundred feet high. Even in October the afternoon sun is hot, so we climb a series of routes following overlapping grooves and arêtes on the shady east face. Here, close to the seaward end of the wall, *pinus thunbergii* – the black pines native to all coastal Japan – grow horizontally from the cliffs, their boughs leaning towards the sea, at once melancholy and enchanting. I thought back to Yokoyama Taikan a century ago, painting these same trees above the moonlit Pacific Ocean at the pagoda in Izura. There is something indefinably evocative about the pines that overhang the ocean at Jogasaki; it reflects the Japanese concept of *mono no aware*, which can be roughly translated as the aesthetic value that comes from not what you see, but what you feel about being in a certain place. It's a very Japanese version of the notion of psychogeography.

The sound of the wind in the pine trees is the thing I'll always associate with the climbing in Japan. There's something about the sound and smell of a Japanese pine forest that's quite different from anything remotely similar – just like so much else in this country.

I woke early the next day to the sound of the wind in the tall bamboo outside the house. Over breakfast, our host Takemune told us about his life in Tokyo running an executive headhunting agency. He'd sold the business in the late 1990s, and moved to Izu with his wife in search of a more relaxed pace of life.

That morning we drove across the mountainous interior of the Izu Peninsula on winding rural roads, and Takemune's wife dropped us off in a glade of Japanese cedars near the summit of Joyama, a mountain in the northern part of Izu, and an example of the phenomenon for which the peninsula is famous. Geologists call these formations a 'volcanic neck', which means the magma supply-path of an extinct submarine volcano. It's too hot to attempt the multipitch climbs on the mountain's impressive south face. Following directions over the phone from our ex-pat British friend Alex, we hike through the cedar forest to the Wild Boar Gorge, where on one side of a shady ravine is a collection of superb single pitch climbs on solid volcanic rock weirdly reminiscent of the fine-grained gritstone of the Peak District.

After climbing half a dozen routes, we pack up and hike back over the summit of Joyama, then follow a precipitous trail down to the valley by headtorch, weaving through forests of cedar, pine, and giant bamboo, passing several massive sink holes formed by joints in the magma that created this mountain.

We emerge on the road looking as if we've been on a jungle warfare training exercise, blinking in the evening light, and covered head-to-toe in luminous green dust. Because we're in Japan – which has the best rail network in the world – there's a station a twenty minute walk from the base of the mountain, and three hours later we're back in central Tokyo. I don't think there's another major city in the world where you can reach a crag in such a wild location so easily by public transport. The extraordinarily effective rail network in the Land of the Rising Sun all boils down to the fact that in the 1950s, when the rest of the developed world was busy building airports, Japan invested in trains.

We leave Yokohama at dawn on Saturday morning to reach the Okuchichibu Mountains. I first heard of Mizugaki from Yuji Hiriyama, Japan's greatest climber of modern times, who told me he thought it was probably the best crag in the country. It's mid-October, and the forests of the Chichibu Tama Kai National Park are beginning to turn to gold, and scattered with the red sparks of *momiji* (Japanese maple) as they fire up into full autumn colour.

Mizugaki is home to Japan's hardest multipitch trad route, *Senjitsu no Ruri* (A Thousand Days of Lapis Lazuli), which takes a line through the centre of the huge Moai face. It was established by Keita Kurakami and Yusake Sato in April 2016 after an epic siege. It would rate at E11 on the British scale, which puts this route up there with the world's hardest multipitch trad climbs of the present day.

I have somewhat more modest ambitions. My friend and guide to the area, Shuntaro Suzaki, suggests we climb at Kamanboron, a sector high on Mizugaki's north slopes renowned for its steep, tough routes on solid granite. The trail weaves up through the mountain's ever-steepening talus slopes – which are also home to some of Japan's best bouldering – until the cedar forest gives way to stunted pines and rhododendron thickets. Suddenly, from between the trees, a cluster of monoliths looms down on us; it reminds me of some of the bigger crags in Germany's Frankenjura, but on much more impressive scale. The trail ends on a raised plinth commanding a stunning vista across the autumnal forest that falls away on either side. The mountain air is cold today, and the Japanese climbers make miso soup on pocket stoves as they rest between routes.

I climbed only four routes at Mizugaki, all brilliant climbs on compact, intricately-featured granite. By four o'clock the evening chill had set in, strands of mist began to rise over the tallest trees in the forest canopy, and the climbers began to head back down the trail. This evening, we will stay with our Japanese friends at the *ryokan* (a traditional inn) of Sizennomura, which is a fifteen minute drive down the hill. With simple, unfurnished tatami rooms, and *onsen* (communal baths), *ryokan* originate from the Edo period (1603-1868) when they were the only places to stay for travellers moving around the country.

That evening, we gather around the low tables, sitting cross-legged on tatami mats, talking and drinking *sake* late into the night. The endless curiosity, laughter, and effortless courtesy of our Japanese friends is a humbling experience. After that evening at Sizennomura, I quickly understood why so many foreigners fall in love with Japanese culture, and why so many who move here either end up staying for years, or never leave at all.

Just a few miles to the northeast of Mount Mizugaki, the granite crags, spires, and towers of Ogawayama are arguably Japan's most famous climbing area. Despite having a strong culture of mountaineering and big wall aid climbing, technical rock climbing didn't take off in Japan until remarkably late – well into the 1970s – when Japanese climbers returning from Yosemite began to use the free climbing style they'd seen in America on their local crags around Tokyo. Ogawayama, with its abundance of splitter granite cracks, was ideal for this purpose, and became one of the primary forcing grounds for Japan's leading climbers in the 1980s and 90s.

It's a glorious autumn morning when we cross the river and hike up through the pines towards Mara Iwa, a two hundred foot spire that stands tall above the forest and is home to some of the best single pitch routes in the area, including the stunning fifty-metre face climb *Excellent Power*. The first route of the 5.13 grade established in Japan, it was famously onsighted by a visiting Stefan Glowacz in 1992 during his famous world tour. In the years since Glowacz's visit put Japanese rock on the world map, climbing here has become increasingly popular, and it's not unusual to see hundreds of climbers on the more popular crags at the weekend.

As soon as a veil of cloud obscures the sun in the early afternoon, we're reminded that these crags are at altitude (circa 1400m) as the temperature falls like a stone. Observing Japanese climbers in action at Ogawayama before we have to leave for Tokyo, it's clear how everything they do is informed by the spirit of gambaru (the verb 'to work hard') so central to the Japanese psyche. Their whole approach is refreshingly different from the way most Western climbers approach the sport. The Japanese appear to invest less importance on getting to the top of a route than they do in showing determination and diligence ('gambari') in facing a difficult task. As a result, many Japanese climbers tend to try climbs and projects that are very close to, if not beyond, their physical limit. They seem unaffected by the psychological issues surrounding repeated failure on the same route that many Western climbers have; for the Japanese, it's all about trying as hard as you can. This reveals a great deal, I think, about what's so intriguing about this great country and its people.

Our final climbing destination in Japan would be the escarpment of beautifully weathered, solid limestone that encircles the summit peaks of Futagoyama (meaning 'twin peaks'), high in the rural upcountry of the Chichibu mountains, yet only a three hour drive from central Tokyo. It's all thanks to our ex-pat friend and nuclear physicist extraordinaire Alex that we make it up to Futagoyama; although late October is normally the end of the climbing season, the whole crag is still in good condition. The location of the cliff, hidden from view below a narrow col between the mountain's twin peaks, seems to emphasize its unique status as Japan's solitary answer to the limestone supercrags of southern Europe.

On my very last day in Tokyo, I catch the train to Kamakura, a small city in Kanagawa prefecture, about thirty miles south of the capital. It was the capital of Japan and the seat of the Hojo Regency in the late medieval period, from 1185 to 1333. Of all the complex components in Japanese aesthetic life, it is the vernacular garden design I find most appealing; the basic concept is to create a miniature natural world in a highly stylised form, usually based around a central area of water such as a lake or a series of engineered streams. In the same way time dissolves when our attention is rapt by climbing, the hours of a chilly late October afternoon disappear as I wander the narrow streets past hidden shrines, through giant glades of green bamboo, and along narrow pathways that lead through the mossy alcoves of secret courtyards watched over in silence by stone lanterns.

In three weeks here, I've only climbed for eight days; it's been a journey through Japanese culture as much as a climbing trip. Sometimes, even when you're on a climbing trip, it's good not to put climbing first. All committed climbers tend to prioritise climbing over all else in life. Going somewhere as culturally rich and socially interesting as Japan is a good way to achieve a perspective on whether this is really such a good idea.

As soon as you arrive in Japan, though, you may quickly find you needed no excuse to come here at all. I know I'll be returning sometime to search for more of that quiet inspiration captured in Yokoyama Taikan's paintings; for that feeling distilled by the wind in the pines leaning towards the Pacific, and the desire for movement and stillness we might imagine there.

Climb magazine, 2017

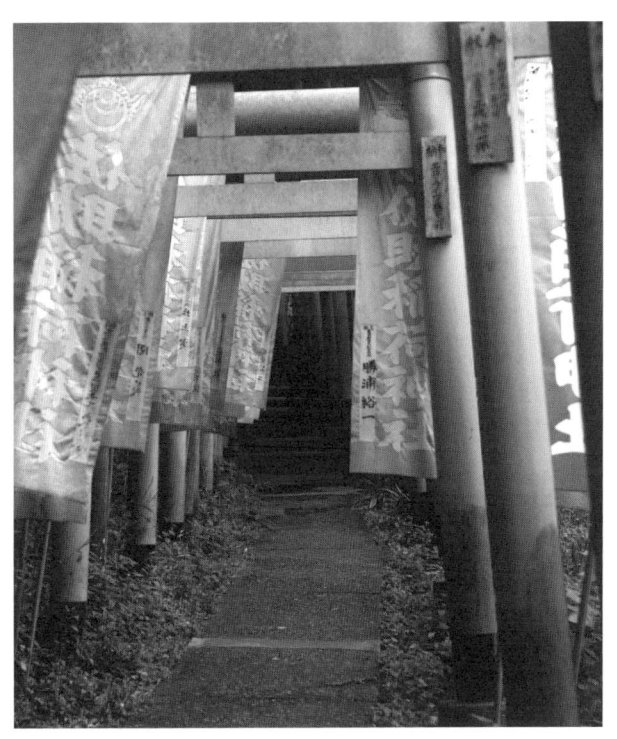

Alleyway in Kamakura, Honshu.

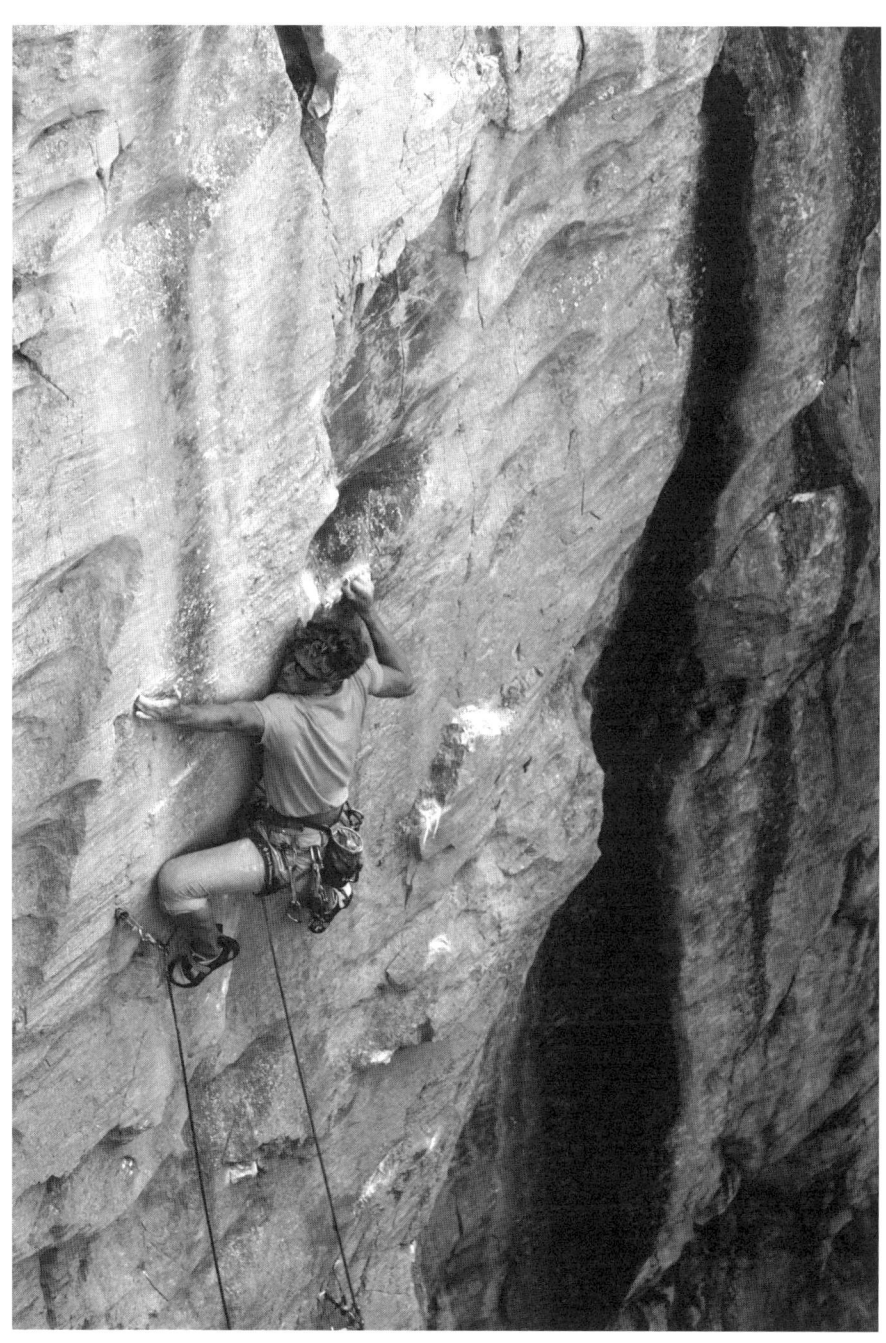

The author adrift on the west wall of Huntsman's Leap on the direct start to his route Dusk Till Dawn (E7).

Rapture of the Deep

The visionary darkness of Huntsman's Leap

The horse stamps his hooves and snorts as the rider pulls him to a halt. Surveying the clifftop, the huntsman traces its contours for signs. The low winter sun flashes off the steel barrel on his rifle as he steadies the horse. Out to sea, squalls scud across the sky and darken the waves. Lundy appears and vanishes through the mist. A pair of choughs quarter the open ground between the horse and the cliff's edge. As they pass, the rider notices a long, low depression in the ground about two hundred yards ahead. Like the shadow of a submerged whale, it reclines among the thickets of gorse and heather. To landward, the shadow widens to suggest the presence of a chasm below, presenting what appears to the huntsman as the opportunity for an epic show-jump. A wild-eyed young daredevil, he can't resist the temptation to attempt the leap. Acknowledging the rider's proposition, the horse senses his boldness and makes a single, powerful snort. At the exact moment of that signal, he digs his heels hard into the animal's flanks, and the pair of them surge forward into the sun. Clods of ripped earth fly into the air, the thunder of the horse's hooves increasing until it overwhelms the din of the sea. An immense black shape now fills the air. High above the sea, weightless for a moment, and with the horse's charge bearing him forward headlong, the huntsman is filled with an exalted freedom. A lone rider against the pale sky. White horses beyond.

It was late one summer Sunday afternoon by the time Mike Robertson and I were racking up on the flat grass above Huntsman's Leap. I was seventeen, and felt as if I'd just landed on another planet. Down in the depths of the chasm, climbers were making their way up some of the classic routes: S*hape Up, Bloody Sunday, Beast From The Undergrowth*. Chalked holds made unlikely connections in the gloom and wove patterns across the coloured stone, white against red, merging into complex hieroglyphs at certain cruxes. The spooky names of the harder routes echoed like ghostly voices in my head: *Minotaur, Headhunter, Snake Charmer, The Witching Hour.*

As you descend into the Leap there is a distinct drop in temperature as the sun only touches its sandy, boulder-strewn floor for a few hours each day. Once immersed in its depths, you enter a transitory, tidal space of exquisite beauty and mystery.

I set off up *Minotaur*, a statuesque climb that pierces the brooding heart of the west wall, weaving through its welded conglomerate channels and weird parabolic bulges like monstrous ice cream scoops. Sinuous cracks threaded through frozen serpentine structures of red and orange stone. Reflected light from the pools of seawater left by the falling tide glinted off the wall, washing the boulders with oily liquid. Right then, a big Atlantic grey seal with a scar on its nose swam in on the tide. He stared straight at me, and then up at Mike with intense curiosity. Above him, a host of awesome climbs waited in prescient silence, like sleeping giants frozen in ice. Little did I know that this introspective chasm would shape my climbing ambitions for years to come.

The more I climbed in Huntsman's Leap, something like a sense of rapture with the place started to form. High on the bold arête of *Compulsion*, far from the tiny wires I'd placed twenty feet above the boulders, I stared into the smooth looking-glass of the wall. Onsight climbing is often about following threads, hoping they will be long or strong enough to eventually tie together. Human beings are neither wizards nor angels, so our conjuring tricks seldom work out the way we'd like them to. I managed to climb *Compulsion* that day, but other routes got away.

Two years later, on a balmy Sunday morning in mid-April, I'm definitely not feeling like climbing a big, bold route in the Leap after an all-night party at Trevor Massiah's house in nearby Stackpole. Unfortunately for me, my partner that day and the most enthusiastic person in British climbing at the time, Tim Emmett, had other ideas. Spurred on by his enormous psyche, soon I was cleaning the barnacles off my shoes and chalking up below the ultra-intimidating line of *Nothing To Fear*, a fearsome testpiece on the left side of the west wall. As it had before, and as it would so many times again, the Leap cast its spell. I suddenly found myself at the crux; a weird, strung-out sequence on bad sidepulls and insecure footholds. With more gas in the tank than I'd dared to imagine, and Tim's voice encouraging from the darkness beneath, I floated into the deep runnel marking the end of the hard climbing.

An hour later, my hangover returns as I find myself holding Tim's rope on a bizarre climb named *Woeful*, which lies underneath the boulder choke in the depths of the Leap. Having long cut off any hope of escape, the tide is lapping at my feet as Tim fights the beast in the black space above. The rock in the Leap's innermost depths has a strong greenish pigment, and colours the water in dizzying jade and emerald light. Moving through this secret region as the reflected light off the sea illuminates the darkness all around is one of the truly great experiences in British climbing – even if you get a full soaking.

As the years flew on, I found new territories in the cool air of the Leap's darkest reaches. In 2010, I teamed up with Dan McManus for an attempt at the gorgeous-looking unclimbed headwall to the right of Pat Littlejohn's masterpiece *Terminal Twilight*. Swinging around on the abseil rope, I discovered a pair of strange twin pockets that allowed access to the headwall. The crafty passage through the water-sculpted flutings that create the link to the top offered some of the best climbing I'd yet done in Pembrokeshire, and *Dusk Till Dawn* was finally complete. As Dan neared the top, darkness was falling, and a big grey seal swam in towards the beach on the rising tide, just like the one who came in on my very first evening in the Leap in 1998. The seal's sudden presence articulated what I already knew about this place, and everything else I didn't know. Mysteries remain undeciphered on these shadowlit walls. Some could produce climbs so futuristic that when completed they will become testpieces for future generations, as the Leap's hardest lines are today for ours.

Perhaps that's the measure of a truly great cliff – somewhere to find inspiration and potential at any stage of your life, and in any era. Or perhaps it's more the measure of the Leap's distinctive kind of magic, that rapture of the deep you always find down there in the darkness of a summer's afternoon.

Climb magazine, 2016

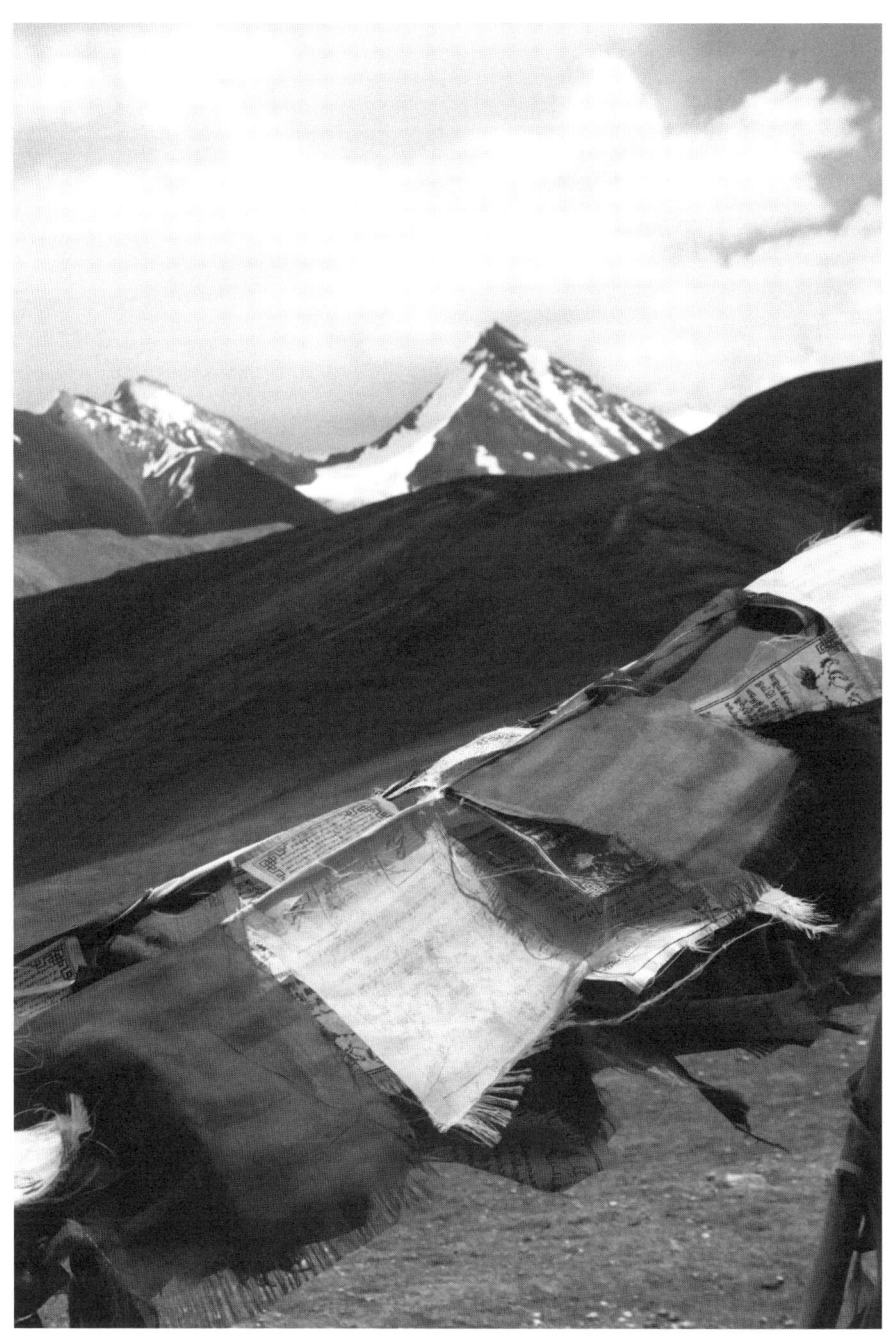

Prayer flags flying on the Zalung Karpo La (circa 5100 metres), Zanskar Range, Indian Himalaya.

The Jade-Star Kingdom

Exploring Zanskar's mountains of ice & fire

Late summer 2006, I'm back in Delhi after a month of trekking over two hundred miles across the heart of the Zanskar Range, from Darcha in the south to Lamayuru in the north, crossing four 5000 metre passes. Pink Floyd's *Wish You Were Here* crackles in my headphones and I stare at the fan on the ceiling. All I need to do is close my eyes, and I'm straight back to those mountains in my mind.

I gaze down from the roof of a 5th century monastery to a quilt of barley and maize that spreads across the floodplain of a powerful river. Above, crumbling chorten monuments cling to scree cones strafed by storms. Wayward goats graze under prayer flags bleached to antique shades by the strong high altitude sun. Evening light drains through poplar groves and highlights the spines of distant ridges, serrated and stark against the black shadows below.

A solitary raven's croak breaks the silence of our camp as I wake to a cold dawn at 4,500 metres below the Anuma La, where a light dusting of snow has fallen overnight, turning the arid mountains ghostly pale. An hour later, on the way up to the pass, I find fresh snow leopard tracks. The leopard, that most elusive of Himalayan creatures, has passed this way overnight or in the last few days. I sense the powerful, secretive animal up there, somewhere above the snow line, watching us.

Two days later, the tiny, otherworldly village of Photoksar clings to the cliff like a medieval fortress above the confluence of two gorges. From the summit of the Sirsir La, the last high pass we will cross before reaching the Indus Valley, higher mountains erupt from the horizon. To the north, towards China, the Saser Kangri group spears the sky in a series of gothic spires. To the west the twin peaks of Nun and Kun, Zanskar's only 7000 metre giants, stand proudly together, anticipating the Karakoram rising beyond them. More possibilities for mountaineering adventures than anyone could fit into one lifetime surround us here. Just staring at these peaks feels like opening a rift. Who could we become here? Here are the reasons the Himalaya has been the refuge of monks and mystics for thousands of years.

Later that night, as I board the plane back to London, I realise that in a month among them, these mountains have worked their way into my skin. I'll be coming back to Zanskar, I hope, as often as time and chance allows.

As the most accessible and straightforward 6000 metre peak in the Zanskar Range, Stok Kangri is one of the most popular mountains in the Greater Ranges. After a long, steady ascent through the moraine, a series of snowfields leads to a narrow col on the impressive west ridge, with spectacular views of the imposing Kang Yatze (6400m) to the south, and north towards the even higher peaks above the Nubra Valley. From the col, the straightforward but often exposed west ridge leads to the true summit. Accessible mountains like Stok Kangri are a great introduction to high altitude climbing, but the true attraction of mountaineering in Zanskar lies in getting off the beaten track. The Korzuk Range in the Tso Mori area of Ladakh, around a hundred miles east of the Indus Valley, isn't just off the beaten track. There really aren't any tracks as such around there at all.

When the expedition team I was leading arrived in Leh, the bustling capital of Ladakh, in mid-August 2014, I was informed by our agent that Lungser Kangri, the 6,600 metre peak we intended to climb in the Korzuk Range, had been closed by the Indian military until further notice. Due to its proximity to both China and Pakistan, Zanskar & Ladakh is a politically sensitive region, and such closures are not uncommon.

Searching for an alternative, challenging yet achievable objective, I came across an online report by a recent French expedition to the Mentok group of peaks in the Korzuk Range, which lies just to the south west of Lungser Kangri, and most importantly was still open for climbing. Comprising a dramatic chain of summits linked by a ridge-line extending for over twenty miles from north to south, dotted with 6000 metre peaks, the French report described ascents of the two most striking summits in the range, Mentok I and II. Our new goal was therefore set.

After a week's acclimatisation in the Markha Valley, then across the 5100m Zalung Karpo La and down the other side through the most spectacular system of canyons I've seen anywhere on Earth, we arrived in a nomadic Tibetan settlement at 4700 metres that was one of the loneliest and wildest inhabited places I have ever seen. Here we were picked up by our agents for the drive across the high plateau of northeast Ladakh to Tso Mori. We were finally ready for our revised objective: Mentok II.

Past a narrow col at six thousand metres, the ridge sweeps northwards. From the 6,150m summit of Mentok II, it looks like the vapour trail of a banking plane: a thin white line stretched across the sky. To the west, the highest summits of the Indian Himalaya's Zanskar range shimmer against the cold blue air. To the northeast, the bulk of Lungser Kangri (6,662m) looms vast and silent, an icy giant watching over the silent waters of Tso Mori lake 2000 metres below.

We'd been on the go since two in the morning. Endless moraine, steepening icefields, a rocky spur, and a final snowfield led us to the summit ridge. To descend, we must make a long traverse of the horseshoe-shaped ridge, two miles at over 6000 metres, before we can descend an easier angled spur to the moraine. The next few hours on the ridge are amongst the best I've ever spent in the mountains. The air is an impossible, crystalline blue. The upper world stretches beyond the horizon in every direction: north into the Chinese Karakoram, east to Himachal Pradesh, west northwest to the lonely summits of Kashmir. Like an early aviator, I trace the unknown liminal between the Earth and the air.

It's just after six in the evening before my team reaches the moraine. It'll be dark in less than an hour. Here, at 5,800 metres, the temperature will fall like a stone. A series of colossal scree-spines extend out from the glacier for several miles, separating us from base camp. Like the defensive structure of an ancient city on an unimaginable scale, the scree seems impregnable. Dark falls fast as we traverse the scree-spines; it's like walking up and down several 100 metre hills covered in boulders the size of footballs. After two hours, the lights of base camp are still invisible. Finally, we reach a tiny stream I recognise by the light of my head torch from our ascent. After a while, lanterns appear and low voices murmur from the night and the distance: our base camp crew come to meet us on the terminal moraine with hot mugs of steaming chai and lemon squash. We'd been on the go for just over nineteen hours.

Some places inscribe themselves on memory more than others, and I've been fortunate to lead several mountaineering expeditions to Zanskar over the past fifteen years. Those mountains are a highly-charged place, with endless interlocking ridges of black and orange rock against the gold-blue sky, pale green poplar trees growing tall beside surging rivers of glacial meltwater, ice-capped summits shining like polished marble in the morning sun, the moonlight beaming back into the atmosphere at night, and the ultra tough, bright-faced locals. As the clasp of winter fastens every year, the mountains of Zanskar and Ladakh fall silent, the frozen domain of the bharal, the lammergeier, and the snow leopard. And when spring returns and the snow melts once more, they are still there in all their glory.

Our climb of Mentok II in 2014 wasn't a proper epic in mountaineering terms. It was just a long day on the hill, on which everything went more or less according to plan, and we climbed well within the weather-window we needed. But had something happened that fell slightly outside the expected parameters – a twisted or broken ankle, say, or sudden poor visibility on that final slog back across the moraine – then a real epic might have ensued. It sometimes takes just one minor event to swing a controllable situation into a much more serious one.

Epics are at the heart of the climbing life, at least for some people. To become an accomplished alpinist, in particular, going climbing means having a certain number of epics that you can learn from and improve. You certainly don't have to go on expeditions to remote places to have epics; they can take place almost anywhere, even on a single pitch crag close to your home. In a psychological sense, part of what it means to be a climber is the capacity to deal with epic situations. To be truly competent, you need to be able to deal with stuff that's outside your comfort zone.

Epics matter because climbing is often at its most rewarding when you manage to pull something off without having an epic that could have very easily turned into one. If I learnt anything from that glorious day on Mentok II in Zanskar, it was that any adventurous high mountain climb must tread the uncertain ground between everything going to plan, and the very real possibility that it doesn't.

Climb magazine, 2016

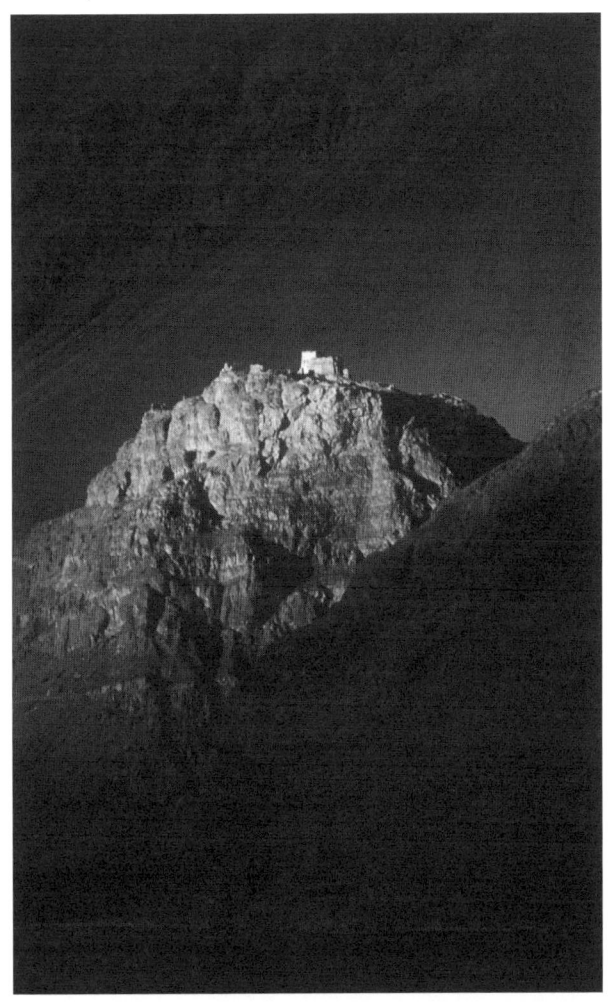

*High above the surging waters of the Zanskar River,
Zangla Gompa is highlighted by a shaft of sunlight.*

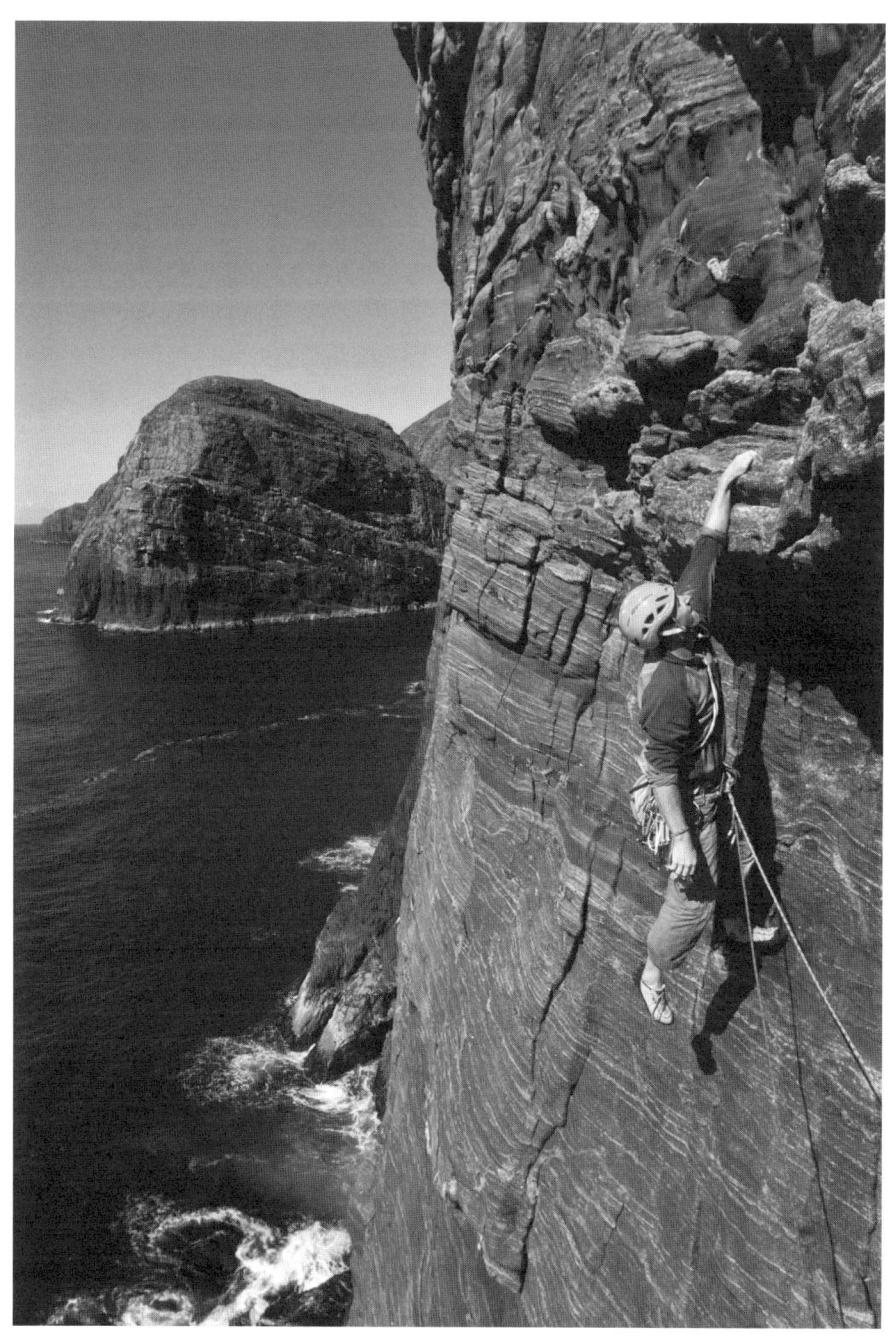

Charlie Woodburn on pitch 2 of Call of the Sea (E3 5c) on Mingulay, Outer Hebrides, Scotland.

Crusoe and the Witch

A stormbound sojourn on Mingulay

The air over Guarsay Mor suddenly turns dark with shadows as we approach. The shadows dart and whoosh and wheel like huge bats through a threatening sky. Then a rush of displaced air hits Charlie and I almost simultaneously from a blindspot on the left. Two of the vicious terrordactyls swoop so low their spiny feet skim our ears.

"What the hell was that?" Charlie spins around, blinking into the sky as another pair set their sights on us with swept-back wings.

"Bonxies!" Garth exclaims. "Keep to the ridge, away from the nests."

"Aaarrrggghhh" erupts from Neil as another airborn attacker sweeps in, and we make haste for the sanctuary of the clifftop. We defend ourselves by swinging our climbing helmets around our heads. It's just like *Top Gun*, but with massive birds instead of F15s and MiG 29s.

We're on the uninhabited island of Mingulay at the southern extremity of Scotland's Outer Hebrides. It lies some twelve miles southwest of Barra, the nearest inhabited place. It's one of the most remote of all the British islands. Although continuously inhabited for around two thousand years, evacuations began in 1907 and by 1912 the island had been completely abandoned by its human residents. As with the even more remote St. Kilda, there were numerous reasons for the evacuation.

It seems likely that the sheer hardship of life on the island – particularly during winter when it became inaccessible for months at a time – was a major factor in the final exodus, when the prospect of an easier life on the larger islands to the north became possible.

Through the rest of the twentieth century, Mingulay was mainly left alone to its resident populations of seabirds, mainly razorbills, guillemots, kittiwakes, and puffins, plus a large colony of grey seals. It was not until the early 1990's that climbers discovered the Lewisean Gneiss crags surrounding the island that would harbour some of the most spectacular sea cliff climbing in Britain. The igneous rock of Mingulay is part of one of the earth's oldest geological groups: it was already about four hundred million years old when the Himalayas were being formed. As you'd think given its antiquity, it is generally extremely solid, with gritstone-like friction and frequent opportunities to place natural protection. The first recorded routes on Mingulay were climbed by Mick Fowler and Chris Bonington, a partnership that indicates climbing of high quality and real adventure was to be found here. New climbs have been pioneered by visitors every summer from 1995 onwards, and the hundred-metre northwest wall of Dun Mingulay, an imposing promontory that harbours the remains of an Iron Age fort, is now one of Britain's best sea cliff crags.

The exciting idea of sailing to Mingulay from the Scottish mainland came about through a plan hatched on Lundy in southwest England, and another showpiece of offshore British climbing. Enter Patrick Trust, our very own Captain Ahab, skipper of *Hecate*, a forty-five foot twin masted cruiser. Patrick has sailed the turbulent seas around the Inner and Outer Hebrides for close to fifty years. The plan was simple: we would sail out to Mingulay from Ardfern on Loch Craignish, where Patrick would leave us to make course north for Barra, Eriskay and Harris, returning in a week's time to pick us up.

The voyage out on midsummer's night couldn't have been better timed, as our departure coincided with a ridge of high pressure over northwest Scotland. The Sound of Jura was almost flat calm as we turned Craignish Point, and the ink-dark water sluiced under the hull. The perfect sea conditions were fortunate, since our plan was to make a short cut out into the Firth of Lorn through the notorious Gulf of Corryvreckan. This narrow stretch of water separating the Isle of Jura from Scarba is one of the most turbulent stretches of inshore water in the northern hemisphere. Formerly classified as "unnavigable", the Admiralty's *West Coast of Scotland Pilot's*

Guide To Inshore Waters still calls it "very violent and dangerous" and says "no vessel should attempt this passage without local knowledge". In certain conditions, the world's third largest tidal whirlpool forms here above a submerged rock pinnacle. Even on the quiet night on which we sailed, standing waves almost a metre high formed at Corryvreckan's seaward reach. The sun rose as we passed through the Sound of Iona off the western tip of Mull at around four a.m., and we headed out into the Minch on a perfect summer's morning.

The panorama that now lay before us was enough to put any thoughts of seasickness at bay. A chain of blue-green islands lazed on an idle sea to the west. Barra to the north, then Vatersay and Sandray; Pabbay lay a little closer still, and then on our course almost dead ahead was Mingulay itself.

We dropped anchor off the sandy beach on the east coast, one of only two landing spots on the island. After a delicious slap-up lunch of shepherd's pie on the aft deck, we had work to do. The dinghy was launched and after at least a dozen runs to shore and back, we'd landed our gargantuan stash of kit and food on the beach with no need, remarkably, to jettison anything. Our skipper, Patrick, rowed back out to *Hecate* and pulled her anchor, giving us a final wave as he motored out before finally disappearing around the point to the north, bound for the sheltered sanctuary of Castlebay on Barra. Finally, we were the castaways on Mingulay we'd first imagined in the smoky back room of Lundy's Marisco Tavern almost two years ago.

Tents were quickly pitched, and before long we'd shouldered packs and climbed to the wide col on the ridge that forms the undulating spine of the island. Dodging the dreaded Bonxies on the higher slopes of Guarsay Mor, we eventually made it down to the clifftop. Despite the ominous signs of a front pressing in, with streams of cirrus darkening and building across the sky to the south, we managed to get a couple of climbs in that evening.

Sunday dawned grey and chilly as the full force of a northerly gale began to make its presence felt. An old Scottish myth tells that the hag goddess of winter, Cailleach Bheur, uses the Gulf of Corryvreckan to wash her plaid, and this then ushers the turn of the seasons from autumn to winter. I hoped we hadn't stirred her from her summer sleep on Friday night as we crossed the Gulf. *Hecate*, after all, is a ringleader of witches in European mythology. Casting spurious thoughts of witchcraft aside over another round of fresh coffee, we put the poor conditions to our advantage and made a reconnaissance of the northern half of the island.

That night the weather cleared, and tomorrow's sights were set on the great precipice of Dun Mingulay, reportedly the island's best climbable cliff. Sure enough, the next morning was fine and breakfast quickly merged into the organisation of gear and ropes, and we struck up the hill to the south west. To gain the promontory of Sron An Duin you must cross a narrow col, and the view down into the chasm below is truly awe-inspiring: the eye plummets more than four hundred feet into a dark cauldron filled with circling gulls.

Throwing the entire length of a hundred metre static line out over a sea cliff is a relatively rare experience; there are few crags in Britain that require more than three hundred feet of abseiling to reach their base. Most of them, like Carn Gowla's America Buttress, are notorious for the committing nature of their climbs. Dun Mingulay is certainly no exception to this rule. Once you're at the base, the only way out is up.

The crag was still wrapped in the morning's shadow. Beyond the point gannets dived for fish, falling like thrown spears into the sea. Seals welcomed us with their familiar inquisitions, appearing and vanishing, then reappearing even closer to shore. Even in calm conditions, a slow Atlantic groundswell sucked along the base of the cliff, reminding us that the next dry land to the west would be the beaches and cliffs of Newfoundland.

It was a perfect day to be here. Just as I topped out on the spectacular forty-five metre final pitch of *Call of the Sea*, the sun swung round and lit up the entire northwest wall of Dun Mingulay. Veins of quartz suddenly sparked like silver fish in the rising light. By six o'clock the tide was low and with another six hours of light to play with, we moved the rope over to the huge arch at the centre of Sron An Duin, home to the most impressive climbs on the island. As I belayed the others on the superb second roof pitch of *The Great Shark Hunt*, a Minke whale surfaced for air just beyond the point, and the evening sun glistened off its back before it plunged again into the deep. I watched Neil making steady progress up the final pitch of *Ray of Light*, a crazy line on which some of the holds stick out like gigantic tusks almost a metre from the main wall. As we sorted gear and coiled ropes that evening, still blessed with sunlight at ten o'clock, we found it difficult to come to terms with just how good the climbing here really was. Walking back over the hill, St Kilda, the most remote of all the British islands, appeared unmistakably on the northwest horizon.

The weather report crackled on the VHF radio early on Tuesday morning, and the possibility of being stranded on the island dawned: "This is Clyde Coastguard. There are gale warnings in force for Rockall, Malin, Hebrides… Now the general situation at 0900 hours. Ardnamurchan Point including the Outer Hebrides: easterly gale force 8, backing south easterly and increasing severe gale 9, imminent. Rain then showers. Sea state: very rough or high."

That night we were battered by the full force of a highly unusual easterly gale howling off The Minch, and the double-overhead surf breaking on the east-facing landing beach didn't make our prospects of escape look any better. Luckily, the wind swung around to the south west later, and settled to a more reasonable level on Thursday. When we woke on Friday – the date of our intended departure – the sea was as calm as when we arrived. I scanned the murky horizon for signs of a mast. As our final supplies of fresh coffee bubbled in the percolator, a familiar voice came in on the radio: "Mingulay, Mingulay, Mingulay. This is yacht *Hecate*. Do you read me? Over."

Our able captain had weathered some of the worst June storms in recent memory and had made it down to collect us in perfect time. A muted cheer drifted over the beach as *Hecate* appeared through the mist. There was a mingled sense of relief at catching the weather-lull and sadness at having to leave the island. A couple of hours out into the Minch, a pod of bottlenose dolphins joined us, diving across the bow and plunging under the hull for almost an hour, leaving us enchanted by their free companionship.

That evening, we waited for the flood tide in Tobermory's sheltered harbour at the northern entrance to the Sound of Mull. Standing with shaky legs on inhabited land for the first time in a week, the wildness of Mingulay vividly returned as I looked out across the cinder sky over Ardnamurchan, knowing I'd have to set sail sometime again across that dolphin-torn sea.

Climber magazine, 2008

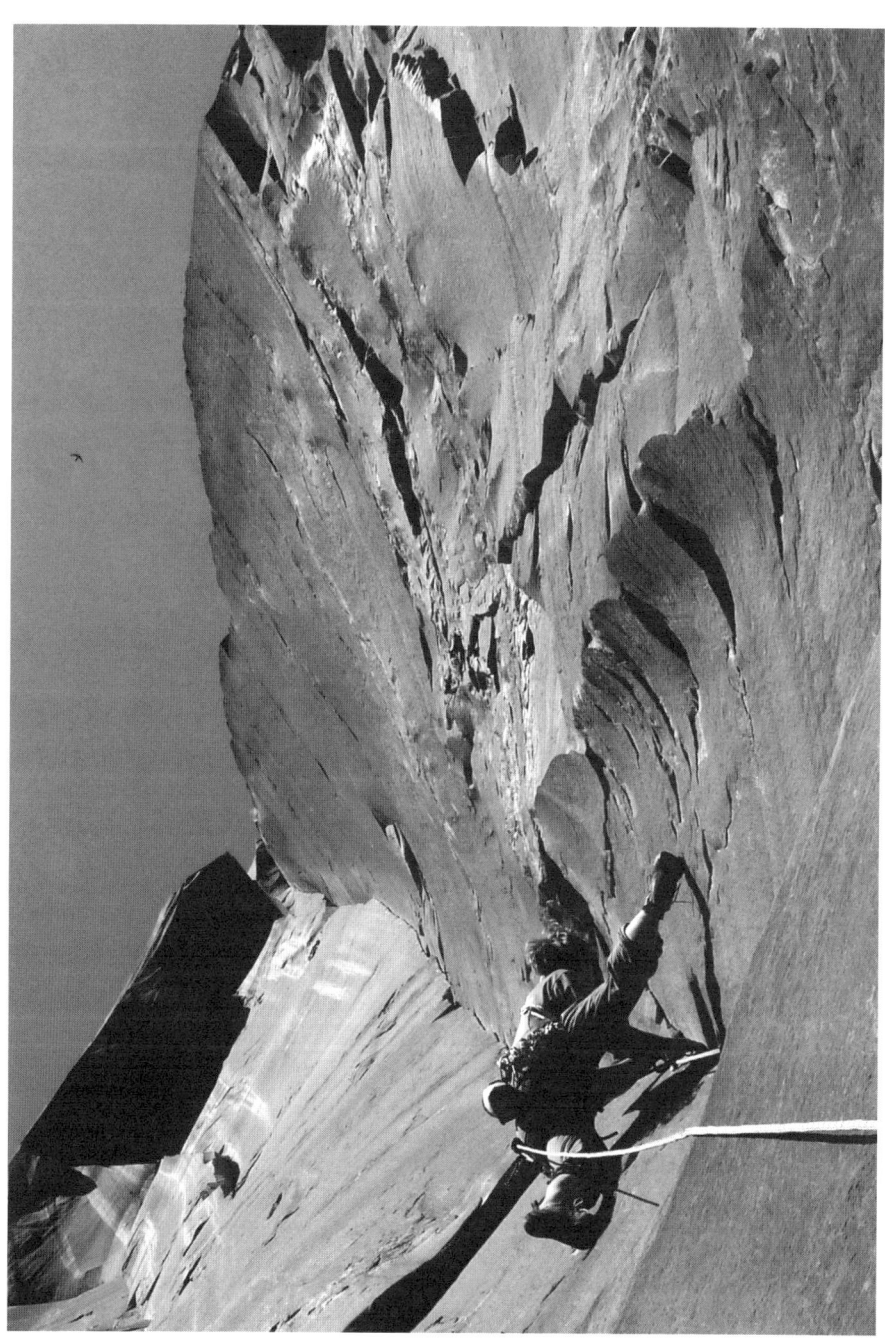

The author leading pitch 4 of Moonlight Buttress (5.12c), Zion, Utah, during a free ascent.

Wind, Sandstone and Stars

The sky-imagined climbs of the American desert

It was getting late when I turned west off Highway One Ninety-One. Low in the sky, a blood-orange sun began to sink into the labyrinths of Canyonlands. Sudden gusts of wind stalked the country, and the surrounding mesa darkened as I drove west. The air grew colder as the road dropped off the plateau. Just past the petroglyphs of Newspaper Rock, through the corner of the windscreen, I glimpsed a buttress of coloured sandstone fired by the last few minutes of evening light. The rough rock shimmered red, chased by the encroaching shade. Through the gloom, I could just make out the two-tone shadow of a colossal dihedral spearing the cliff. Jet black lines of splitter cracks scored the walls to either side.

Instinctively, I pulled over and cut the engine. In the middle distance, the slender towers of the North and South Six Shooter broke the horizon. There was no sound at all except the faint hush of the wind in the cottonwood trees. In every direction, high-walled canyons extended into the sandstone tableland, and I came to terms with where I was as night fell across the Colorado Plateau. After years of waiting, I'd arrived in Indian Creek, the heart and soul of rock climbing in the American desert.

Nothing you have heard about crack climbing on Windgate sandstone can fully prepare you for the first time you climb on it. American climbers often say that if you think you know how to climb cracks, but haven't been to Indian Creek, then you're in for a surprise. This is broadly correct. For your first trip – for the first few days at least – it's a good idea to bear this in mind, and reign in your climbing ambitions accordingly. Before Ray Jardine invented the first active cams in the 1970's, only a few routes had been climbed in the Utah desert by a daring band of pioneers. The reason for this is obvious: virtually all the routes follow perfectly parallel-sided splitter cracks, which nothing other than camming devices can effectively protect. The evolution of climbing at Indian Creek therefore mirrors the development of active cam protection itself.

Getting used to the unique movement of climbing here is a bit like acclimatisation for altitude: if you get it right, things start slowly but surely to feel easier. If you get it wrong, it'll feel more like you've been doing a few rounds with a heavyweight champion wearing a pair of washing-up gloves. At some point, you'll cross a watershed and feel much more comfortable cramming fingers, hands, fists, elbows, feet, shoulders and knees (or a combination of all the above) into the world's most perfect and laser-cut cracks. Many seasoned Creek climbers cherish their memory of the day when it all started making sense, and routes which previously had seemed desperate became relatively approachable.

After a few weeks of perfect spring weather in April 2010, the Creek started to feel like home. I climbed many of the classics. When I finally succeeded on my first 5.13 crack, a classic line called *Ruby's Café*, I finally felt like I was making real progress with the rubix cube of Windgate sandstone.

After seasoning ourselves in the Creek, my climbing partner Matty Rawlinson and I drove across to Zion National Park with just one route on our minds. It had been suggested to me by various reliable sources this great sandstone canyon might be home to North America's best multipitch climb: *Moonlight Buttress*. It is the definition of a perfect line, following a single crack that bisects a slender pillar of marmalade-orange sandstone a thousand feet high. It's still popular as an aid route, but our intention was to free climb it.

We crossed the Zion River early in the morning, the icy chill of the water sending shockwaves through feet and legs. But what a way to wake up, with that awesome climb waiting directly above. We got stuck into the first couple of pitches without a moment to lose. Skittish, sandy holds and an interesting rising traverse up to the right lands us on the ledge known as the Rocker Block at the base of the first dihedral: this is the beginning of the 5.12 climbing that continues for the next 700 feet. We had arrived, as they say Stateside, at the real deal. Right on cue, the morning sun hit the ledge as we prepared for the challenge ahead. I head off up the first dihedral pitch, and after a short tussle with its tricky boulder problem start, I make it to the belay. Matty joins me and we rearrange the rack before he heads on upwards into the corner. The crack is just wide enough here for fingertips; that's it. He arrives at the belay and I follow the pitch, climbing quickly and confidently now. After a while, we make it through the dihedral crux pitches to arrive at the notorious 'slot' pitch, a kind of flared overhanging chimney. Matty leads this one in full fight mode, arriving on the belay ledge hyperventilating. It's easier to follow as the ropes don't get in the way of your feet, and I lead on up what turns out to be my favourite pitch of the whole route, pitch eight. It's a stunning finger crack in what feels like the very centre of *Moonlight Buttress*, the exposure now overwhelming with the river over a thousand feet below us. The rest of the route passes in a dehydrated haze as our water runs out, and Matty leads on through the last pitches. There are hugs and cheers as we coil the ropes on the apex of the buttress. It's been, we decide, a very fine day of rock climbing.

After *Moonlight Buttress*, we returned to the Creek for a few days. Still flushed with the afterglow of the climb, the place seemed even wilder and more peaceful than before, with the spring break crowds now back home. I climbed a few more classics, and explored the stone maze that is Canyonlands. The day before I had to leave, a friend and I climbed *North Face Route* on Castleton Tower, the most famous of the freestanding spires in the Castle Valley just north of Moab. From the summit, I kept having to remind myself that I wasn't on an aeroplane: the otherworldly expanse of the Colorado Plateau stretched the horizon's limit in every direction. It was only then that I began to understand the most painful thing about climbing in Indian Creek – much more acute than the sting of broken skin from the rough sandstone or aching muscles the morning after a long, hard day – is having to leave the place behind.

I drove up Highway Six early the next morning, heading for Salt Lake City and my plane. As I turned north off the Interstate and across the empty desert of Carbon County, I understood that in climbing, as with so many other things, we can only know paradise by knowing what it is like to lose it.

Climb magazine, 2010

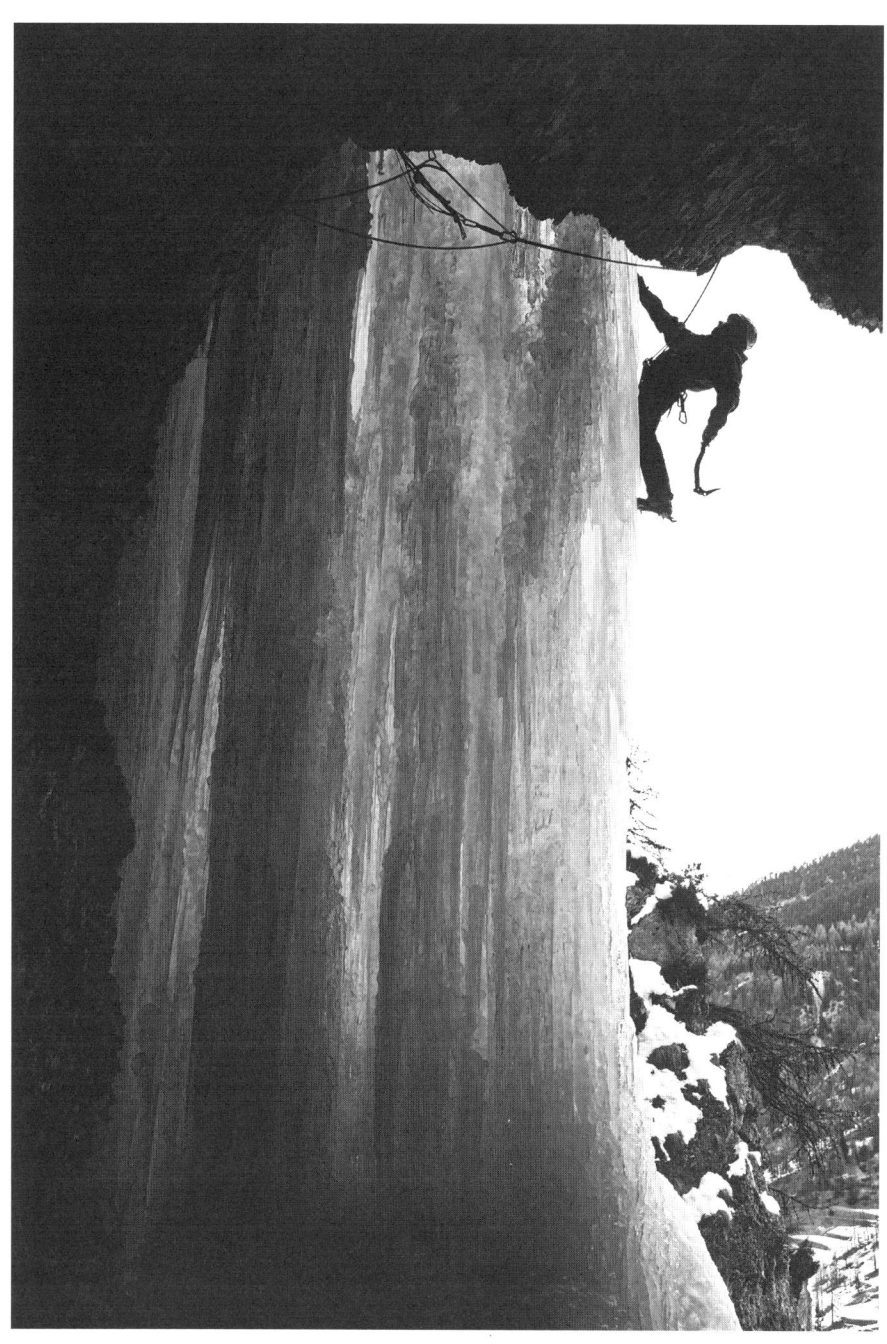
Giles Cornah on the big icefall of Sombre Heroes via the mixed line Prends Moi Sec (M7), Ceillac, France.

The Magic of Falling Mercury

The lure of the frozen world

My axe makes a dull thunk as it breaks the translucent gleam of the plastic ice, and I pull out from the security of the stance in a small cave and on to the main icicle. I'm on the final pitch of *Nuit Blanche*, one of Europe's classic icefalls, a proud cascade that spills almost three hundred feet down to Mont Blanc's Argentière Glacier from the rim of the Grandes Montets ski area.

This climb has everything you could want from a pure ice route: steep, continually challenging climbing with massive exposure, commanding one of the more outrageous positions in the Alps, yet just an hour away from the car. Somewhat uniquely, you don't even have to walk in to *Nuit Blanche*; instead, you ski down to it, and then ski out. As I near the top of the final column, I place a screw, taking the time to glance out over my left shoulder and up the glacier towards the outline of the Droites; a twisting, serrated slice in the sky like the superimposed spine of a prehistoric reptile.

I then turn and stare across the valley towards the crenellated granite ridges above Flégère, and glimpse the black dart of a lone skier plunging from the shadows of a high couloir. Finally, I look back down the pitch to see my partner, Ben, hanging out from the belay. We're both grinning from ear to ear. It was many years ago, but I still remember reaching the top of *Nuit Blanche* as clearly as yesterday. These electric moments are at the lifeblood of climbing, and winter climbing creates them in abundance; I think it's got something to do with the cold, the adrenaline, and the big effort required to do it at all. I got interested in ice and mixed climbing many years after I started out on rock; I was thrilled to find a whole new dimension of the vertical world, complete with its own corpus of specialist knowledge, secret techniques, and hidden treasure-maps. The feeling of leading my first steep icefall, the fat pillar of *Sombre Heroes* in Ceillac, France, felt like a big deal in the same way that leading my first extremes as a teenager felt like a big deal. At the same time, a whole new region of experience was revealed by the magic world anticipated by December's falling mercury.

On April 1st 2013, I fulfilled a long-standing ambition to climb *Central Icefall Direct* in Llanberis Pass towards the end of the astonishing cold snap that particular spring brought to the UK. Finding this classic testpiece in prime condition so late in the season sounds like an April Fool. It nearly was, since the route collapsed less than a week after my ascent with my friend and ice climber extraordinaire, Tim Emmett.

At the same time, the profoundly ephemeral nature of ice climbing in a place like North Wales, with its small mountains so close to the sea and the warm waters of the Gulf Stream, is a reminder that mountain activities in general, and something like ice climbing in particular, depend to a great extent on certain climatic factors. At the time of writing, a decade since I climbed *Central Icefall Direct*, I understand from my local contacts in the area that the route has not fully formed into a climbable icefall since that Arctic spring I climbed it in 2013. Climate change is transforming mountain landscapes and ecosystems across the world; glaciers are melting, some classic climbs in the Alps are disintegrating due to a breakdown in permafrost, and many European ski areas at lower altitudes are threatened with closure. Mountain sports, and particularly winter sports, are conditional on the climatic factors that bring ice and snow; once they go, so do these activities eventually. For my part, climbing *Central Icefall Direct* on a perfect spring day with a great friend reminded me of that revelation I'd had at the top of *Nuit Blanche* three years before. Winter climbing distills all the real reasons we started climbing in the first place: the search for pure adventure, the desire to explore hidden realms, and the joy of forging strong partnerships.

Climb magazine, 2013

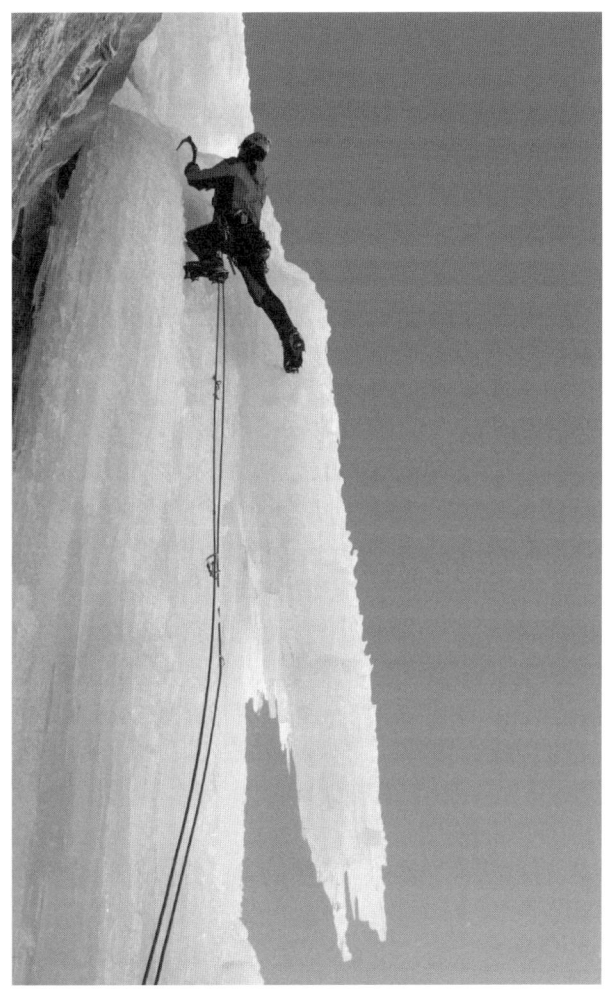

*The author leading pitch two of Nuit Blanche (WI6),
Mont Blanc massif, Chamonix, France.*

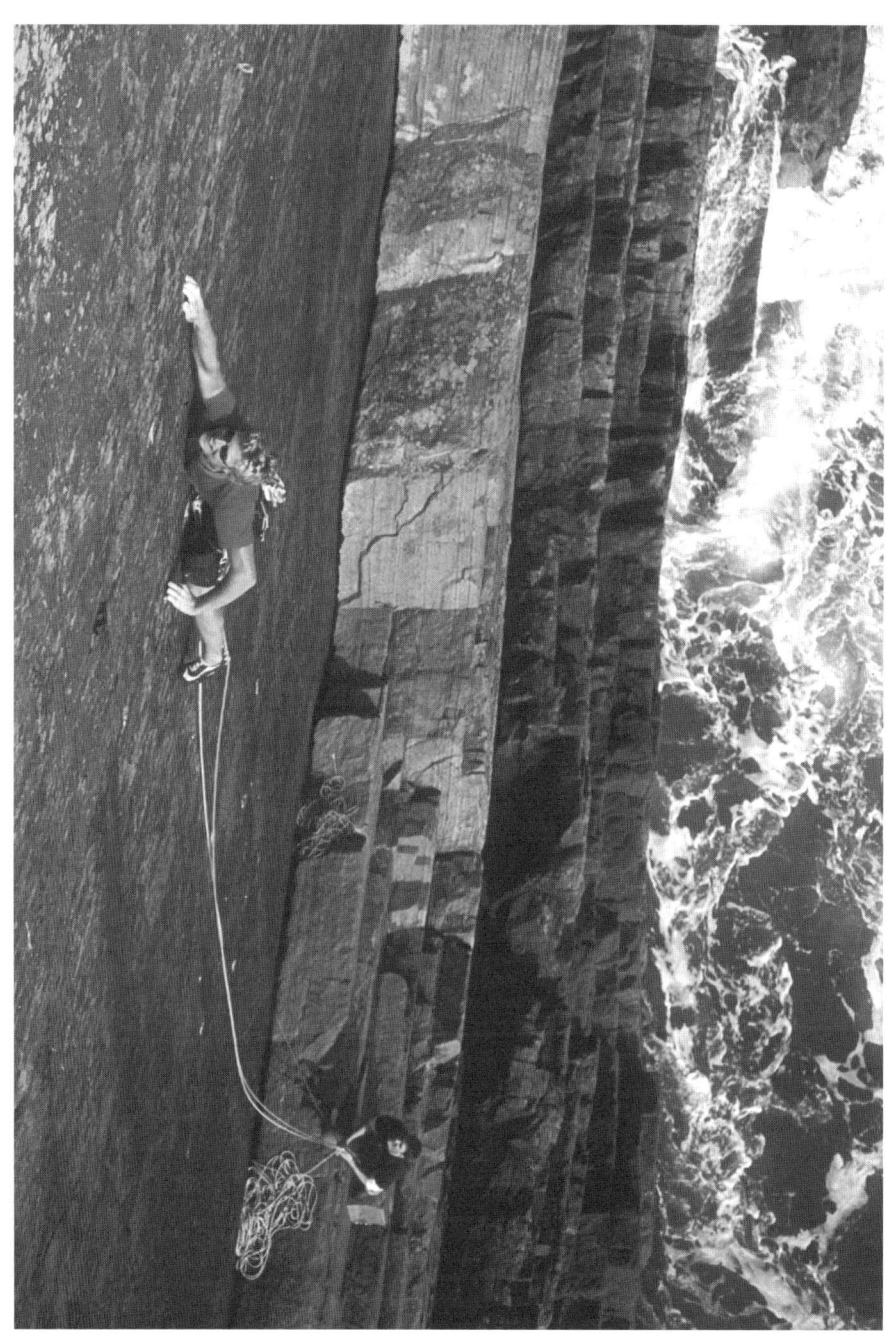

The author on the first ascent of Daddy Cool (E8 6b) at Carreg y Barcud, Pembrokeshire, in 2005.

Flow in Climbing

Towards optimal experience in the vertical

How might we describe the experience of climbing on top form? And what exactly lies at the heart of peak performance? Canadian climber Peter Croft, one of the most successful free soloists in the history of rock climbing, gives one perspective: "When you're out soloing a long route, you get into this rhythm that is beyond exhilarating. You get this super clarity, and you can see things really perfectly."

Croft's observation distills the key ingredients of confidence, control, and focus that are so critical to the most rewarding moments on the crags and in the mountains. Sports psychologists have a useful way of describing this: it's a signal of an athlete experiencing a so-called 'flow state'. Flow – a concept first developed by the famous Hungarian-American psychologist Mihaly Csikszentmihalyi – is a condition of complete engagement in which a person becomes totally focused on the task or challenge they are facing. At one level, flow is very good way of describing the psychic state an athlete enters when they achieve peak performance. At another level, it is a useful way of understanding *how* to achieve optimal performance itself.

In her book *Flow In Sports*, Susan Jackson writes that "flow offers something more than just a successful outcome; flow lifts experience from the ordinary to the optimal." In other words, flow can be clearly linked to the most successful and rewarding climbing performances, whether we are on an 8000 metre peak or doing an hour of indoor bouldering after work. But what does the flow idea actually tell us about the things we can do to climb at our best? All climbers can learn something from the flow concept about the ingredients that make for optimal climbing experiences.

The so-called 'Challenge-Skills Balance' refers how an athlete feels about their skill level in relation to the challenge they are facing. The key point is what you believe you can do will determine your experience more than your actual abilities. There are countless cases of climbers facing and succeeding on what seemed to be an impossible challenge. Henry Ford famously said that "whether you think you can, or whether you think you can't, you are probably right". This paradigm is as true for a beginner racking up for their first lead as it is for a world-class climber.

Susan Jackson writes that "to enter flow, goals should be clearly set in advance, so that the athlete knows exactly what he or she has to do". It is also equally important to be flexible and creative about your goals, in my view. You will be make your climbing more enjoyable, and stand a better chance of success, if you can 'goal swap' on the spur of the moment; choose another route, go somewhere else, and so on.

Feedback in sport refers to the many signals an athlete receives about their own performance, both from their own body and also from the external environment. If you receive positive signals about your performance you will be more likely to enter flow. Feedback is linked, I'd suggest, to a sensation of control. Having a sense that you are in complete control of what you are doing is very important for optimal performance, and in any climbing situation involving risk, a sense of control is even more important, since being out of control has major implications for your safety. You need to feel calm and in control, ideally, even whilst operating close to your physical limit. Control in climbing is about being able to focus and also to relax at the same time. It's not easy, and it takes a lot of practice.

Feeling self-conscious is also something to be avoided. Self-consciousness and distraction are the opposite of focus. If you find it difficult to focus when the cliff is noisy or there are a lot of people around, then climb somewhere else. Susan Jackson observes that "athletes report that time seems to both slow down and speed up in flow. 'Having time to think' is one way athletes describe the altered perception." It might be worth thinking about your experience of time after climbing well.

The concept of so-called 'autotelic experience' is arguably the most interesting point of all in terms of finding flow in climbing. An autotelic experience is something that is intrinsically rewarding, and which is sought after for its own sake. As John Redhead once said, "if you climb to get to the top, then you should question why you do so".

Broadening the discussion of flow, Haruki Murakami, Japan's most celebrated living writer, also has a taste for running marathons. In his brilliant book *What I Talk About When I Talk About Running*, Murakami recalls running his first non-stop triple marathon, aged forty-seven, around Lake Saroma in Hokkaido. He experiences a remarkable breakthrough at the forty-seventh mile, becoming certain that he will finish the race and moving from a state of exhaustion into one of physical energy and psychic clarity. He describes having a cathartic feeling that "[now] I didn't have to think anymore. All I had to do was go with the flow and I'd get there automatically. If I gave myself up to it, some sort of power would naturally push me forward."

Murakami is talking about finding flow on mile forty-seven of the ultra-marathon. In my experience, you sometimes can only recognise that you've experienced flow by the feeling of it passing. In this respect, a flow state is similar to the transcendental experiences that religious people and mystics claim to have when they engage in deep meditation. At a psychological level, flow is perhaps similar to one of the goals of meditation: a complete sense of being in the present, the mind completely clear of any distracting thoughts. One of the commonly-observed consequences of flow is that time is transformed in the space during which it occurs – so the period can seem much longer or shorter than it actually was. Against a cultural space in which everyone is increasingly distracted, the pursuit of flow seems like a more useful occupation than ever.

There is a point in any discussion of flow where ideas about spirituality take over. Regardless of whether you're an athlete or a mystic, though, the flow state emerges not from external ambitions, but from consistent, focused attention and complete engagement with the task or challenge in hand. If you can find a way of accessing and harnessing this elusive mind-body duality, then you're on to a winner.

Climber magazine, 2010

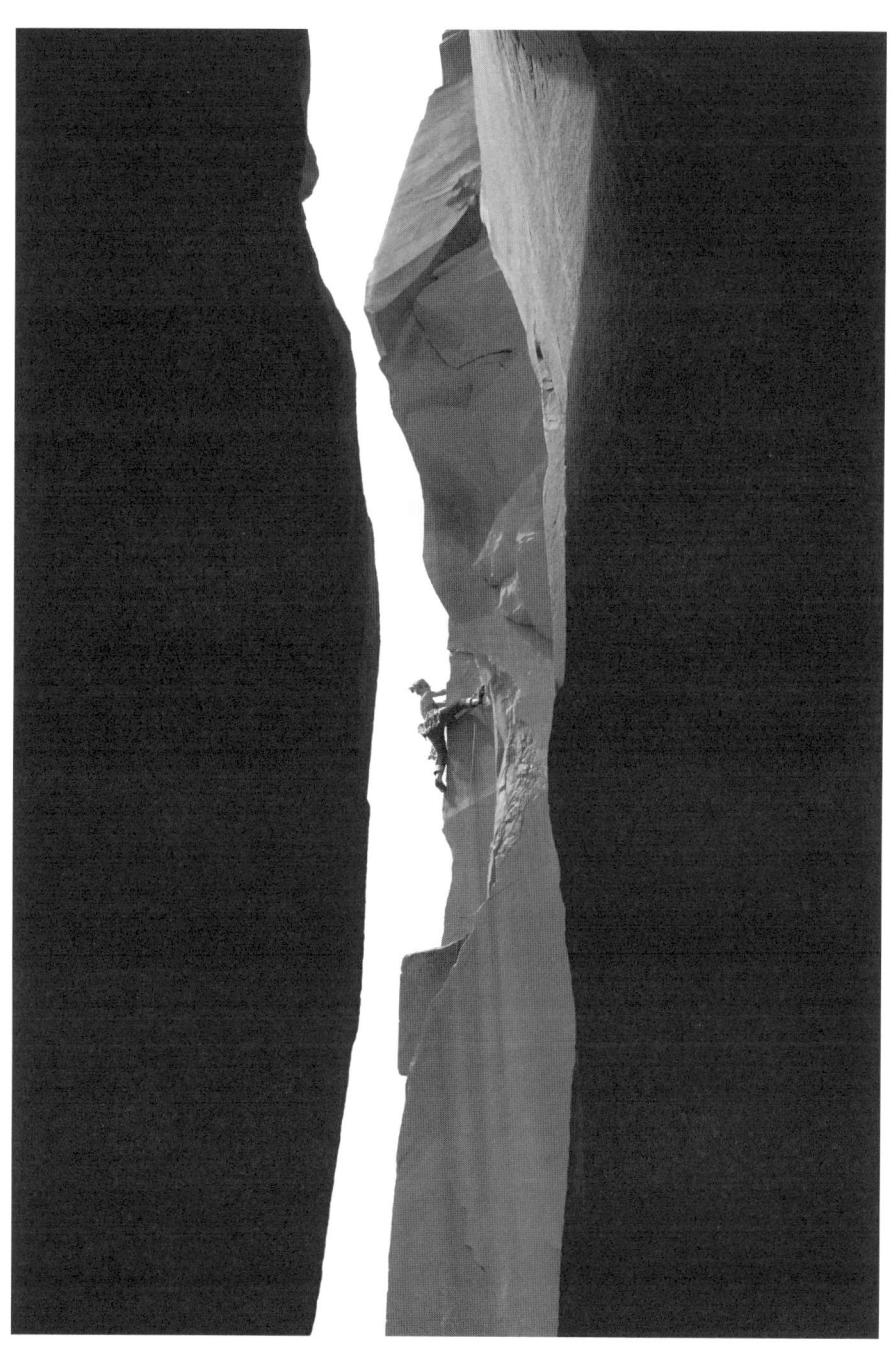
Kate Rutherford moving up to the headwall of Family Home Night (5.12c), Indian Creek, Utah.

High Exposure

A very short history of climbing photography

We live in a visual world on a scale unimagined by previous generations. The confluence of digital photography and the internet have blown the doors of photography wide open; anyone can capture an image and send it around the world in seconds. Yet for climbers, photographs have always been dream-catchers, and the camera has become an essential tool for the interpretation of experience. Climbing photography now pervades all levels of the sport, and particularly powerful images have become part of the history of climbing itself.

The relationship between climbing and photography goes back to the early nineteenth century – well before any photographic technology existed that could be taken into the mountains. Back then, climbing literature was the chosen media for the transmission of experience. In the 1820's, at around the same time the French inventor Nicéphore Niépce made the world's first permanent photograph, the English Romantic poet Samuel Taylor Coleridge wrote a remarkable account of a descent of Broad Stand – the long bastion of slabs below the summit of Scafell – in one of the earliest pieces of identifiable mountaineering literature.

Coleridge became stranded on a ledge, and recalled: "When the sight of the Crags above me on each side, and the impetuous Clouds just over them, posting so luridly and rapidly northward, overawed me. I lay in a state of almost prophetic Trance and Delight".

Coleridge's 'Trance and Delight' is a consummate statement of the Romantic condition, although it's also pretty likely that he was in the grip of an opium trip, being the nineteenth century Hunter S. Thompson that he was. In their strongly visual quality, these evocative lines also anticipate the rise of climbing and mountaineering photography. Coleridge had no means by which to take a photograph of Broad Stand. Instead, he created a kind of textual negative of his experience. His description is filled with the sense of the unknown that every adventurous climber understands. It's a very modern description, and far more in tune with the heady reality of adventurous experience in the mountains than many expedition reports written a century later. In Coleridge's eyes, climbing is a vivid interaction between the self and the world. This is also what the very best climbing photographs achieve, in their balance of space and subject, of light and shadow, and of action and reflection.

Without photography, you needed to write a decent account of your climb to be acknowledged, and this is precisely what nineteenth-century pioneers of alpinism such as Mummery and Whymper did. They simply couldn't carry the heavy Victorian box cameras of the day around with them in the Alps. The germination of climbing photography had to wait for the development of the first portable camera: George Eastman's Kodak Brownie box camera, introduced in 1900. The 'box brownie' remained popular for decades – many of the earliest climbing snapshots were taken on these – but their relative bulk still made them impractical for use on serious adventures and expeditions, and they certainly couldn't produce the same quality negatives as the larger-format cameras of the era.

It was German optical scientist Oskar Barnack's extraordinary breakthrough in 1913 that heralded the beginning of climbing photography as we know it. Barnack had found an ingenious way of using 35mm cine film to produce a still image, and his 1913 prototype, the famous Ur-Leica, was the first truly compact camera capable of producing high quality exposures. The First World War stalled his research, but by 1925 he put the Leica I on sale. The world's first production 35mm camera was received with immediate popular acclaim, and produced an almost instant photographic revolution: quality images could now be obtained anywhere, any time.

Eric Shipton took a Leica I with him to Nepal in 1933 on his legendary Everest reconnaissance expedition. Interestingly, though, Shipton and H. W. Tillman didn't take any photographs from the actual summit of Nanda Devi after their first ascent in 1934 – even though they had a camera on the trip.

By the 1940s, in the face of ever-advancing photographic technology, it became clear that that no account of a climb or expedition would ever be complete again without the addition of a few photographs. So when Edmund Hillary stood on the summit of Everest for the first time in 1953 – nineteen years after the first ascent of Nanda Devi – it was unthinkable that he wouldn't take a photograph. His striking image of Tenzing Norgay on the summit instantly became – and still remains – the world's most famous climbing image. It was splashed on the front of newspapers across the Western world and further afield. That epoch-making photograph was to herald an approaching revolution in both climbing and photography, as the technical standards of both climbers and cameras rocketed to previously unimagined levels in the post-war era.

The 50s were a dramatic decade in Himalayan mountaineering, with twelve of the fourteen eight-thousand metre giants seeing first ascents; only Dhaulagiri and Shishapangma had to wait until the 60s. They were also hugely important in the development of European and North American climbing, with major first ascents such as *The Nose* on El Capitan in Yosemite, the *South West Pillar* of the Dru in the Alps, and some of the biggest lines on British crags like *Cenotaph Corner* and *Coronation Street*.

These huge leaps forward were matched in the world of photography, where the pace of technological development in compact cameras was driven by the Western media's insatiable thirst for images, and by the new prosperity of post-war consumers. In 1959, Nikon introduced their groundbreaking 'Nikon F' – the world's first single lens reflex (SLR) camera to use 35mm film. This enabled photographers to capture images of a quality previously associated with far heavier cameras, and the legendary 'F' would be used to great effect by the leading climbing photographers of the 60s and 70s. Perhaps the most influential of a small group of British post-war climbing photographers was John Cleare, an adventure photographer and filmmaker, whose iconic book *Rock Climbers In Action In Snowdonia* laid the foundations for a new wave of climbing photography, and was, in many ways, decades ahead of its time.

Cleare's ambition in his book was extraordinary: he set out to capture significant ascents in a 'fly-on-the-wall' manner, using the previously unknown technique of taking shots whilst hanging from a rope. This opened a new world of photographic possibility. Cleare also sought to illuminate the climber as a kind of visionary figure, moving through a landscape of dreams more mysterious than the climbs themselves. Cleare reveals this in the caption to his unusual shot of Peter Crew on *Pellagra* at Tremadog: "I had this dream, see, and I was falling upwards in a shaft of light."

Cleare's stunning images presented a huge challenge to future photographers. You can see the influence of his work in all the best modern climbing photographs by people like Galen Rowell, Heinz Zak, Simon Carter, Andrew Burr, and many others. Today, even the most experimental shots still retain many of the compositional qualities first seen in Cleare's book: the dramatic silhouette; the monumental object highlighted by a small figure; the reflection of a shadow on the wall; a shallow depth-of-field drawing the eye directly to that of the climber; the presence of changing weather at the edge of the frame.

But what special ingredient makes a good climbing photograph into a great one? Often, historical achievement is a major catalyst. Hillary's photo of Tenzing on top of Everest in 1953, for example, is exceptional only due to the event it records. Yet some climbing photographs also weave a moment in history into the frame of a strong image. A great example is Doug Scott's striking image of Pete Boardman and Joe Tasker, high on the West Ridge of Kanchenjunga in 1979. Two grinning figures in bright red down suits lean on a tiny snow bank as the ridge steepens. In the background, it falls away for thousands of metres below before vanishing into a sea of cloud drifting over the Sikkim foothills. It is an iconic shot, and defines the zeitgeist of seventies Himalayan mountaineering. Scott's photo also gained an unnatural poignancy, since just a few years later, Boardman and Tasker would themselves vanish into the cloud high on the North East Ridge of Everest.

Rock climbing photography, too, can capture that elusive spirit-of-the-age. Bernard Newman's beautiful black and white composition of Steve Bancroft making the first ascent of *Strapodictomy* at Froggatt Edge in the Peak District in 1976 is a strong contender for the greatest British cragging shot of the 'seventies. It so perfectly highlights the attitude and equipment of that era. With Bancroft in his rolled-up Levi's and EB climbing shoes, the photograph is so steeped in the atmosphere of its time you can almost smell the dope smoke hanging in the air.

By the 80s, 35mm colour slide film had become the norm for the world's leading climbing photographers, allowing richly-saturated and finely-grained images to be produced even in the most difficult conditions. Cameras had developed extensively since the sixties, too. In 1978, Polaroid introduced their innovative SX-70, the world's first SLR autofocus camera. Pentax, Nikon, Minolta and Canon all soon followed suit: by the late 80s, the proliferation of autofocus systems allowed climbers to be frozen, mid-move, on the very hardest climbs. This coincided with a dramatic leap forward in international climbing standards.

Climbing magazines were evolving fast from the black-and-white pamphlets of the early 60s into the glossy format we're familiar with today. And, at the same time, an all-new and intriguing phenomenon arose: the climbing photo-book. Although mountaineering books containing extensive photography had been around for some time, three significant and highly influential books appeared in the early 1990s that confirmed the new power of the photograph as a climber's dream-catcher: David Jones's *The Power Of Climbing* (1991), Doug Scott's *Himalayan Climber* (1992), and Uli Weismeier's *Rocks Around The World* (1996).

Although completely different in style and subject, the books shared a single theme: the power of documentary photography within the climbing world. Scott's book is a kind of photographic autobiography, recounting the best tales from dozens of major expeditions across the world. *The Power of Climbing* is a remarkable historical record of the British forcing grounds of the Peak District, Yorkshire and North Wales in the early 90s. Shot exclusively on black and white film, it remains a unique account of an unforgettable era. *Rocks Around The World* is a very different kind of story: a photographic record of a globetrotting road-trip by one of the world's best climbers of the era, Stefan Glowacz.

When these books first hit the shelves, few could have predicted the tsunami that was about to sweep through the photographic world. In the late 90s, a secretive photographic revolution was bubbling in the research laboratories of the Japanese camera manufacturers. In 1999, Nikon released their D1, a 2.74 megapixel digital SLR model. Selling at a hefty £4,000, it was affordable only to pros with a taste for the new, or keen amateurs with deep pockets. It's an odd thought that this camera, with far poorer resolution than today's smartphones, was sold for the same price in 1999 as something like the Canon R5 in 2022. One of the world's best mirrorless cameras, it has a resolution of 45 megapixels and shoots at 20 frames per second.

However, by 2003 Canon launched their groundbreaking 300D model for just £600, confirming that digital SLR technology had entered the mass market. Many professional photographers were initially skeptical of the benefits of digital. This was understandable since those early digital cameras produced far poorer images than a sharp 35mm slide. However, when Hasselblad released their prototype medium-format digital body (the H1D) in 2002, with a resolution of 22 million pixels, the jaws of the professional photography world hit the floor. By 2005, it had become fairly clear that the days of 35mm film would soon be numbered.

What did this technological whirlwind mean for climbing photographers? Well, it improved the odds for the photographer: exposure can be tested and compositions reviewed before the climber reached the crucial move, and a virtually unlimited number of photographs can be taken. One of the most obvious consequences is that the number of climbing photographs in circulation has multiplied at an astonishing rate. Some of these images are mediocre, but a great many of them, including many shots by amateur photographers, are absolutely brilliant.

Drone photography and mirrorless cameras have been two of the major innovations for climbing and mountaineering photographers in the 2020s, both of them in different ways enabling extraordinary images to be taken from some very unlikely places. One of the tremendous positive influences of digital technology is the way in which it has sparked a booming interest in photography in general, and the basic standards of climbing and mountaineering photography have been pushed considerably higher than they ever were in the analogue era. Increasingly spectacular, beautiful, and unusual climbing photographs are being taken all the time, amplifying and enhancing our global climbing culture.

Yet even within the whirlwind of imagery that surrounds our lives, we can still isolate those moments of clarity that are the lifeblood of our experience on the cliffs and mountains. Any one of us can rediscover Coleridge's "state of prophetic trance and delight" on Scafell, or John Cleare's visionary dream at Tremadog, and reach for the camera. We'll encounter it when that sudden rainbow falls through a clearing squall, when our partner's rope flashes darkly against a setting sun, or when the snow sparks with moonlight as we approach the shadowlit wall.

Summit magazine, 2008

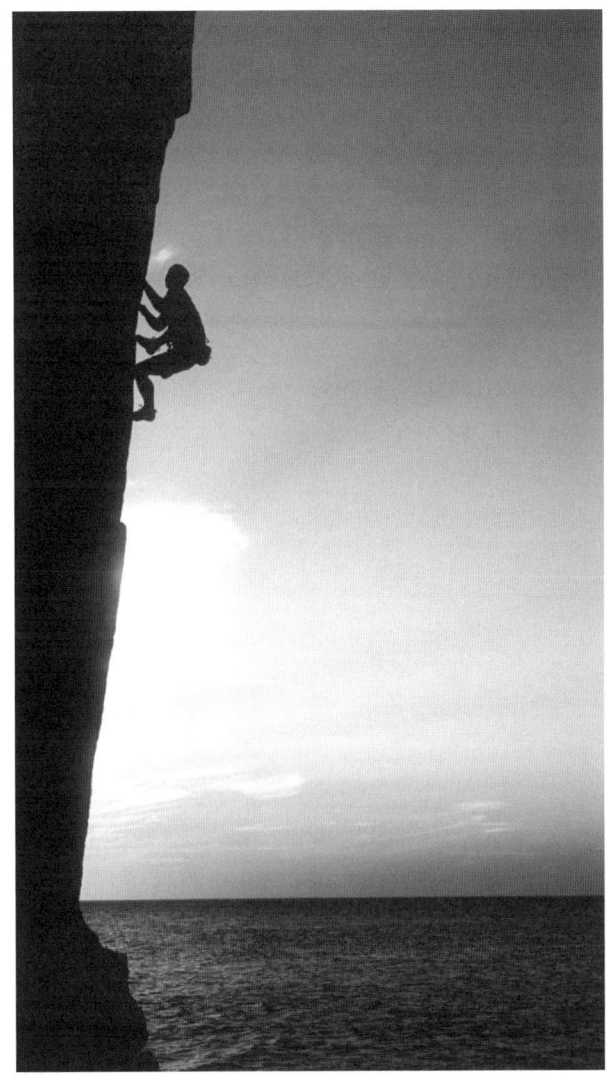

Neil Gresham deep water soloing pitch 1 of The Flying Dutchman (E7 or 7b+/S2) on Lundy Island in 2006.

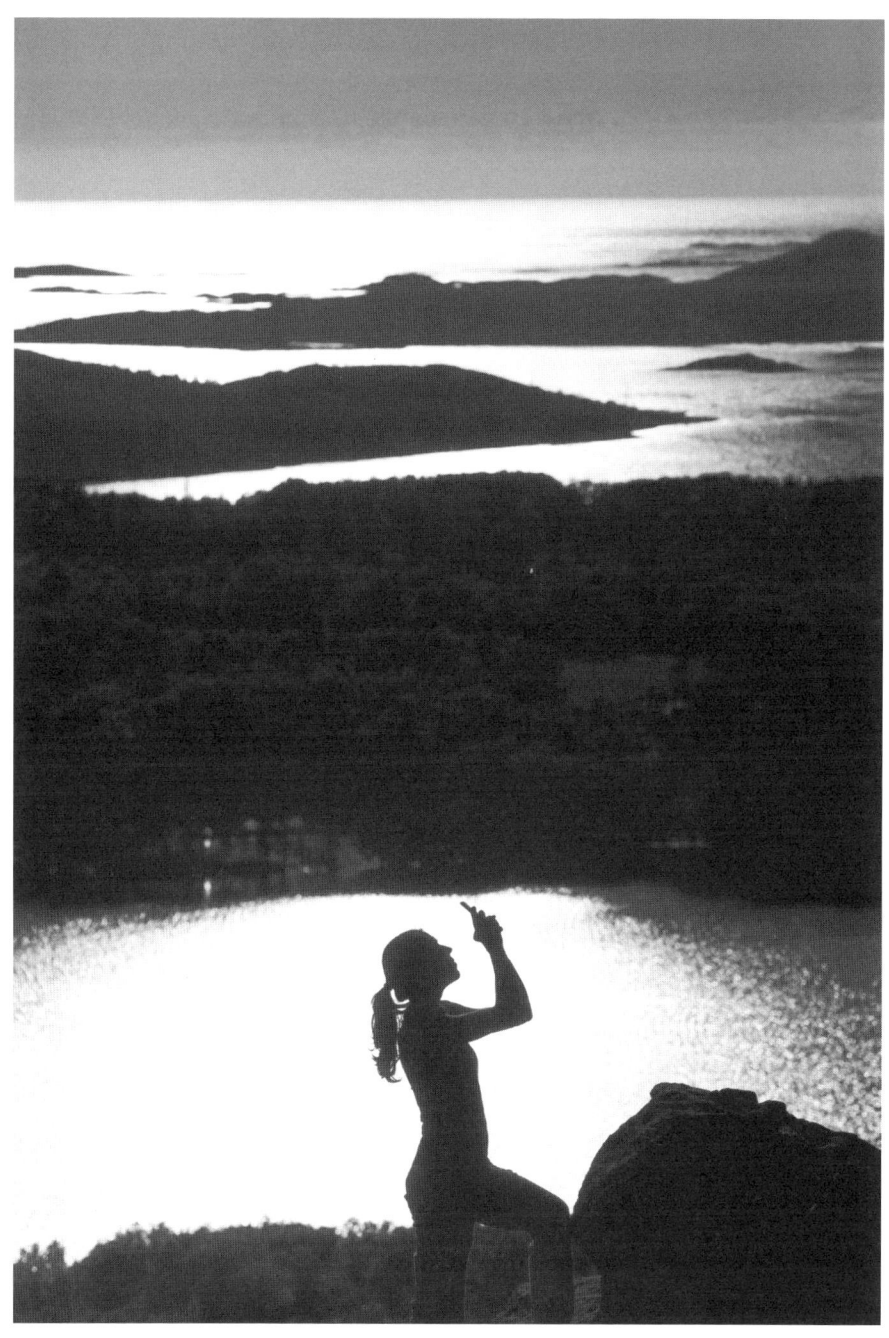
The future of guidebooks? A climber cross-references a route at Flatanger, Norway, with a topo on her phone.

Totem Masters

Climbing guidebooks through the looking glass

"Conveniently situated near the graveyard of Morwenstowe church, Henna is the highest precipice on the Culm coast, and the most frightening…" Thus begins the magnificent description of Mick Fowler and Mike Morrison's esoteric five hundred foot monster route, *Breakaway,* in Iain Peters' masterful 1988 guidebook to North Devon and Cornwall. Peters pulls no punches in the route description itself:

"Terrordactyls, ice screws, stakes, and a strong belief in one's immortality are the main requirements for an ascent… After pitch 2, retreat would be at best difficult or at worst terminally easy." Written a generation ago, it still inspires awe and fascination. It remains to this day my all-time favourite description of any rock climb in any guidebook.

Reading it again now, over thirty years after that guide was first published, the power of the route description is more striking than ever, partly because it is difficult to imagine a modern guidebook writer coming up with anything like it, or a guidebook editor including such a description in a finished book. Peters revels in the fact that *Breakaway* is an absurdly loose and extremely dangerous challenge, and celebrates the route as an example of the eccentric, iconoclastic, and inherently hazardous nature of climbing.

His description is deeply characteristic of the maverick, punk-influenced nihilism and its parallel hedonism in late twentieth century British climbing which John Porter identifies in his brilliant book *One Day As A Tiger,* and that would last at least a few years more after Peters' guide was published. I'm glad I caught just the tail-end of this madcap tradition as a young climber in the late 1990s and early 'noughties.

What's perhaps even more important about the description of *Breakaway* in this guidebook is that there's no photograph or top of the cliff whatsoever; nothing at all to give any real idea of the exact line of the route. There's an OS grid reference (this was the pre-GPS era) for the crag, and just that single page of ominous text. A laconic, almost throwaway sentence is all we have from which to work out where the five hundred foot route possibly goes:

"The climb takes the central fault gained from the left." This tradition of deliberately vague route descriptions has a long and complex history, part of which is rooted in the idea that too much information about a climb reduces the adventure it might offer. Some Norwegian climbers today still frown upon recording first ascents in certain areas, believing that route descriptions are incompatible with the pursuit of true adventure. It's hard to dispute this position from an objective standpoint: grades and guidebooks limit the possibility of authentic adventure, but they're also extremely useful.

From the perspective of practicality, the description of *Breakaway* in Peters' 1988 guidebook is sub-optimal, containing scant information about the line, the location of the belays, nor any visual references of the cliff itself. Anyone setting out to climb this route, the author has correctly concluded, will either already be a very experienced and proficient adventure climber, or completely insane, or even conceivably both. They will therefore have the necessary skill to find and follow the line roughly described in the text. Since we're dealing with an adventure climb of the highest order, Peters' description doesn't dilute it by providing too much information about the route.

Examining a guidebook like *North Devon and Cornwall* (1988) alongside a raft of modern guidebooks, as I have done recently, an intriguing paradox quickly becomes obvious. Whilst modern guides have become enormously better tools as guidebooks, I wonder if they have also become inferior to some of their predecessors in certain ways? A great climbing guidebook is not just a tool. It should also be a source of inspiration, a looking glass, a kaleidoscope of dreams through which we might project and realise our own most cherished ambitions.

As a guidebook attempts to condense ever more helpful information into its pages, it necessarily loses some of its potential power as an inspirational text. If you put one of the better British or American guidebooks from the 80s or 90s on your kitchen table next to a more visually dense modern guidebook, you'll immediately notice this process taking place. I should stress here that I'm not suggesting our modern guidebooks are inferior to those of the past; clearly they're far better at doing what they say on the tin. But perhaps they have lost something in the process of becoming such good tools for climbing. Many modern guidebooks have abandoned their historic role as inspirational texts. And perhaps this is because the current generation get most of their climbing inspiration online.

For a generation of young climbers for whom inspiration doesn't come from books but from Instagram feeds, guidebooks may not matter much. The concept of the book itself, for the tech-savvy youngster, is an artefact of the analogue age of human history.

The transition of books from printed into digital form raises some further important questions about climbing guidebooks. Whilst there has been a big drive over the last few years from major guidebook publishers to develop versions of their guides accessible through a bespoke app, as the technology has progressed the limitations of the digital medium for climbing guidebooks, and therefore the current practical advantages of printed versions, have become clearer.

At the moment, there are perhaps three key reasons for the superiority of printed guidebooks over digital ones. First, colour screens are still not yet properly readable in strong sunlight, rendering a screen-based guidebook app useless in the common scenario of squinting up at a crag, cross-referencing a visual topo with the wall above in bright sunshine.

Second, screens run on batteries, and batteries run flat. When you've just hiked two hours up a Scottish mountain to approach a climb, a flat battery would be more than inconvenient – it might ruin what would have been a perfect day's climbing.

Third, and perhaps most importantly right now, digital guidebooks work best on a screen somewhat larger than a normal smartphone. This means that to get the best from your digital guidebook you'll have to take a tablet to the crag, and the possibility of damaging an expensive piece of tech in a dirty, dusty crag environment is extremely high. The conclusion we might draw from the above is that until colour tablets exist that are low cost, weather-sealed and virtually unbreakable, readable in direct sunlight, and have batteries that last for 40 hours-plus (even in cold conditions) digital app-based climbing guidebooks are unlikely to take off in a big way.

If print guidebooks are to remain with us, at least for now, then what might the most successful guidebooks of the next decade look like? Like many climbers who've been involved with the sport for many years, I own a lot of guidebooks. I've given away the majority of my old, out-of-date guides, not wishing to clutter my already heaving shelves with books that have been made obsolete by newer, better, and more accurate versions. But I have kept a few old guides, such as Iain Peters' 1988 *North Devon and Cornwall* mentioned earlier. Surely it's this – the act of keeping a guidebook after it's been replaced by a newer version – that defines a truly great piece of guidebook writing by an author and their publisher.

How many modern guidebooks will fall into the category of classic climbing texts like *North Devon and Cornwall (1988)*? Much of the culture of adventure climbing and mountaineering, interestingly, is still constructed through written narratives, whereas the culture of contemporary sport climbing and bouldering is largely defined by visual narratives in the form of photography and videography. Whilst writing this, on the table next to my laptop there's a copy of a slim guidebook to one of Norway's most famous crags. There's hardly any text, lots of full page topos, and plenty of decent action shots. It does everything you require from a guidebook for a mainly single-pitch area. It details how to get to the cliffs, where the routes go, how hard they are, and any other important information such as if trad gear is required. It doesn't include information we don't need to know, such as where to buy groceries or how far the crag is from the nearest airport. People smart enough to climb should be able to work these things out for themselves, and guidebook publishers who choose to include such details are kidding themselves that they're improving their products by doing so.

Too much information and too many intrusive instructions can seriously detract from the quality of lived experience. More specifically, the wrong kind of information in a guidebook can negatively impact our enjoyment of climbing. A good topo should show the line of a route accurately, but there's a critical threshold beyond which more information no longer necessarily improves a guidebook's basic function.

Personally, I see the perfect guidebook of the future as something of a hybrid, incorporating the accuracy and detail of modern guides with the flair and maverick originality of the best guidebooks of the past. This would ideally be achieved within a volume as inspiring and carefully curated as a large format hardback book, striking the balance between delivering all the information a climber might need – as most modern guides do – but not including any details that are not really needed.

At some point in the future, all climbing guidebooks may become digital and updated in real time. Moderators could add new routes and crag information to digital guidebooks constantly, removing the need for updated editions.

As of 2023, though, the technology isn't quite there. Until there's a cheap, unbreakable and 100% waterproof and dust-proof tablet device with near-endless battery life and a colour screen readable in full sunlight, old-school printed guidebooks will be the defacto guidebook format of choice for most climbers around the world. In a wild environment, the simplicity of printed paper is sometimes better than the complexity of tech.

You can always take a photo of the page or pages you need on a mobile phone, too, which works well for multipitch routes, and many climbers do just this to translate print editions into instant digital form. The most successful climbing guidebooks of the future, I think, will probably find a way of combining modern information-delivery systems that are already largely perfected with more of the contextual, qualitative elements that certain historic guidebooks excelled at through original and charismatic descriptive text.

Most of the best examples of design evolution in history have incorporated older forms or iterations with newer elements to produce something far better than either the original or the most recent versions. As baseball player Yogi Berra very wisely noted, the future ain't what it used to be.

www.monographmedia.com, 2023

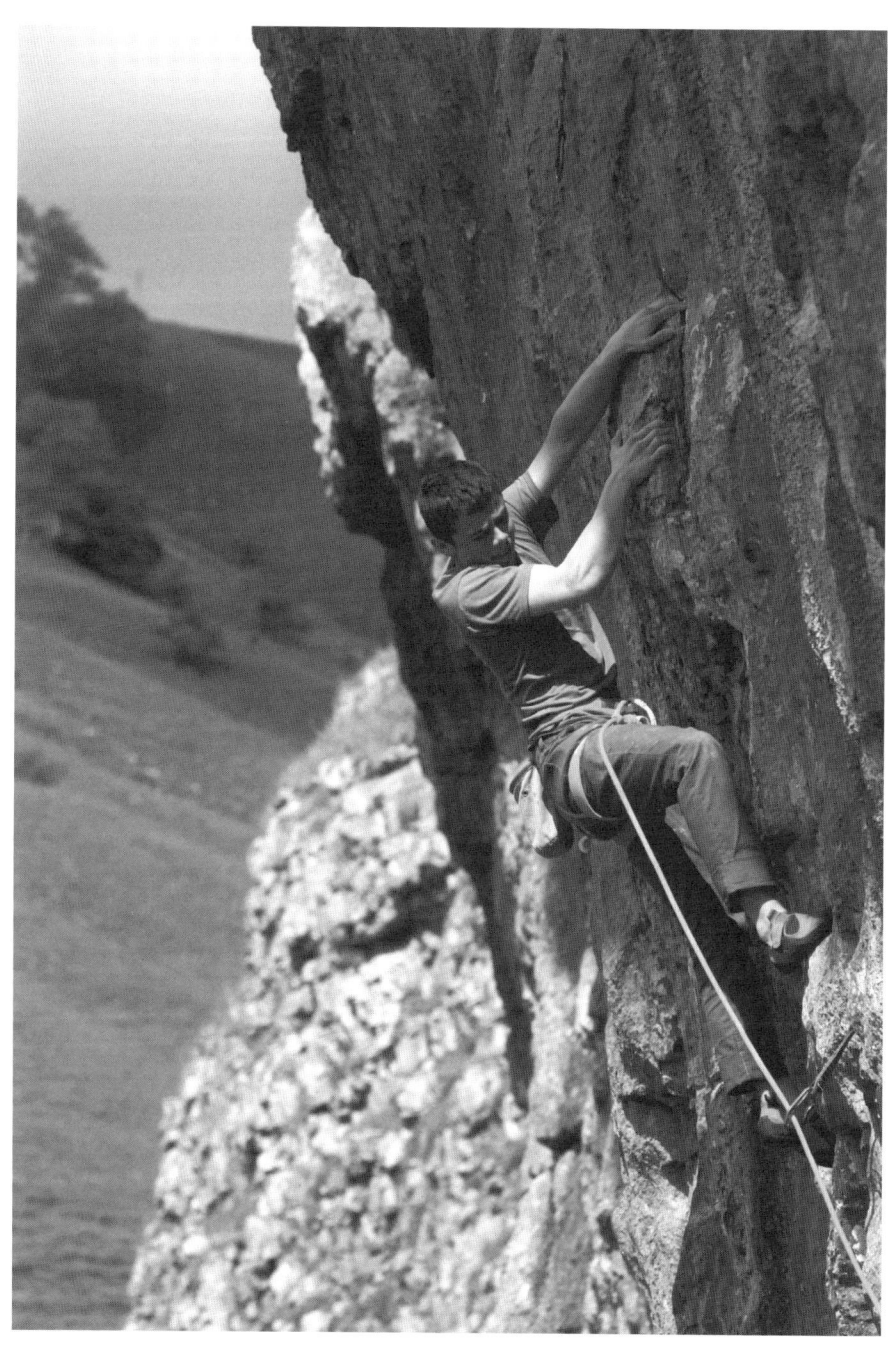

Rising star of British climbing, Jim Pope, aged 15, on Love Sculpture (8a+), Yew Cogar, North Yorkshire.

The Sporting Contract

Passing the torch to the next generation

It's a chilly November afternoon at Stanage Edge in the dusk days of 1995.

A wide-eyed kid pulls on his rock shoes below the steep southern wall of High Neb Buttress. He glances nervously upwards on the first few moves of John Allen's bold face climb, *Old Friends*. There's a musty tang of peat and wet leaves in the wind as he sets off up the first section, and fixes some small wires – which he doesn't really know how to place – in the shallow flake before the crux. After a moment of supercharged adrenaline, he makes the moves as all the gear falls out below him. The rest of the climb floats by in a haze. After a while, he's standing on the flat gritstone plinth above, his face pressed to the wind. Southwest, over Abney Moor, the horizon is fading under an advancing line of mist that smudges the edges of the fading light. Suddenly moved by a strangeness in the air, the boy wonders how his desire to climb might direct the unknown path of his adult life. Where, exactly, might climbing take him? He doesn't know. The mist begins to swirl around him, thickening the light.

If you began climbing as a teenager on the gritstone edges of northern England, like I did, there were always swathes of cirrostratus scudding across the moors. You might have shivered in a cheap cagoule, fumbling with gear you didn't know how to place, wondering how soon it was going to start to rain.

It was all a bit grim, but in an exciting sort of way. There was something intangibly romantic about the dank smell of peat in the wintry air and the bitter tang of lichen on your fingers, and about the way you went home with a head full of dreams of crags and mountains beyond the horizon after talking to the older generation. When I first heard about the climbing in North Wales from old hands on the gritstone edges of Derbyshire as a thirteen year old, I simply couldn't wait to get there. In a few short years, I would discover the place for myself, just as they did once.

When I began climbing in the early 1990s, climbing walls and sport climbing barely existed. Learning how to climb outdoors, therefore, involved a considerable element of trial and error, much of which involved making all sorts of dangerous mistakes and learning from them. You had to be smart and learn fast to survive. This was not necessarily a very good environment for learning, but it was certainly an exciting one.

Today, learning to climb is much easier and far safer than it once was, as there are now thousands of indoor walls across the developed world. This is obviously a very good thing. But given the far more complex nature of outdoor climbing compared to its indoor counterpart, most of those who are now introduced to climbing indoors quite understandably never venture outdoors. Why go to the trouble of nicking a car for a late night joyride if you can play Grand Theft Auto?

In recent years, indoor and competition climbing has become not just a separate genre of climbing, but something much more like a separate sport altogether. This fact, along with a host of other factors, will not only shape the future of climbing as we currently know it, but also as we can barely imagine it.

The 18th century Irish political philosopher Edmund Burke wrote that "[Society] is a partnership in all science, in all art... [and] in every virtue and in all perfection. As the ends of such a partnership cannot be obtained in many generations, it becomes a partnership not only between those who are living, but between those who are living, those who are dead, and those who are yet to be born."

If Burke's far-reaching insight is even partially true, one of the most striking features of modern Western society is the partial collapse of the social contract he highlights. This is partly due to the failure of elites to properly address its causes, such as the fact that property is unaffordable for a great many young people in rich countries. At the same time, and despite the imminent problems this process anticipates, self-organised adventures are one way the young can assert their independence from the generation preceding them. At a biological level, teenagers rebel to demonstrate their genetic independence from their parents, and adventure sports and other outdoor activities can channel this instinct in an extraordinary way.

When the Burkian social contract is translated into the realm of climbing, it would suggest that the generation currently active is obliged to pass on everything they know to the young, who can then use that knowledge to reinvent the vertical world in whatever way they wish to. One of the defining strengths of climbing and mountaineering is that the rules of the game, such as they exist, are not concocted by a self-interested gerontocracy, but by the key players themselves – the current generation. That's the point missed by ageing figures who complain about why things were "better in my day". Things were never better, of course; they were just different.

The leading Slovenian alpinist Marko Prezelj has observed that "paradoxically, as young [climbers] live in an era of information, they have lost real contact between each other, with different generations, and with different climbing cultures." It is the vital duty, therefore, of climbing's various institutions to promote the kind of contact Prezelj is talking about. To give the major national climbing and mountaineering organisations credit, they generally do try to achieve this through international meets and so on. Such events are important; without the fertile mingling of different generations and cultures, climbing's social contract begins to break down, as Prezelj has suggested.

At the same time, and as a saving grace, the crags will always be out there for another generation to explore and discover anew. The gritstone edges of the Peak District on which I first cut my teeth as a climber are just one such proving ground. I hope there's another wide-eyed kid at High Neb Buttress on a misty, windswept autumn afternoon, embarking on an epic journey through climbing just as I did one November day far away and long ago.

Climb magazine, 2016

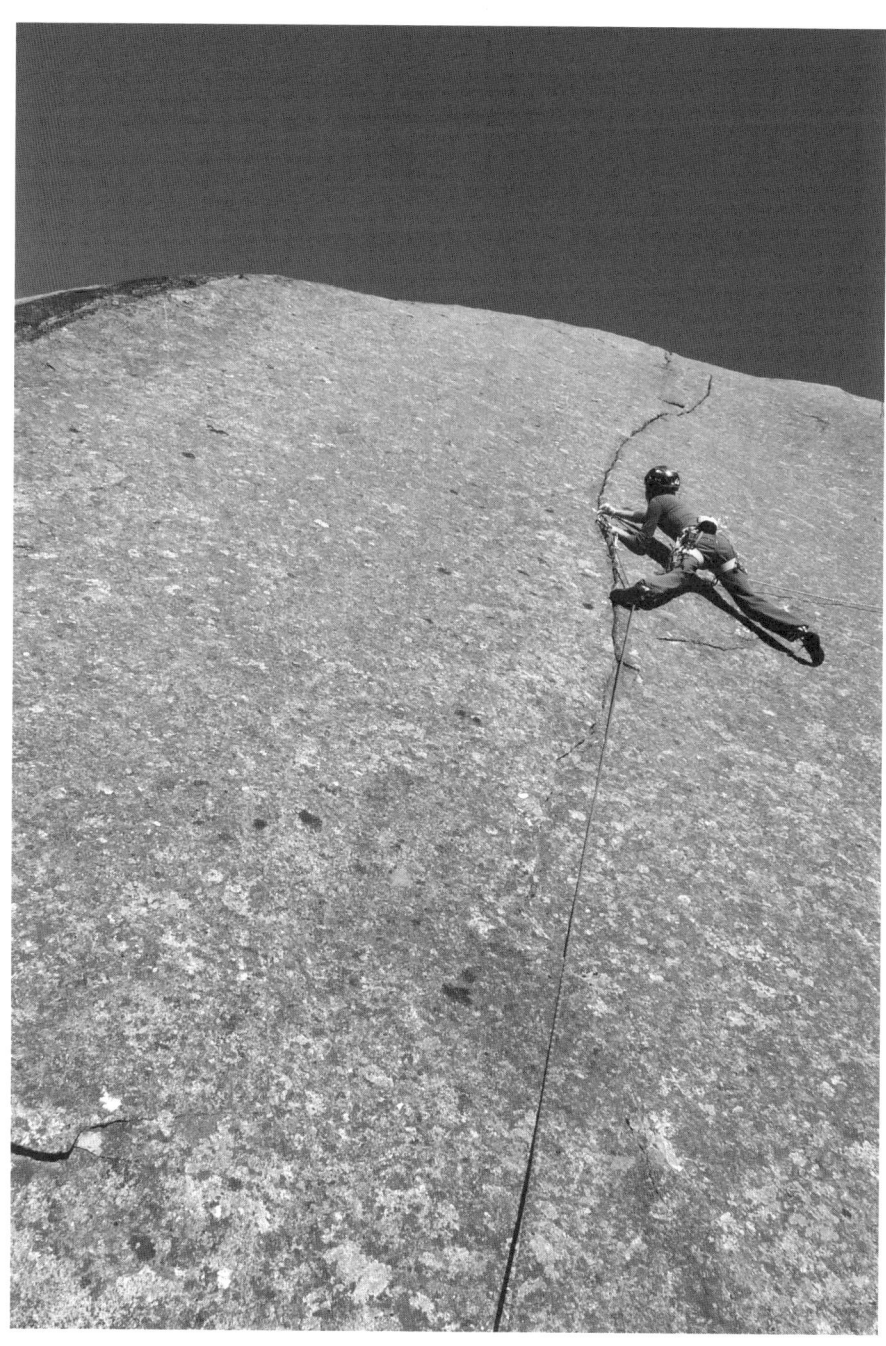

Malin Holmberg expertly leading the crux pitch of The Lady of the Lake (N9), Lofoten, on the first ascent in 2011.

The Skilful Climber

On encounters with experts

On a cold spring day of fast-moving clouds, a nine year old boy is riding his bicycle along an English country lane. Enchanted by the wind in his hair and the freedom of cycling, he doesn't notice the flat rear tyre until it forces him to swerve off the road and into the long grass. He's miles from home, and wonders what to do.

Faced with the choice of either pushing the bike home or fixing the tyre, he decides to try and repair it. He doesn't know how, but this doesn't deter him. After a while, he manages to get the wheel off the bike. Just as he's wondering what to do next, a middle-aged man cycles down the lane towards him.

"Sorry, could you help me out" the boy asks the man. "I've got a flat tyre."

The man doesn't answer back, but nods his head and smiles as he stops beside the boy. He reaches into the saddlebag of his own bike and pulls out the tools the boy doesn't have – tyre levers, a pump, and a spare tube. Momentarily confused, the boy simply says "Thanks a lot". The man smiles back as he nods at the boy, pointing to his ears and mouth whilst shaking his head. The boy suddenly understands what the man wants to explain to him through these gestures: that he can neither hear nor speak.

Quickly and carefully the man changes the tyre on the boy's bike, showing the boy the best way to do it, using the tyre levers together, working the bead off the rim, and inflating the new inner tube a little before fitting it. The boy is awestruck at the man's seemingly effortless skill. In ten minutes, the boy is back on the road, cycling off into the cloud-shadowed spring day, and the unknown path of his life.

I never knew the man's name, or what he did, or where he lived. I have a feeling he may have been a professional mechanic from the way he used his tools. Whilst anyone with an iota of mechanical competence can show a kid how to change a bike tyre, there's something about that encounter that's stayed with me. It was the way he showed me how to perform the simple repair: he never hurried, but the tyre was fixed quickly. Although the man was profoundly hard of hearing and speech, he articulated what it means to do something with skill in a way that a nine year-old remembered forever.

Performing with skill, fitness and talent is what we strive for in climbing or any other adventure sport. I'd like to suggest that the truly skilful climber isn't the one who concentrates purely on achieving their personal best as an athlete, or on having the biggest adventure they can have. Instead, it seems to me that the genuinely skilful climber is the one who's as keen to help out others as to do their own thing, just like that unknown man who showed me how to change a bike tyre more than thirty years ago. His skill as a mechanic was self-evident, but his greater skill was in that selfless, spontaneous act of being in a position to help someone out, and then doing so freely. Sometimes, people with disabilities have profound empathy with others in difficult circumstances due to their own lived experience, and this was certainly true in this case.

When I interviewed Yuji Hirayama, Japan's greatest climber of all time, he talked about how he wanted to help Hans Florine in his quest to set *The Nose* speed record on El Cap, because Hans had supported him on his 1997 near-onsight of the *Salathé Wall*. That, it strikes me, is a truly skilful kind of partnership.

A search for balance can be important in the climbing life, because climbing is an inherently selfish activity. But how to achieve this in a sport which places individual attainment above all else? Part of it could be about being an attentive belayer, a good partner, and a trusted friend to the people you climb with. I've been lucky enough to climb a lot of very good routes on five continents. Nonetheless, some of my best climbing experiences have been holding the ropes of partners. Things like hiking up to the summit of Djupfjord Wall in the Lofoten Islands for the third time to help Malin Holmberg complete the desperate final crux pitch of our big new route, *The Lady Of The Lake*; or shouting encouragement to Matty Rawlinson as he led the last hard pitch of *Moonlight Buttress* in Zion. And on each of those climbs we had a laugh whilst doing something challenging.

Alex Lowe said that "the best climber is the one having the most fun." He's right, because climbing becomes dramatically more enjoyable as soon as you reach the point that it's not all just about you and your routes, but also about helping out your partners with theirs.

This is the path of the truly skilful climber; a path I was guided towards more than thirty years ago, by a wonderful deaf man who I met completely by chance whilst out cycling on a country lane.

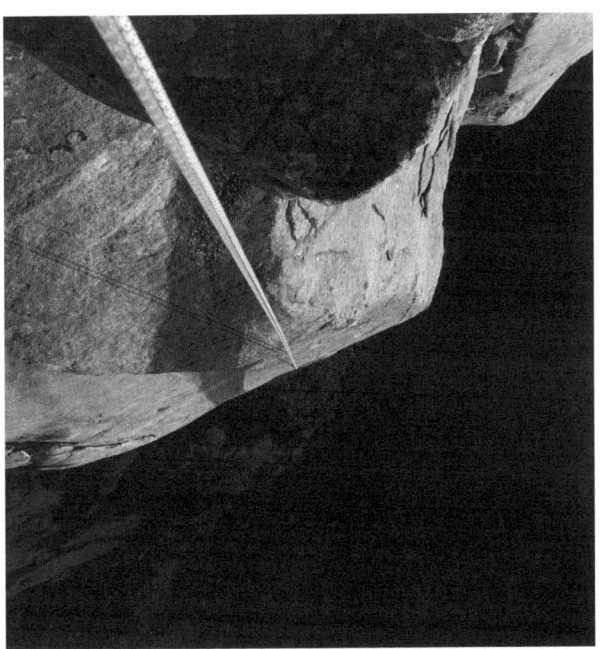

The fellowship of the rope: my partner, Malin, somewhere in the void below high on Djupfjord Wall, Lofoten.

Climb magazine, 2015

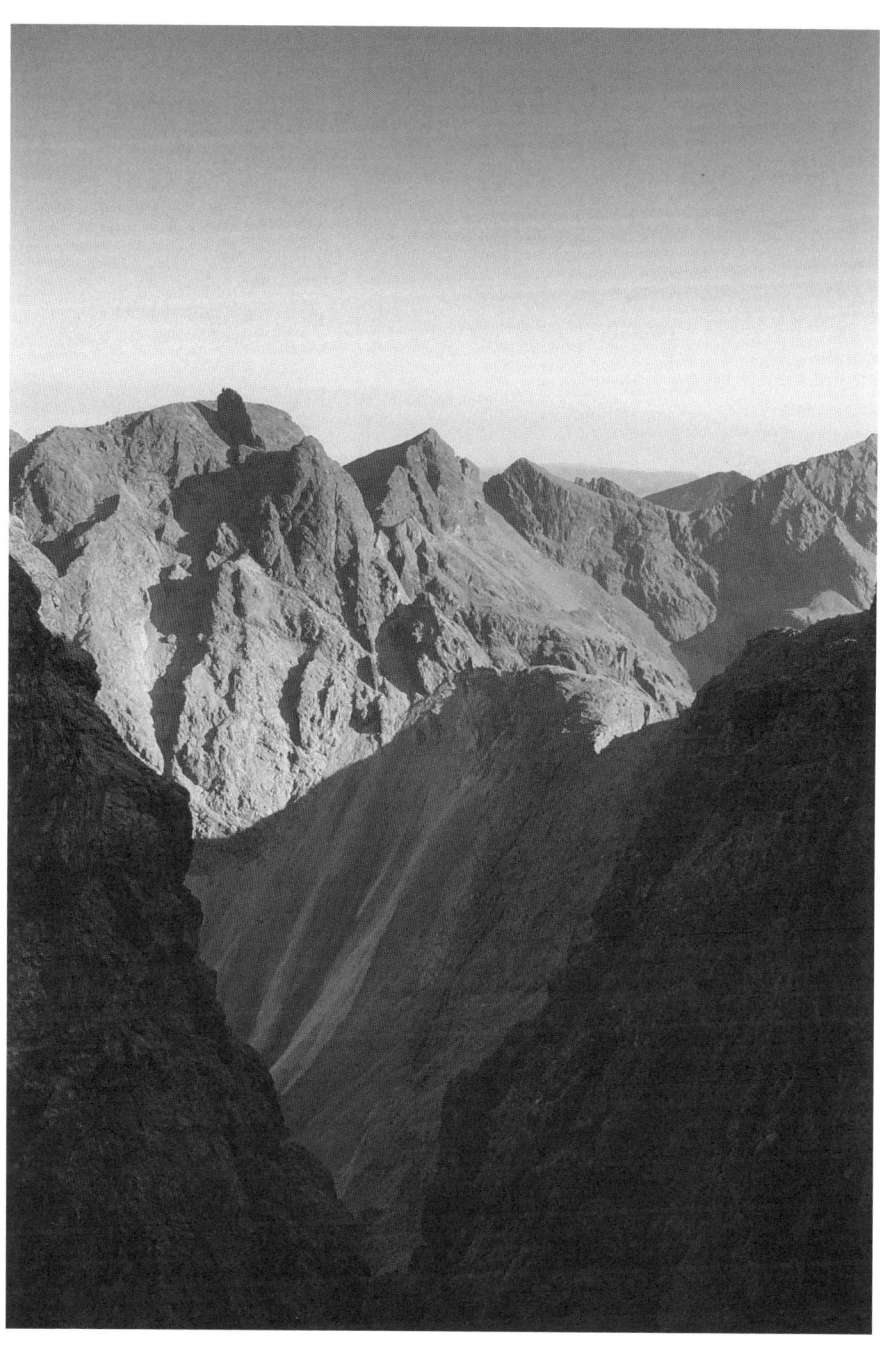

The Inaccessible Pinnacle and the central Cuillin seen from the top of the Great Stone Shoot on Sgùrr Alasdair.

Heaven's Highway

The precipitous marvel of the Black Cuillin

The Cuillin Ridge on the Isle of Skye is the showpiece of the Scottish mountains; it's the best of the best. Rising from the clear waters of Loch Scavaig and carving around the Coruisk basin in the shape of an Arabian scimitar, this striking line of interconnected summits is one of the only truly alpine features in Britain. Despite their relatively low elevation, their jagged verticality makes these peaks appear far higher than they really are. From most places on the southern part of the Isle of Skye, and from almost anywhere on the adjacent mainland, the Black Cuillin utterly dominates the skyline. To any serious climber, this place is heaven's highway.

Traversing the Cuillin Ridge from end to end is the most compelling mountaineering challenge in Britain. It is approximately twelve kilometres from the southernmost summit of the ridge, Gars-bheinn, to the most northerly peak, Sgùrr nan Gillean. This includes roughly 3000 metres of ascent and descent, a great deal of scrambling, plus a few sections of relatively easy but very exposed rock climbing. Most people, therefore, take a rope, climbing equipment, and overnight gear to complete the traverse of the ridge over a couple of days, with a bivouac at a convenient halfway point.

The really big challenge, though, is to traverse the ridge in a continuous, one-day push. I did the Cuillin Ridge in this style in early June 2023, taking about three times as long as Finlay Wild's quite remarkable 2013 record of two hours and fifty-nine minutes, and finally completing an ambition I'd had for a very long time. My overall strategy was to take my time and enjoy myself up there rather than to go as fast as possible. I settled on a pretty minimalist approach to equipment, taking three litres of water in a twenty-litre running pack, a windproof top, sunglasses, a mobile phone, two tins of sardines, some Highland oatcakes, four energy bars, and a bag of jelly beans. I thought about about taking rock shoes for the more technical climbing sections, but decided it wasn't worth the extra weight.

I woke five minutes before the alarm at 3.55 a.m. It was already light. In midsummer in the north of Scotland, the night doesn't last very long at all. I was already packed and ready, but in my haste to get going I made a rookie error, knocking my coffee percolator off the stove and dispersing its boiling contents on the toes of my left foot. Ouch. My lightweight Scarpa trail running shoes offered zero protection from the scalding liquid, and I realised I could have scuppered the whole mission before the starting line. Never being one to let minor setbacks get in the way, I took a large dose of Ibuprofen, put on a fresh pair of running socks, and set off hoping for the best.

Heading up into the Black Cuillin in the early hours of a perfect summer's morning is one of the great experiences in the British mountains. The waters of Loch Scavaig glinted darkly far below as I weaved my way along the narrow trail that leads to the southern end of the ridge. By six o'clock I was on the summit of Sgùrr nan Eag, the first and most southerly Munro of the Cuillin, basking in the glorious rays of the rising sun. It felt fantastic to be on the ridge with a pack that weighed less than five kilos; it's quite often the case that the less stuff you bring, the more you can enjoy yourself in the mountains. Even better, the Ibuprofen seemed to have worked and the pain in my toes had somewhat subsided.

On reaching the TD Gap, the first technical climbing section, a pair of Americans were already roped up and engrossed in the pitch that leads up from the narrow col towards the summit of Sgùrr Alasdair, having bivvied on the ridge overnight. I scampered down the tricky, balancy down-climb of the abseil route into the Gap, and they kindly let me passed on the awkward climb out of it. On top of Sgùrr Alasdair, the highest Munro of the Cuillin and one of the most spectacular mountains in Britain, I took a short break to take in the surroundings. The tents and camper vans in Glenbrittle far below were nothing but minute specs against the hazy sand dunes. To the south, the islands of Eigg, Rum and Canna floated lazily on the still waters of the

Minch. To the west, the long profile of South Uist created a strange mirage on the distant horizon. The Cuillin itself stretched away northwards before arcing to the northeast; a jagged, jet-black line of gabbro spires superimposed on the morning sky.

The descent from Sgùrr Alasdair's satellite summit of Sgùrr Thearlaich is complicated, and I was glad of some route-finding assistance here in the form of Tom Prentice's ultralight topo guide. I wound my way down the natural spiral staircase and then up towards King's Chimney, which I bypassed via Collie's Ledge, a kind of narrow horizontal walkway across the side of the mountain. Suddenly I was at the Inaccessible Pinnacle, one of the highlights of the entire ridge. Clambering up the dinosaur-spine of the pinnacle, I realised its fame was well-deserved: it might well be the best low-grade rock climb in the UK. I met some more climbers on top, busily faffing with ropes and gear, and nipped passed them to make the surprisingly tricky down-climb of the abseil line, which is probably the technical crux of the ridge if you don't take a rope.

As the morning pressed on, more summits came and went, and it became quite hard to keep track of which mountain I was on. The Three Peaks and the Three Tops all blurred into one as I fell into the groove of unencumbered movement across the Black Cuillin. By the time I'd reached Bidean Druim nan Ramh, the mountain that marks the start of the northern part of the ridge, I felt glad to have made such an early start as the sun's heat was building. I still had a litre of water left, and gobbled some jelly beans for a sugar hit to help me along the final section.

Reaching the Bastier Tooth not long after noon, the final bit of proper rock climbing on the ridge raised its head in the form of *Naismith's Route*, the classic Severe that leads to the apex of the Tooth itself. This is actually the climbing highlight of the ridge; it's a marvellous little climb up one of the most distinctive geological features in Britain. I arrived on top of Sgùrr nan Gillean, the final Munro and the end-point of the ridge traverse, at around one-thirty, and had a celebratory lunch of tinned sardines by the summit cairn.

As I descended towards Sligachan in the heat of the afternoon sun, I reflected that traversing the Cuillin Ridge in a single push had been by far the best thing I'd ever done in the British mountains. Whether you do it as I did or take the classic two-day approach, this is the greatest mountaineering route in Britain by a country mile.

www.monographmedia.com, 2023

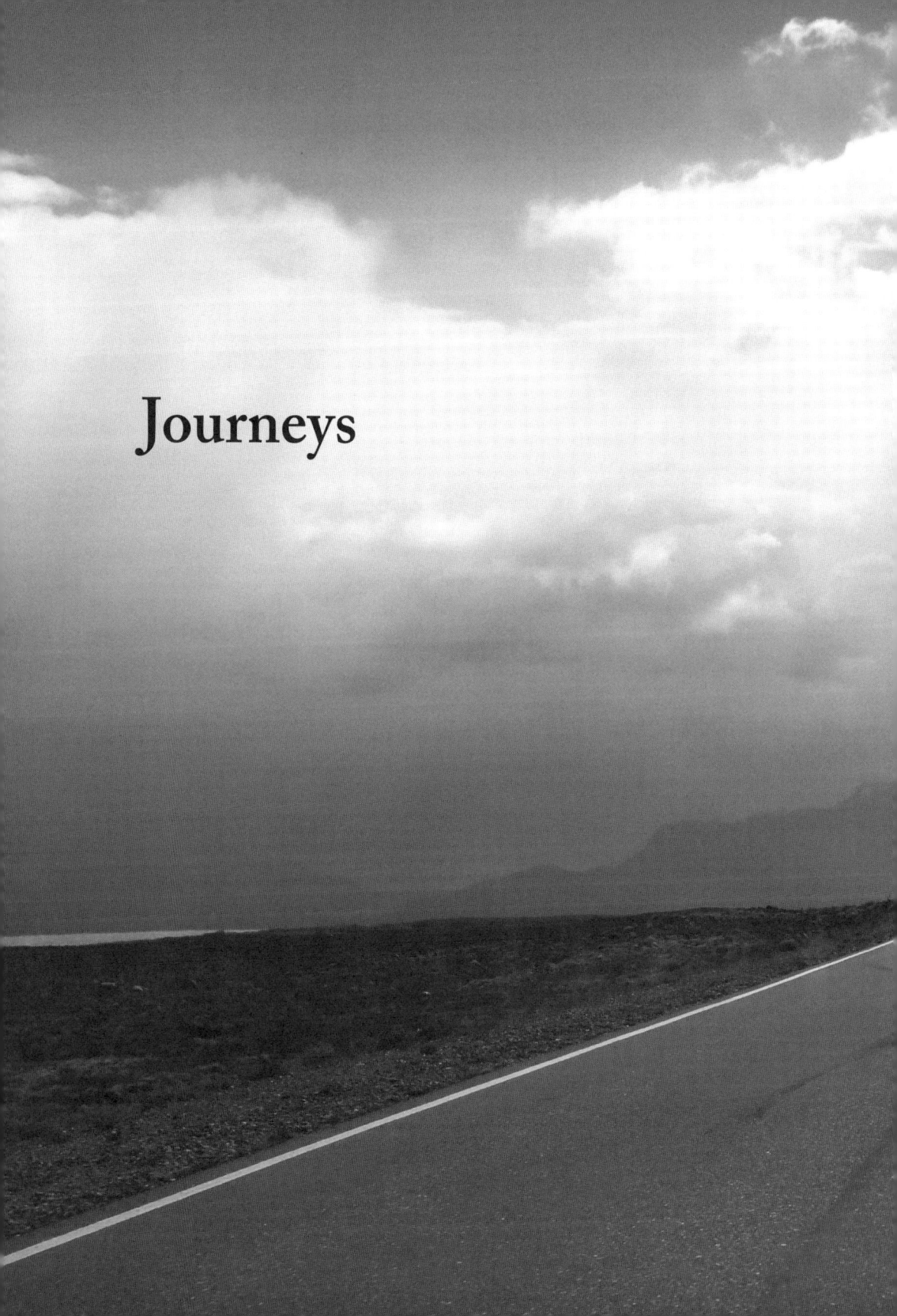
Journeys

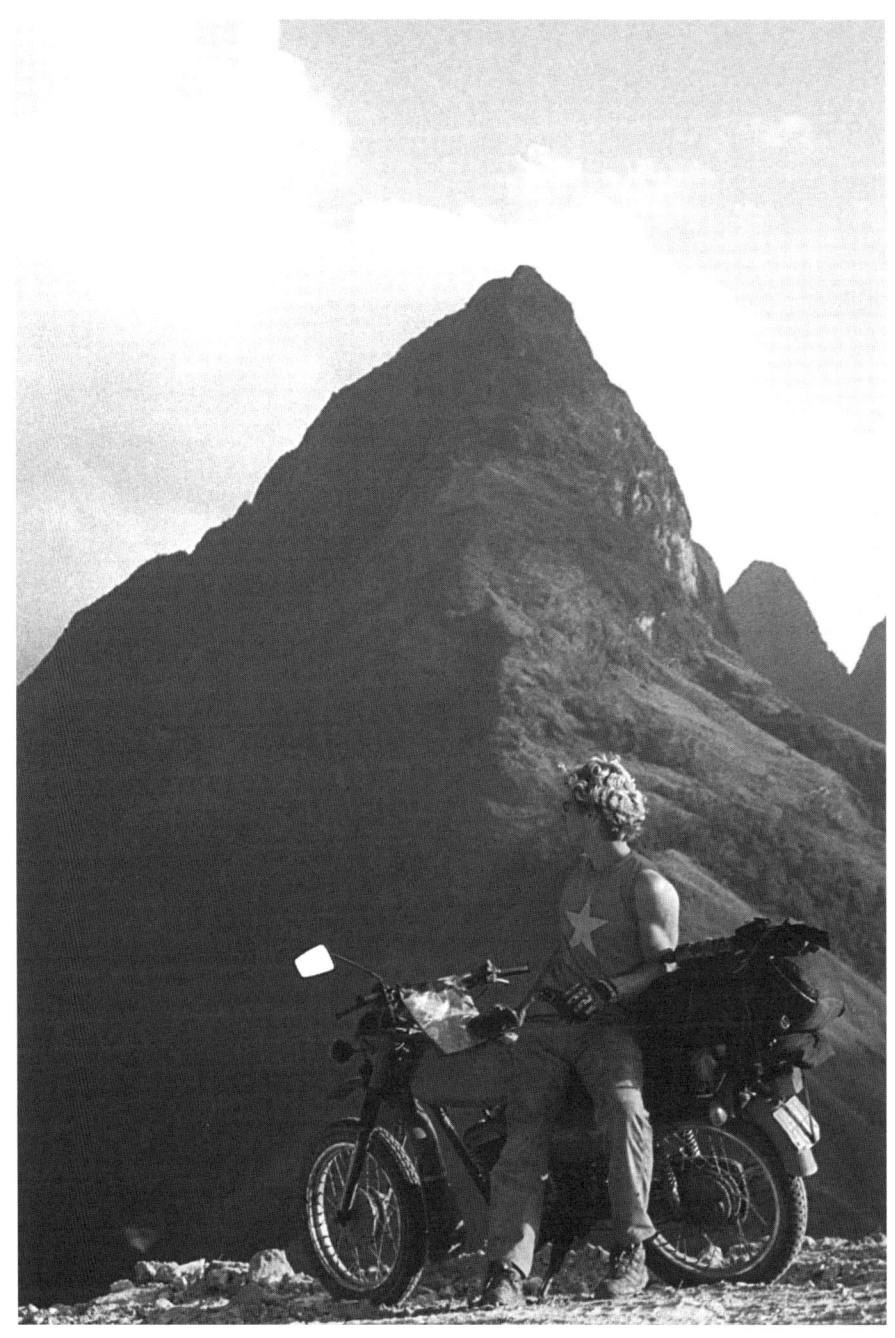

The author and the Minsk on the road in the mountains of northern Vietnam in November 2003.

Facing East

Six months by motorcycle across Southeast Asia

It was late afternoon when the plane hit the runway at Noi Bai airport, and my blood was running hard. The very first time you arrive, Southeast Asia hits you like the gust of an incoming typhoon. The yellow sky has a sudden magnitude. The evening air smells of dust and smoke and drains. And people do everything differently here.

I'd ridden dirt bikes as a teenager, but I learnt to ride a motorcycle for real in the morning rush hour on the outskirts of Hanoi's French Quarter. My two options of sink or swim couldn't have been clearer.

"You must look both ways, every time you go", Mr Cuong had told me in the refuge of his tiny workshop off Rue N'Oc Quen. I'd found out about Mr Cuong, the capital's celebrated dealer of Minsk two-stroke motorcycles, via an intricate series of tip-offs from Mr Hoi at the Bau Long Hotel, a glorious old place in the heart of the French Quarter. It would become my second home between two journeys through the mountains of northern Vietnam.

After a long conversation charting the entire Hoi family tree, Mr Hoi's brother-in-law finally gave me a lift to Mr Cuong's workshop on the back of a scooter. We wound through the tiny lanes closely packed with artisans' stores and overhung by balconies of cascading creepers before making a sharp turn down a slightly wider alley. The heady, slightly sweet smell of petrol fumes mixed with two-stroke oil wafted from the open workshop. Motorcycle components were scattered like spilt beans on the street outside. Mr Cuong greeted us with a big smile that spread across his lean, intelligent face.

"Ah, Mr David, we have been expecting you. We are preparing strong Russian motorcycle for you!"

"That's great, thank you. Do you think I'll be able to take it later today?"

"Yes, today. Or maybe tomorrow. Before that you can be visiting our city."

"Okay, great... so I could collect the bike tomorrow then?"

"Certainly, very soon, maybe tomorrow. Or Wednesday, early morning. We call you."

Mr Cuong's 'very soon' captured the unique entity sometimes referred to as 'Asian time'. His business was not organised around deadlines and targets. Work was simply done. It could be today, or tomorrow, or the next day. The elegant logic of this system is infuriating to many Westerners, but the enforced delay allowed me to explore Vietnam's enchanting capital at leisure.

Later that evening, I walked back to the Bau Long along the northern edge of Hoan Kiem. The city lights reflected off the dark water in a mixture of smudged neon and black ink. The narrow reception appeared to be deserted until I picked up the metronomic tones of Mr Hoi's ancient grandmother snoring quietly behind the desk. At the same time, I spotted a note in my name pinned to the board above the desk: *Minsk ready. See you morning time. Mr Cuong.*

I crept upstairs to my room on the roof and walked out on the balcony. The night air had a welcome chill after the humidity of the day, a reminder of the imminent approach of winter. I inhaled deeply, contemplating the unknown roads that lay ahead. The lights of Hanoi's French Quarter flickered like magic lanterns in the growing dark. I'd finally arrived at the beginning of my journey.

I rode north out of the city in the coolness of dawn, passing the shadow-boxers on the edge of Hoan Kiem Lake and women with coloured garlands making their way to the early morning flower market. It was more than five years since I'd ridden a motorbike, and the unique Vietnamese system of traffic flow was something of a baptism of fire. Weaving through the multi-directional chaos of the main intersection before the bridge over the Red River, my new found confidence on two wheels was chastised as I watched a blind man make his way with a slow assuredness across at least seven lanes of traffic. It is said of Vietnam that every time you walk into the street, you'll meet a hero. It didn't take me long to find out that heroes come in a thousand disguises in this country, and that few of them wear medals.

After leaving Hoa Bin and the coastal plain, the legendary Highway Six winds its way northwest from Hanoi into the mountains of Lai Chau province, the most northwesterly part of the country, which borders Laos to the west and China to the north. The potholed tarmac turned to dirt as the road wound into the hills above Hoa Bin. At a narrow col before a steep descent, I stopped close to some road-workers, who immediately invited me over to drink tea under their tarpaulin. The Vietnamese taste for green tea so strong and thick you can stand a chopstick in it is wasted on me, but I shared some nonetheless, glad of their hospitality. Afterwards, they passed me their huge tobacco pipe. I took a drag and spluttered, unused to the acrid smoke and fierce nicotine hit. This produced convulsions of laughter among my companions, and bright smiles flashed through the smoky air.

That night I slept in the makeshift bamboo shack of another group of road-workers. They had been toiling all day to clear a landslide, a common problem on the roads of northern Vietnam. It was late by the time they'd blasted the last of the rubble away, and they were adamant that I should stay with them instead of continuing on in the dark to Thao Nguyen, fifty kilometres to the north. We stayed up late that night, drinking their home-brewed sake and playing cards. When I said goodbye and made ready to leave early the next day, the wife of one of the workmen, Mai, refused to let me go without a bowl of hot noodle soup. I was hungry, and it was one of the best breakfasts I've ever had. Mai's simple, sacred gesture in the cool dawn air would stay with me more vividly than any good luck charm for the next six months.

I arrived in S'on La at sundown, covered in the thick red dust characteristic of Vietnam's northern roads. It had been an exhilarating day. The air was cold as the wind came down off the mountains and contoured through the deep limestone valley. Some of the women here wore the brightly coloured headscarves of the White Thai, their teeth jet black from years of chewing beetle nut. As I travelled north, the Honda scooters of Hanoi and the lowlands were quickly losing favour to the more robust, archaic Minsk. I was now in familiar company on these mountain roads.

The next day I had a flat tyre on the long descent into the valley north of the small town of Tuan Giao. Just west of here lies the old garrison of Dien Bien Phu where the Viet Minh forces finally ended the French occupation of Indo-China in 1954.

I've changed many bike tyres over the years, but find it hard to recall a more laid-back place to have a puncture than on that long hill into the Luan Chao valley. High overhead, two eagles circled on a thermal. The green thread of the valley twisted north like a piece of twine spun between the mountains. I had only been there for a few minutes, finding the necessary tools and the spare tube, when a local man stopped to see what I was doing. It quickly became obvious he was an expert Minsk mechanic, and gave me an impromptu lesson on the rear drive train and brake assembly of this vintage Russian machine in beautifully concise broken English.

"See, this one go this way" he said as he dismantled the hub.

"Two bolt. Like this. Then sprocket, quick fix. No problem."

"Thanks very much, I see now" I replied, enthusiastically testing about half of my Vietnamese vocabulary in one sentence. My companion burst into friendly laughter. Vietnamese is a pretty complex, tonal language, and later that evening I worked out that I'd probably told him something like "my enormous duck is grateful".

This episode, like my impromptu evening with the road-workers above Mai Chau, left me with an overpowering sense of the communal spirit of the north Vietnamese. It is underscored by an intensely energetic practicality, and an unselfish willingness to help others. Both the French, and later and more disastrously, the Americans, gravely underestimated the power of this communality in the form of the north Vietnamese fighting spirit.

What the Viet Minh forces accomplished at Dien Bien Phu in 1954 should possibly have made U.S. President Lyndon Johnson think twice when he said, on 24th November 1963 "the battle against communism must be joined… with strength and determination". This marked the point of serious escalation in the Vietnam War, the most deadly conflict of the late twentieth century. Johnson might have thought twice, perhaps, about his hawkish stance on Vietnam in general, and the Viet Cong (who were a reinvention of the Viet Minh) in particular, in the mid 1960s, had he remembered how the Viet Minh foot soldiers encircled and then cut off the massively better armed and equipped French garrison at Dien Bien Phu under the command of the extraordinary Vietnamese general, Võ Nguyên Giáp. He should have remembered the utter disbelief of the French when the Viet Minh commenced artillery fire from the mountains high above the town: Giáp's soldiers had hauled their antique weaponry up the mountains around the garrison by hand, under the cover of night, rendering the French garrison a gigantic sitting duck.

I'd lost an hour fixing the puncture, and it was already late in the day. Beyond Ban Nam Nen, the broken asphalt turned back to dirt, and I rode on hard towards the head of the valley. The Minsk's dust-cloud sank fast in the cooling evening air. When biking in hot weather at this time of day, swarms of insects suddenly fill the air and glasses or a visor are essential to keep them from your eyes. I had to stop several times to clear my lenses of squashed flies. Each time, I was glad of the chance to take in the red and green mountains that had begun to rise high on both sides of the valley. I stopped for a final time on the col before the endless switchbacks down to Lai Cao, as the evening mist began to thicken on the rice fields a thousand metres below.

A few days later, I attempted to cross a high pass in the province of Th'a Nugen and over to the valley of the Song Da, the Black River. The speculative route I took was defined by the key on my large scale Nelles map as an 'unclassified road/cart track'. I had travelled down such roads before, and expected the Minsk would take such a route in its stride. As I gained height in the lengthening shadows of evening, the road deteriorated into single track too narrow for a horse, let alone a vehicle, and I began to distrust both my judgement and the cartography of my 1:1,500,000 scale map. At dusk, I arrived in the remote village of Th'on Ban Doc, its bamboo houses clinging on stilts to the mountainside which neither roads nor electricity had yet reached. In 2003, this region was still populated almost exclusively by the White Thai people, one of the numerous indigenous tribes of northern Vietnam that the Vietnamese government somewhat euphemistically calls "the ethnic minorities of Vietnam". This statement ignores the fact that outside the larger towns, relatively few ethnic Vietnamese live in these remote northern hills.

Before long I was surrounded by several dozen children, all of whom watched me intently, silently fascinated by my northern European features and blond hair. I was invited to stay in the village that night by the couple who ran the village school, who were of Vietnamese ethnicity and, amazingly, spoke some English. They told me that only a few old people in Th'on Ban Doc had seen a white person before, and that none had ever visited the village. I must have seemed a strange, unexpected guest that night in this remote river valley high in the mountains of Vietnam.

Saying goodbye to my hosts the next morning, I felt fortunate to have been their first foreign visitor, and curious as to what might have changed in Th'on Ban Doc were I to return ten or twenty years from now. Would a metaled road have replaced that precarious trail, clinging to the mountain slopes by the narrowest of margins?

The subsistence-based agricultural system and traditional way of life here is made possible, at least in part, by its relative inaccessibility, which is a natural barrier to both tourism and outward migration.

I turned the Minsk to the south and out of sight of the Da valley as smoke rose from brushwood fires on the edge of the steep, terraced fields where buffalo and horses grazed; this is an aesthetic familiar to anyone who has travelled in the high country of northern Vietnam.

The ride back to Hanoi that day was long but incredibly exhilarating, and would remain one of the best days of riding on the whole trip. I left Thon Ban Doc at six-thirty that morning, and arrived back in the Bau Long Hotel just after ten in the evening. At Bao Ha, I took the dirt road that follows the west bank of the Red River downstream to Van Yen, where a tiny ferry – just large enough to squeeze a small motorcycle on – allows a better road on the east bank to be picked up. I followed it to the larger town of Yen Bai, from where a glorious ride up into the hills took me over to Don Hung and Highway 2. The final few hours of the day, battling across Viet Tri and then on down the Noi Bai freeway (Vietnam's only dual-carriageway in 2003, and one of the country's busiest roads) in the dark was particularly hard going.

I shadowed the slipstreams of the bigger trucks for protection from other, mainly unlit vehicles. I'd ride like this until the dust got too much to bear, then seek respite in the slow lane. Finding my way back into central Hanoi was easier than expected. City lights and well-lit roads are a blessing after night riding. I crashed out almost instantly after inhaling a bowl of noodle soup and slept solidly for eleven hours.

I woke to the mingled sounds of motorcycle horns and exhausts, and the clanging of the gigantic metal spatulas the noodle-stall women use to turn the contents of their woks, and the rattle of vegetable-carts rolling on the cobbles. My body was deliciously tired from yesterday's fourteen-hour ride, but I felt a sharp clarity of thought and mind distilled by my first journey into the mountains of Asia.

I left Hanoi for the final time the following afternoon, on the notorious Highway One that joins the capital with Ho Chi Minh City (formerly Saigon) in the south. This route is also known as 'The Highway of Death' on account of the disproportionate number of fatalities it sees – hardly surprising considering the volume, speed and anarchy of its traffic. All vehicles, from buffalo-carts to trucks, few of which are likely to possess either lights or brakes, must use and compete for the same unmarked, undivided strip of tarmac. When running this gauntlet on a motorcycle, you can quickly sympathise with the great traveller Wilfred Thesiger's sentiment that "the most destructive invention of the twentieth century [was] the internal combustion engine". The combined effect of high speed chaos is amplified by the presence of Vietnam's main north-south rail line, which in some places runs very close to the road. The madness of this motorised version of Russian Roulette was illustrated by the hapless scooter rider who recklessly swerved off the road to my left and was promptly crushed by a passing train. I didn't actually witness the moment of his death, but heard a sickening crunch over the devilish din of the road.

I stopped a short distance ahead for a much-needed drink of water to quell the rising surge of nausea in my stomach. In response to what was evidently a regular event, a woman tending the noodle stall opposite quickly scurried across the road with a white sheet, in which she wrapped the scooter rider's remains. She neatly, calmly and quickly sealed each end with strong tape, as if she were sealing a parcel: it looked like a procedure she was extremely familiar with. She then scurried back across the road at the first opportunity, to attend to her unwatched wok.

Very slowly and very, very carefully, I steered the Minsk back into the multidirectional chaos of Highway One. At Hanam, I saw a dusty sign to Chi Ne, a village on the much smaller road to the west. I swung out of the traffic of the Highway of Death with deep gratitude, and headed out of Hanam on an empty, leveled dirt road heading west. I could smell the damp freshness of the air coming down off the green hills in the distance, and took a deep breath. Keen to make good progress south, I rode on late into the evening that day, through glorious open country criss-crossed by rice fields that were interrupted occasionally by towers of limestone rising erratically from the plains.

The next day I made much better progress south, enjoying the peace and freedom of open, dusty roads after the mayhem of Highway One. I arrived at Du Lu'ong, a small garrison town on the Ca river, just after sunset. Tomorrow I would head west over the Cau Treo pass into Laos. That night, over a glass of local sake, I reflected on the miles I'd travelled through the wild northern hills of Vietnam over the past month. A kaleidoscope flashed and turned across the garish walls of the cheap diner where I'd ordered a noodle soup and salad. I felt completely at ease on the road, in the world, and with my own company at the same time. And I felt an inexplicable affinity with the rugged hills, red rivers and tough, close-packed villages of northern Vietnam.

It had been a voyage of self-discovery, too. Ted Simon, one of the greatest adventure motorcyclists of all time, rode around the world on a Triumph Tiger between 1974-78. He pointed out that "To be worth making at all, a journey has to be made in the mind as much as in the world of objects". That night I dreamt of a bright country of green mountains and red deltas, and of being on the road once again with the rising sun.

I left Vietnam the following morning at the Cau Treo pass. Riding across the long, deserted bridge over the Ca river at dawn, a dusty sun rose over the wide plain to the east. To the south lay Vinh, the sprawling new city built from the rubble of fighting between the French and the Viet Minh in the 1950's, and the American bombing in the early 70's. The badly potholed road twists out of Du'c Tho and makes a dramatic swing west, disappearing headlong into the dense jungle of the Annam Highlands. I reached the summit of Cau Treo just before eleven. The misty sky above the jungle changed colour ceaselessly. Blue-grey cloud thickened to black, then cleared. The thick-tangled trees vanished again under curtains of veiled mist. Across the border, Vietnam faded to the east.

After Cau Treo, the road plunges through a steep-sided ravine. Soon the densely wooded hills of Annam break and the land stretches out across a wide flood plain. At the new bridge where the Nhuong and Kading rivers meet, I stopped for a while. The merging waters swirl fiercely here, with great turbulence from their combined strength. Alone on the empty plain, in the middle of that bridge in the heat of the day, I leaned heavily on the Minsk and felt the vibration of the moving water judder through the bike's heavy frame. The hum of that mountain water through the old machine seemed to anticipate the long road ahead, like thunder before rain. I took a last, long look down into the swirling current, then a deep swig of cold water from my flask, and kicked the Minsk into reluctant life. I rode on west into a dazzling blue afternoon, towards Mueng Pakxan, the Mekong river, and another country.

Laos's long, lonely Highway 13 is the only road connecting Pakse and the south of the country with its capital, Vientiane, and the mountainous north. It shadows the Mekong's east bank, following the great river north for almost a thousand kilometres, from the remote frontier-post of Veun Kham on the Cambodian border. From Vientiane, it twists into the mountains of the Xiangkhoang Plateau, eventually descending back to the Mekong at Luang Prabang.

Vientiane in late autumn might be the closest thing to a perfect capital city you can find, if you like the laid-back vibe. A French traveller in the 1950's called it 'Asia's Timbuktu', and with good reason. The city is now more commonly known by its local name, Viangchan, but I actually prefer the exotic, sultry-sounding beauty of the French version. The traffic here is slow and lazy, reflecting the pace of life. The interminable bustle and noise of most Asian cities is conspicuous only by its absence; there was no real rush-hour in Vientiane in late 2003.

The presence of the Mekong immediately to the south gives the place an implacable timelessness. On the southern fringe of the central quarter the sidewalk merges imperceptibly with the huge sandbanks of the great river. In the strong midday sun, the heat swirls above it and the river floats on curtains of shimmering air. With the coming evening, the reedy, languid smell of the river suffuses the streets. In the frequent lulls between passing cars and rickshaws, you can hear the brown water turning and gliding between the sandbars. The current is still strong here, even though the river is almost half a mile wide. It is easy to forget that it's still more than a thousand miles from where the Mekong meets the South China Sea.

I stayed in Vientiane for a week, and the city was a perfect opportunity to get my bike in the best possible mechanical shape for the road ahead. One morning, on the way to the Chinese embassy to collect a visa, the Minsk suffered the first – and most serious – of various mechanical failures that would befall it over our six-month journey. Braking for a junction, I made a normal gear-change down into second, and a sudden, monstrous crack erupted from the engine, soon followed by the crunch of shattering shrapnel inside the bike's single cylinder. It was not a good sound.

If you've ever accidentally left your keys in a trouser pocket and later hear them inside your washing machine on full spin, you'll appreciate the alarming nature of a similar sound coming from the guts of a motorcycle engine. After a few seconds, inevitably, the engine died.

I tried the kick-starter, which was completely stuck, confirming my worst fears. One of the piston rings must have sheared or snapped off, hence the shrapnel sound, and a piece of it had jammed between the piston and the cylinder sleeve. The only solution to this was to find someone who could re-bore the cylinder and replace the piston: I thanked my lucky stars it had happened in the capital. The time it took to find a mechanic competent enough to carry out the repair and get the Minsk on the road again gave me the chance to have some much needed downtime from the road.

I left Vientiane later than I planned, partly because of the need to get the bike in good shape, and partly because of a girl from California. Since the bike would undoubtedly have got jealous if I stayed on – and perhaps do something worse to me than destroy a piston – and also because her way was south and mine was north, I rode out of the city on December 10th as the afternoon heat shimmered on the plain to the west. As the shadows lengthened and the air cooled, the southern edge of the Xiangkhoang Plateau buckled the horizon. To the north, a wall of green mountains flashed in the day's last light. A shiver of excitement went down my spine: I would ride that way tomorrow. I aimed to stop at Vang Vieng for the night before making a dawn start for Luang Prabang.

Rising at dawn and leaving very early was a habit I'd got into in Vietnam. When you're travelling on a bike, it's wise to make good use of all the hours of light available. Riding in the dark is better only as a last resort than a regular habit. Some of the best things about adventure motorcycling in Asia are the things that happen in the early morning: feeling the cool rush of the morning air as you shift into gear and hit the road, the smell of the new day breaking across the hills, greeting farmers on their way to work, and hearing the bright, bubbling chatter of local kids walking to school with the dust on their shoes.

Almost as soon as I left Vang Vieng on Highway 13, the country changed immediately. From the long plains of the Mekong the land rises in sweeps of hills scored with terraced fields: these are the easternmost foothills of the Himalaya. Quickly, without warning, the terraces are engulfed by mountains too steep for cultivation. Thin wands of mist began to streak the highest trees, refracting the sun into columns of green and white light. On the first big climb of the day out of Ban Phatang, the Minsk's rebuilt cylinder hummed with new confidence.

I began the ascent of Highway 13's highest pass in the late afternoon. Leaving Muang Kasi, a windswept one-horse village with a wild-west air, I felt the unmistakable chill of winter for the first time. A cold wind blew down the valley, and odd, stray gusts kicked up dust devils in the dirt. I turned up the collar on my jacket to reduce the chill.

I saw the stray flare of metallic red of a car bonnet amid the ochre leaves and road dust as I passed, almost as an afterthought. Stopping a short distance further on, I walked back to discover the identity of that alien colour in this rugged land. Weeds grew through the buckled wheels and empty windscreen, giant bamboo sprouted from the rust of the big V8 and the paint flaked off like ash under my fingers. It was an early 70's Pontiac Firebird, one of the most desirable American GT cars of its era. It looked like it hadn't moved an inch for twenty five years. What strange events had left it to rust here, on the summit of one of the highest passes on the Xiangkhoang Plateau? Had some flashy U.S. Army Captain been forced to abandon it here in the chaos of the American withdrawal from Vietnam?

Two boys suddenly poked their heads above the driver's side wing, laughing innocently. I couldn't help but do the same. Our laughter broke my reflective mood, and I showed the inquisitive boys my map. In the most rural areas of south east Asia, few people have seen detailed maps of any kind of their home region. These two were particularly captivated by the sight of their village and the mountains around it in two dimensions, creased and held together with sellotape.

"Look, the Mekong" I said, pointing to the blue line of the great river cutting through the darker swathes of the mountains.

"Ah, Meaak-Hong" they said, pronouncing the great river's name as they pointed west towards it with the heavy local emphasis on the second syllable. Their dark eyes lit up as they recognised the line of the river on the map, as if they somehow understood its enormous power.

I stopped for a final time that day on the last high pass of Highway 13, before the road twists down the long hill to Xiang Ngeun and the final stretch back to the Mekong and Luang Prabang. There were blue-grey clouds shrouding the higher mountains to the north. A cold wind blew across the shoulder of the ridge and through the narrow gap in the trees at the summit of the pass. A wave of lonely delight swept over me as I sheltered behind the bike, drinking green tea from my flask, and watching the world turn as the light drained out. I stayed there for a while as the sky steadily darkened in the east. When the air grew colder, I started the bike and rode on into the evening. I could see the lights of distant trucks winding up the hill far below, and flicked the switch on my own headlight as I began the descent.

After travelling four thousand miles through Yunnan and Sichaun in southern China – but without the Minsk, since it was impossible to get it into the People's Republic – at the beginning of January I crossed the Mekong from Ban Huai Sai in Laos to Chiang Khong in Thailand. I'd been on the road just two months, but it already felt like years. From here, my plan was as ambitious as it was loose. I aimed to ride through the mountainous north of Thailand before heading south and east, across the plains of Phitsanulok and Chaiyaphum, and crossing into the remote jungle of northern Cambodia just south of Surin. Then I'd head west for the Gulf coast and take the road south from Bangkok down the Kra Isthmus. After that, improvisation would be my strongest ally. From the very beginning of this journey, I'd sought to travel by instinct. As I rolled the Minsk off a dugout canoe at Chiang Khong in northwest Thailand, I felt that the roads I'd already travelled had been a pretty strong initiation for whatever was going to happen next.

The visible difference between the Mekong's east bank at Ban Huai Sai in Laos to the much more modern and developed country I found on the river's other side captured the gigantic cultural and economic contrasts visible everywhere in the developing world. The striking contrast between the sleepy, dusty, one-street town of Ban Huai Sai and its bustling counterpoint to the west defined the separation of developing Asia from the rapidly Westernising, modern Asia of Thailand in the early twenty-first century.

Rolling into Thailand was like entering a new world. Gone were the makeshift roadside bamboo shacks fastened with twine and coconut leaves; gone were the wide-eyed, filthy street kids with tangled black hair that only half concealed their brilliant smiles; gone were the buffalo meandering unattended through the dust in the evening. The strange sight of more familiar customs replaced them: busy cafés were filled with people working on laptops, reading newspapers, or chatting loudly into mobile phones, and swarms of modern Japanese cars and pickup trucks jostled for position on major intersections.

After stopping briefly in Chiang Khong to change money on the black market (where you always get a better rate, by the way), and picking up a few long-overdue spare parts for the Minsk – a new back tyre, spokes, a chainwheel, clutch plates, and a chain – I set off quickly for Mae Chan and Mae Ai, a small hill-town built on a narrow ridge about a hundred kilometres west of the Mekong and just south of the Burmese border. Just north west of here, the Doi Ang Khang poppy fields stretch into the distance towards that elusive country to the north. These fertile plains ringed by green mountains – the heart of the so-called 'Golden Triangle' of global opium production – were one of the major sources of the notoriously pure heroin that circulated among American GIs in the latter years of the Vietnam war.

I stayed that night in Mae Ai in a tiny room that overhung the steep slope of the mountains falling away to the north. I rose in the half-gloom at five thirty, and left as the dawn was just starting to break across the hills to the east. There were blue and red clouds along the horizon, and as the sun rose on the hill down towards Chiang Dao the entire sky reflected the rising sunlight for a few seconds. It was a moment so luminous and three-dimensional that no photograph could ever have reproduced it, and set the tone for a thrilling day's riding.

With its re-bored cylinder and new piston now bedded in, the Minsk was on the best form of the entire trip. The higher quality of the gasoline in Thailand may have helped too, since much of the fuel I'd been running on until now had been of an ambiguous octane rating and probably contained lots of dodgy dilutions. In much of the developing world, petrol is often diluted with kerosene, paraffin, alcohol, or any other flammable liquid you can think of. After leaving Pai on the Nam Mae Pai river, I took Highway 1095 around Thailand's extreme northwest tip to Mae Hong Song.

This is one of southeast Asia's more spectacular roads, and a motorcyclist's dream: it curls along the Burmese border like a shadow-boxer, shrugging off even the steepest climbs with deep switchbacks, and diving into secret, sharp-edged ravines at a moment's notice. I reached Mae Hong Song, a bustling market town in the country's extreme northwesterly corner, just as the air was cooling and the dusk sky begun to drain the colour from the hills. Green turned to deep blue and gradually faded to grey, with the shadows of the spines of more distant ridges interlocking and vanishing in the darkness beyond.

Before arriving on the outskirts of Mae Sariang itself, the road climbs over a little pass just north of Pang Mu and plunges into a deep valley. The road here is overhung with black crags that seem to float above the highest trees. Half-lost in the thick air of that jungle evening, I felt that this valley might go on into interminable distance, towards some place lost in time beyond the edge of the map, and out of reach.

I found a simple room in a quiet guesthouse on the edge of the river, and remember it as one of the most beautiful places I've ever slept in. There was a bamboo mat on the wooden floor with a candle beside it, a mosquito net, and a table and chair against the wood-panelled wall. That was all. The window was open, and the room smelt of the river outside and the hills beyond it. I sat on the edge of the flat concrete roof, and smoked a cigarette as the creatures of the night began their soporific drone.

In this part of Asia, the frogs and cicadas seem to increase their volume as the last strands of light drain out of the sky and the neon bulbs of street vendors begin to flicker. A bright star rose in the southwest and I ate fresh fish from the river for dinner that a little boy brought over in a box on the back of his bicycle. Served with the staple Thai side-dish of rice and vegetables, it was the most delicious thing I'd eaten for months. Orkney writer George Mackay Brown sprung to mind: "a river fish has been left at my door. / Gifts come like autumn leafage in my doorway."

The next morning I rose at dawn, and left Mae Sariang with unexpected regret: it is the kind of place where time could melt like ice in the sun, where mornings merge with evenings and evenings with nights. Perhaps I should have stayed, but I had really got to get going if I was to make it to Prah Nang Bay in the far southwest of the country by the end of January using my chosen route through Cambodia: I had some serious distance to cover.

It was another truly exhilarating day. After plunging down a series of steep hairpins thirty kilometres south of Mae Sariang, Highway 105 picks up the east bank of the powerful Mae Nam Moei river which rises in the distant highlands of Myanmar. The road shadows the riverbank for over a hundred kilometres, until it reaches the remote border-town of Mae Sot. Eager to make swift progress east, I took a short-cut route over the mountains above the temple of Wat Don Kaeo. The landscape changes as you travel south from the jungle mountains of Chiang Mai and Mae Hong Song towards Sukhothai and the great central plains of Phitsanulok and Chaiyaphum. The intense green of the steep-sided hills is split by the blue and white flashes of rivers. Eventually, these rivers fade into the ochre vastness of the open land that stretches across the heart of the country. In midwinter, the plains lie parched by months without rain, and the grass turns the same colour as the earth itself.

Thailand's Highway One is the main arterial trunk road connecting Bangkok with the country's second city, Chiang Mai. As I crossed the huge intersection at Tak from west to east, the powerful feeling of entering wilder country washed through me. Perhaps it was the way the sky opened out like a fan in the absence of interruptions like towns and other roads, expanding the horizon in every direction. Or perhaps it was the arrow-straight strip of tarmac that vanished into the haze in the distance ahead. Most of all, perhaps, it was the feeling that very soon I would be crossing the border into Cambodia.

I made it to the small plains city of Phitansulok by dusk that day, riding into town under the familiar leer of the neon signs scrawled with tightly-packed Thai characters and the occasional bit of hilarious English. I noticed a sign for 'Chow Mien – Fried

Children' scribbled on a blackboard under a noodle stall: an all-time classic among the perpetual comedy of lost translations in East Asian signage.

That night I stayed in a cheap, old-fashioned hotel on the eastern edge of the city, which I chose because it seemed well-placed for my early departure tomorrow on Highway 12, heading east. It was hot that night and I slept directly underneath the old fan, clunking around slowly like a decrepit windmill. The sweet smell of garlic, ginger, and chilli in smoking wok oil still hung in the air; an unmistakable scent of the provincial towns of Thailand, Cambodia, and Malaysia.

When I left Phitansulok at five thirty the next morning, the air was still dark and cool. The road outside was quiet, but the low drone of a heavy truck pulling away from the intersection around the block reminded me that I was on Thailand's arterial east-west highway. For the first time since the jungle mountains of northern Laos, I had to fasten my leather jacket right to the top to keep out the chill. As I swung out of the hotel courtyard and into the street, my headlight beam glanced off a puddle, like the beam of a lighthouse. I had a long journey ahead today.

Given the Minsk's top speed of eighty kilometres per hour, I covered the seven hundred kilometres from Phitansulok and Buriyam in a respectable twelve hours, with a few stops. In the middle of the vast plains of Nakhon Ratchasima, to the east of Chaiyaphum, the clutch lever warped alarmingly as I changed gear to negotiate some potholes. I then watched in horror as it bent and lazily snapped like a lead pipe. In slow-motion, it fell and vanished into the dust, leaving a useless, sharp stud of cheap Chinese alloy protruding from the lever housing. I thanked my lucky stars that this had happened today in Thailand, where I knew I would find a replacement part, instead of tomorrow in northern Cambodia, where I certainly would not. I put it down to traveller's luck, keeping my fingers crossed that no more serious problems would occur on my journey south for the Gulf of Thailand and the coast.

Sure enough, a mechanic's shop appeared on the outskirts of Muang Yang, a tiny place on the Mae Nam Mun river. As I rolled the bike up the ramp I felt like a thirsty traveller who'd just discovered a spring. It only took a few minutes to replace the broken lever with a new (and far superior) Japanese spare, and I bought another before I left just in case. There wouldn't be any well-equipped workshops like this until Phnom Phen, Cambodia's capital, at least three days' riding from here.

Buriyam is a busy provincial town about sixty kilometres from the Cambodian border, and the sort of place that remains completely off the tourist radar. I wound through the bustling streets as the night-market vendors were opening their stalls, passing a few glum-looking Chinese hotels. The scrum of scooters and bicycles I was in ground to a halt at a crossing. At that moment I noticed a flash of neon from a small café to the right, half-obscured by foliage, with the drawl of Thai pop music trailing out from a ghetto blaster in the yard. It looked like the right kind of place to wind down after my long day of riding across the hot plains. Besides, I was dehydrated and had begun to hallucinate about bottles of cold, sugary liquid touching my lips. I pulled in and cut the engine. My hair was thickly matted with sweat and my skin was covered in a layer of red dust. Dark stains of two-stroke oil ran from the backs of my hands right up my arms. I was completely spent from thirteen hours of riding, and the evening blurred into sleep. My hosts, Toni and Mai, couldn't have been more welcoming and let me sleep on a bamboo mat on the floor after I'd had dinner. I think they could tell just how tired I was from the long ride.

I woke just before dawn on the bamboo floor of the café kitchen to the quiet sounds of the street outside: a lone scooter passing; the distant rumble of a truck; a wok crackling down the alleyway opposite. Toni and Mai, the café's owners and my spontaneous hosts, were still fast asleep on bamboo mats inside. The courtyard was silent in the cool, scented air. I rose quickly and rolled up my sleeping bag, packing my few possessions into the narrow backpack I strapped to the back of the bike. I left a note thanking them for their wonderful hospitality, drank a thick black coffee from a street stall, then rolled the bike down out of the yard. The engine was completely caked in dust from yesterday but started first time. Good on you, Minsk, I thought. I wanted to leave Buriyam as early as possible and make good progress to the border.

An hour before I reached Surin, the last big town before Cambodia, the Minsk threw another of its inexhaustible jokers from the pack of breakdown cards. This time, the rear suspension mount [the piece of metal that attaches the rear shock absorber to the bike] partially sheared. This was far worse than yesterday's clutch lever problem. It strongly suggested that the bike was in no fit state to take on my intended route, which took in some of southeast Asia's worst roads between the border and the Cambodian capital, Phnom Phen.

This time, yet again, my luck held out against the red jester from Belarus. As always in Thailand, I managed to find a place where I could get the bike back into a functional state. I saw the welding rig out of the corner of my eye, and swung into the yard with an urgency that surprised the two men fixing the dismantled scooter that lay strewn across its concrete floor. They looked up with the uniquely rapid and attentive interest of tradesmen who know they've seen something that could quite possibly be a big, profitable job.

If Monty Python's Flying Circus had ever been given a brief to design a motorcycle, they would have come up with something very similar, I thought, to this cantankerous contraption that had somehow, quite miraculously, got me across seven thousand kilometres of southeast Asia so far. The only added comic feature was my stubborn belief in the wretched thing's roadworthiness. An hour later, and with the rear suspension mount now expertly welded back into some sort of shape, I gave the welder a decent tip for his efforts and rolled the Minsk out of the yard with a hasty goodbye. The road from Surin to Kap Choeng, the tiny outpost of Thai civilisation glued to the edge of the Dangrek Mountains, passed without further incident.

As I waited in the small queue of jostling locals, mostly Thai traders taking cheap Chinese domestic goods south to the remote jungle villages, I felt a sudden, inexplicable sense of excitement. Perhaps I watched *Apocalypse Now* at an impressionable age, but Cambodia was somewhere I'd always dreamed of arriving. As a teenager, I'd imagined crossing a river overhung with giant trees, where the far bank was a place no modern communications had yet reached. At Kap Choeng, the nearest river is ten kilometres away and most of the tallest trees have been felled, but my sense of childish excitement still remained.

Once I'd got my visa stamped and cleared the border, the country suddenly expanded in front of me. It felt like a cinema opening into widescreen. To the south, all traces of industrialised, civilised life seemed to vanish, and there was nothing but jungle and swamp stretching beyond the horizon: the heart of darkness.

I couldn't wait to get going. As soon as I'd shaken off the usual coterie of scam-merchants, con-artists, and random space-cadets who predictably appear at border crossings throughout the developing world, I squeezed the Minsk's throttle, stamped into third, and hit the road south. My dust-cloud completely obscured the view in the mirror back towards Thailand, and the steady riding of the past ten days quickly disappeared in a thick, red-brown haze. I knew I was in for a tough day.

As I gained speed, the flow of dirt riding quickly returned. My body loosened up with the increasing feedback from the bike as I stood up on the pedals, and I suddenly noticed the machine had begun to shudder alarmingly with the change in angle at every corner. I was concerned about the morning's improvised repair failing, and dropped my speed, taking each bend with more caution than usual. As the ruts deepened, with great relief I realised that the shuddering wasn't a sign of the bike threatening to disintegrate, but of the deepening corrugations on every change in the road's direction; these are a constant feature of dirt roads in Cambodia.

I made good progress south into the Kampong Highlands, and the road deteriorated with every passing mile. I found myself riding through giant potholes rather than around them; their sheer scale meant that circumnavigation was impossible. It was simply a case of down one side, and up the other. Although the Minsk had begun to produce a new cacophony of shallow wheezes from the dust-choked carburettor, I felt it was on its best form of the entire journey so far.

It was already late in the afternoon. I knew I had no hope of reaching Siam Riep in daylight, but decided to press on: the thought of finding a clean bed in a quiet room in which to collapse for the night was enough to sustain me through the next six hours of combat motorcycling. As night fell, the jungle grew suddenly, impossibly dark in minutes. I thought of Graham Greene's observation in *The Quiet American*, the classic novel about colonialism in Southeast Asia, that "the men grow old here the same way the sun goes down: they are boys and then they are old men".

I was enveloped by a wild expanse of darkness stretching infinitely across the land, shrouding everything in its grasp. Just beyond the faint smudge at the edge of the road where dust merged into trees, the brushwood camp fires of local people lit my way far better than my fading headlight beam. The unmistakable smell of chilli and garlic crackling in hot oil – the signature scent of East Asian streets – mingled with the damp cool of the jungle. By the firelight, I could make out the shapes of houses along the roadside. Shadows of villagers hovered between them, sometimes moving imperceptibly, like dreaming ghosts.

This evening, winding slowly south through the western Cambodian jungle by firelight, I felt I'd turned a new corner in my journey. The creature comforts and perfect tarmac of Thailand were gone, replaced by a harsher reality of barely drivable roads and a complete absence of signposts, petrol stations, or anything else a traveller might find useful. All the infrastructure we normally associate with driving simply didn't exist here. If you wanted to get some petrol in the far north of Cambodia in the first years of the twenty-first century, you'd have had to ask someone in a roadside teahouse, who might have a few plastic bottles under their bamboo table.

The last ninety kilometres to Siam Riep took an interminable five hours. My headlight had failed almost completely as the alternator coil was completely clogged with dust, and I rode in the slipstream of a convoy of big trucks on that last stretch, following the movements of their tail lights, trying to anticipate the biggest of the approaching potholes. When the trucks negotiated the largest of all these craters, their lights would disappear almost completely with the sudden change in angle.

Facing up as the truck plunged down, the tail lights became dim red smudges in the dusty night air. It was the closest thing I've ever experienced to being at sea whilst on land; the complete spatial disorientation of riding without a headlight was amplified by the necessity of using the moving tail lights as my guide. Occasionally, I'd misjudge a pothole and both the back and front shock absorbers would bottom-out with a sickening crunch. Each time, I thought the Minsk would simply give up the ghost, making a spectacular final protest against its abuse with a gargantuan shearing clunk.

I rode like this for almost four hours. When I thought I couldn't go on much further, I stopped at a roadside tea shack. I was completely exhausted, but at the same time strangely energised. There is a reserve energy tank in a person's body that can only be switched on in certain situations, and only when all available blood-sugar and fluid has been used up. Tonight, somewhere in Cambodia, was one such time. "Is it far, Siam Riep"' I enquired in French, which some Cambodians speak well, to the dreamy girl who'd served me tea, with the hope that she'd understand it.

"No, no, not that far" she replied, to my amazement, in perfect French. Her bright, faraway smile and the hot tea she brought gave me a renewed burst of energy, and I set out on the final stretch of my epic day. The convoy of trucks were far ahead by now, rolling east into the night, towards Pnomh Phen and the Mekong Delta.

The air changed as I set out, and a cool dampness blew in off the approaching trees. Western Cambodia was heavily deforested by the Khmer Rouge, but much of the old jungle around the ruins of Angkor – the extraordinarily advanced civilisation that existed here over a millennia ago – remains intact.

I finally arrived in Siam Reap around ten thirty. I felt like I'd been playing a fourteen hour chess match whilst riding a wild horse: I was completely beat. Following my instincts, I found a quiet place to stay on the edge of town. I ate some noodles, struggling with the effort of twirling them around my chopsticks, and drank a beer. When you're really tired from a long journey involving a lot of physical exertion, the feeling of stretching out on freshly washed linen is the most delicious thing in the world. The night was quite cool, so I turned the fan off and listened to the castanet drone of cicadas until I fell into a long, dreamless sleep.

I woke to a shaft of bright sunlight falling through a chink in the shutters. I must have slept in, I thought, reaching for my watch. It showed 7.45: a serious lie-in after a run of pre-dawn starts. In the east, everyone rises with the sun; this morning, waking well after sunrise felt like utter decadence, which was only enhanced when I realised I didn't have to leave Siam Riep until Thursday. It was only Monday; just the prospect of three whole days without riding distance felt like I'd been given a first class ticket to the most luxurious hotel on the planet. All I had here was a simple room with a mango tree outside the window. Luxury, of course, is a relative concept.

Later that afternoon, reading by the pool of an uptown hotel under the shade of a pair of coconut trees, I met Sonya. She taught English and French at a school for Cambodian orphans set up by French missionaries. Born in Marseille, she'd spent most of her life moving. She was forty-one, but looked a decade younger. She had that distinctive beauty unique to some French women in early middle-age; a dusky, sophisticated sensuality. There was also a mysterious sadness in her bright green eyes. Sometimes, when I told her of my journey, it faded and was replaced by a deep luminescence, as if she were trying to reach back to a part of her life locked away in time and imagination.

In the growing dark, we shared a cigarette under the green thickness of the papaya trees in the back garden. The air was very quiet and still, except for the faint drone of cicadas somewhere else, somewhere out there in the jungle. I looked across and saw that she was trying not to breathe, and her eyes were half-closed.

'Shhhhhh', Sonya whispered, so quietly I could barely hear her, as if she were trying not to disturb the sticky, inky silence of the Cambodian night. I felt she wanted us to become part of it, standing there encased in that immense black space stretching into the outer dark of the jungle, into its old stones, its tangled trees, and its endless mysteries. Later, she kissed me as we said goodbye, before disappearing into the night like some sacred spirit who'd come to meet me from another world.

The next morning, I left Siam Riep at dawn as a huge, blood-red sun was rising in the southeast sky. The air felt like it was going to get hot out here, on the empty plains of the interior. As I rode east, the heat built up relentlessly, and the road worsened. The good asphalt around Siam Riep was completely gone, replaced with huge, bone-jarring potholes and giant ruts. These craters were the results of zero maintenance for several decades, coupled with the fact that even in the summer rainy season, big trucks still made their way, painfully slowly, across the country, gouging ever-deeper corrugations in the mud and dirt. I couldn't begin to imagine what the roads would be like during the rainy season.

After a couple of hours, the plain began to wobble under the sun and only the movement of riding would create a tolerable temperature. Having already run out of the insufficient water I'd filled up in Siam Riep, I stopped at a coconut stall somewhere in the middle of Cambodia, in the white heat of the day. Nothing moved. On the road, and on the plain, everything was perfectly still. It felt as if the earth itself was being pressed down by the huge, relentless weight of the heat shimmering above it.

The boy at the coconut stall had a tiny umbrella, which gave him just enough shade to cower under in the presumably long intervals between customers. He cracked open a coconut with a single, expert blow from a big machete, inserted an enormous straw, and handed it to me. There is absolutely nothing that tastes better than cold, fresh coconut juice when you're really hot and really thirsty. I drank the whole thing without stopping. When I'd finished, I asked for another. With a bright smile, the boy grabbed the biggest one in his bamboo basket. I sat under the tiny patch of shade under his umbrella, drinking the second coconut, feeling the life return to my parched body. The effects of dehydration is one of the biggest dangers of adventure motorcycling in hot countries, and today I'd had a full dose of it. Conscious of the inherent dangers of being dehydrated – instant lack of concentration, spatial disorientation, and delayed reaction time – I stopped regularly to drink through the rest of that very hot afternoon.

Approaching Pnomh Phen at sunset, the city emerged like a dream-vision from the far bank of the Mekong. You can smell the great river long before you see it; here, still a few hundred kilometres from the sea, it is almost a kilometre wide. Underneath the new steel, the murky water flows resolutely south towards the South China Sea.

Ramshackle high-rise blocks and old wooden-rafted houses were jumbled together with delightful recklessness. The sudden brightness of washing hanging out to dry transformed the drab squalor into instant, colourful life. In the lull between waves of traffic, the sound of crackling woks spat hot oil into the evening.

That unmistakable east Asian smell of ginger, chilli and garlic frying reminded me that I'd not eaten for twelve hours; I needed to find a place to stay and get some dinner. Arriving in a developing world capital at dusk after riding a motorcycle all day is one of the least relaxing ways of entering a new place it is possible to experience. I turned off the main drag and followed signs for a hotel down a lively side- street packed with stalls that sold everything from cheap, corrupted petrol to deep fried insects. I checked in, and took a cold shower. Standing under the delicious, refreshing rush of water, I felt layer after layer of dust and grime rinsing out of my hair and skin. Having a cold shower with soap is one of the best physical experiences you can have when you've been riding a bike for a whole day in tropical conditions; I had to unclog the plug of sand and silt at least three times.

It was completely dark outside by the time I flung open the wooden shutters and looked out into the street. Men were packing up their wares around the daytime stalls and women were bustling around setting up the night market. As in every city across the world, the streets changed with the coming of night. The bright, busy reality of the day was replaced with something quite different; a smokier, stranger, more mysterious energy.

I went out and ordered some 'special noodles' from the woman with the biggest wok along the entire street, based on the principle that you should try everything once. She said something in Khmer to her small assistant who fetched a pot of live, writhing creatures. My empty stomach turned. As she grabbed the pot, something large and hairy fell out. So that was what 'special noodles' meant in this part of Phnom Phen: Tarantula Noodles. Spidery legs had been plucked and battered, and were now bubbling in hot oil. I was starving, and needed to eat something, spider or no spider. The first bite was crunchy, the second was slightly more crunchy, and on the third my teeth sank into a fat, succulent pod, about the size of a matchbox. Amazingly, it was delicious. And it's not every night that you clasp a deep fried tarantula between your teeth, after all.

Over the next month, I rode down the coast of the Gulf of Thailand and across to the west coast, where I stopped for a while to climb with some friends from back home on the world famous limestone of the Krabi area. We also developed some new climbs on a small island off the coast further south, near Trang. Our twin-engined speedboat gurgled out from the pier into the slow current, then accelerated hard as it cleared the sandbar at the mouth of the river. Limestone towers rose along the horizon. Ko Laoliang, a tiny island in the Surin Archipelago about twenty miles off

Thailand's south west coast, was exactly the break from the road I'd been searching for intently, but never quite found. In the few weeks I spent living and climbing there, the dusty hills of East Asia that I'd travelled faded into a dream-sequence, locked up with the battered Minsk in the chandler's shed by the pier.

On rest days from climbing, we'd head out in longtail boats to catch fish and swim. We'd play chess and cards on the beach by firelight until late in the evening. It was easy to imagine staying there indefinitely, casting adrift on the complex currents that twist between the islands. One morning, I peered over the edge of a longtail beyond the point of the reef, where the seabed plunges into darkness. Shadows of sharks spiralled under us in pursuit of tuna, headed for other islands to the south and west. Laoliang was a great recharging place just beyond the mid-point of my journey, and without the respite it offered I would never have made it as far as I did on the second half of the expedition. But I couldn't stay there for too long, I realised, or else my momentum would be lost. When my friends left for home, I jumped back on the bike and headed south for Malaysia.

For the last six weeks of my journey, a current swept through me stronger than any I'd yet encountered. Each day, I'd be on the bike by dawn, and travel by instinct. I rode south for two days straight and arrived in downtown Kuala Lumpur, my clothes almost black with oil and dust. But there was no time to lose. With an Indonesian general election less than two months away, and the known reluctance of their authorities to grant tourist visas at politically sensitive times, I had to cross the Malacca Straits as soon as I could.

Two days later, I rode the Minsk up a plank on the Malacca docks and on to a boat for Sumatra. Night fell as we approached the low-lying, murky, mangrove-enchanted coast. A dim fog of shore fires blew across the dense jungle beyond the black water of the Straits, and the shadows of big trees loomed beyond it. The sea smelt of oil and salt and heat as the lights of fishermen crossed our wake. I'd reached the final stage of my journey, and I could already feel it might be the most thrilling chapter of the entire trip.

A brief logistical inconvenience delayed my Indonesian adventure: the Minsk was impounded on arrival in Dumai by Indonesian customs police because I had no Carnet [an internationally recognised vehicle passport]. I was, it seemed, faced with no other option than to steal my own bike back before dawn from their compound.

My plan sort of worked: the compound was guarded only by a fat, snoring policeman and a cat. Flushed with success, I rode into the jungle bound for Pekanbaru at high speed. Success didn't last long, though; I was stopped five hours later by a heavily armed Indonesian military police roadblock. Surrounded by sullen soldiers with assault rifles, I was quickly bundled into a military truck heading back for a police cell in Dumai. What should one do in such circumstances?

My only remaining option now was to befriend – and then to bribe – the customs officer in charge into making a temporary Carnet for the Minsk. He was actually really helpful, and I saw the bribe more like a tip for sorting everything out. With all the paperwork done, the next morning I hit the road north for the volcanic highlands of central Sumatra.

Most days, I'd ride on into the dark after the evening mist had swirled low over the rice fields. I rode northwards into the province of Aceh, then troubled by a long-running civil war. A week later, I waited on the quay at Sibolga for the night boat to Pulau Nias, a small island off the west coast of Sumatra.

The old ferry rumbled slowly out into the tide as lightning hit Musala Island to the south. It was still dark when I rolled off the boat at Gunungsitoli, with the air full of thunder and mosquitoes. But the storm passed, and in the afternoon I made good time down the east coast to Sorake Bay. I stayed there for a couple of days – my only respite from the road in a month of continuous travelling in Sumatra and Java – in a small wooden shack on stilts by the low stone breakwater under the first line of coconut trees. The first morning, I woke to an oily sun recoiling off the perfect surf that broke hard on the outer reef. By evening, the sound of the sea was as loud as a low-flying plane as the wind picked up. The village kids raced around with sticks, whistling and laughing, trying to capture fugitive coconuts. I could have stayed for weeks here – it's one of the best surfing spots in Asia – but my visa only lasted one month.

I had to get back on the road if I was to complete what I had in mind. In ten days of hard riding south and east, I made it as far as Wonosoba, a beautiful highland town on the edge of the Dieng Plateau in central Java. Once there, I had to acknowledge that I'd come far enough: there was just no way I'd get back to Dumai to return to Malaysia before my visa ran out if I carried on eastwards.

Midnight. The road trembled like the surface of a beaten drum. Big trucks bear down on my battered Minsk, sometimes with their headlights blown out, headed east for Surabaya, Yogyakarta, or who knows where. An hour from the eastern suburbs of Jakarta, I'm eighteen hours deep on my longest day of riding on the entire trip. In mountaineering, the most dangerous part of a climb is the descent. Likewise, the most dangerous part of adventure motorcycling is the last few hours of a very long day. I was dehydrated and needed to find a place to sleep. I saw a hotel sign flickering amid a cluster of shabby neon and broken glass on the road's far side, checked my mirrors, and turned right.

In a split-second, the air-blast hit me from behind as it passed, and a smoke-machine of oil and diesel fumes knocked me sideways. The huge, unlit timber truck howled its horn like a death knell and snarled off into the thick Indonesian night, a black rider on the dark. Shaken, I pulled into the sanctuary of a concrete yard stalked by scowling dogs and hunkering shadows. I hit the kill switch to cut the engine and noticed that my left mirror wasn't there: the truck had clipped it off as it passed. As near-misses go, as Captain Willard says of Colonel Kurtz in *Apocalypse Now*: "He was close. He was real close."

I parked up, checked into a filthy one-dollar room, smoked a Gudang Garam, took a cold shower, and slept the sleep of the dead till dawn. The only two survivors from the mission to kill Colonel Kurtz in *Apocalypse Now* are the battle-hardened and pragmatic Captain Willard, and the soul-searching Californian surfer, Lance Johnson. Willard makes it by thinking he knows what he's doing and improvising when he needs to. Lance Johnson makes it by understanding he doesn't know what he's doing, and believing in what he improvises. My journey had been a combination of the two.

Rising the next morning and feeling as lucky to be alive as I'd ever been, I hit the road west for Indonesia's sprawling capital city, Jakarta. Beyond it lay the ferry back to Sumatra, a long, hot ride through the eastern Sumatran jungle, and on to Dumai and the boat back across the Malacca Straits. From there, I'd head south to reach the Minsk's final resting place in an Indian motorcycle collector's garage in Johur Bahru, right at the southern tip of Peninsula Malaysia.

A few days before the very end of the journey in Malaysia, I was riding towards Danau Ranu, a crater lake lost in the volcanic mountains of southern Sumatra. It's one of those places you see on the map, and it just kind of pulls you in with a lonely, exotic allure. The rainy season was in full swing by now; piled thunderheads built up over the jungle hills to the south as I approached the lake on the tiny, winding road from Pekonbalak to the southeast. Women were busy working in the rice fields, moving softly through the luminous green surrounding them. A sudden wind blew in from the west, gusty and strange. I could smell the salt of the Indian Ocean, no more than ten miles away, on every gust that rose and fell. The sky darkened almost to squid-ink black, and the rice fields gleamed bright and green. I came up a short crest in the road, and quite suddenly there it was. The lake nestled in a great cradle of volcanic mountains, a thin fringe of coconut palms lining the shores. The surface of the water was very dark, and the wind was blowing patterns across it. I pulled over to the side of the road and cut the engine. The sky was so dark it felt as if it were dusk, yet there was still no rain. It was just me, the bike, the jungle mountains, and the monsoon wind blowing off the Indian Ocean. And right here, directly ahead, was perhaps the most beautiful place I'd seen on the entire journey. Maybe it was the wild weather, maybe it was the cumulative fatigue of the epic journey now largely behind me, or perhaps it was the thought that it would all soon be over, but I suddenly felt tears streaming down my cheeks. I stared down the mountain towards the dark water of the lake, at once joyful and also full of grief for everything I was about to lose.

All that I'd seen and done on the roads of East Asia over the last six months, I realised, would very soon be gone. The expedition had changed my life, but soon it would become nothing more than a memory, a lost rider on the past. But I was sure that it was the right time for the journey to come to an end. The freedom of the open road, after a while, just becomes another way of saying you've got nothing left to lose.

The fact the Minsk made it to the end of the journey is a testament to the stubborn simplicity of Soviet engineering, to the rambunctious roadside mechanics of southeast Asia, and some improvisation on my part. That I made it to the end, on the other hand, might be proof that traveller's luck isn't something you just wait around and wish for, but something you can actively create.

I climbed back on the Minsk, stamped on the kick-starter, and felt the two-stroke motor hum into life yet again. I rode on towards Danau Ranu with the wind in my hair, the thunder beginning to break out even closer now across the mountains around me. As I reached the point where the road met the shore of the lake, the first drops of rain began to fall, like an ending.

'Facing East' is a compilation of extracts from my journal from an unsupported, solo motorcycle expedition in East Asia made between November 2003 and May 2004. I travelled circa 14,500 kilometres through eight countries, beginning in Hanoi, Vietnam and ending in Johur Bahru, Malaysia. The motorcycle I used was the two-stroke 125cc 'Minsk', a Soviet-era two-stroke machine made in Belarus that was still widely used in northern Vietnam at the time. The bike's frame had been welded in seventeen separate places by the end of the expedition. I also used twelve tyres, three chains, two rear sprockets, two chainwheels, four air filters, two electrical coils, three sets of rear shocks, and an unknown quantity of two-stroke oil. I purchased the Minsk from Cuong's Motorcycle Adventures in Rue N'oc Quen, Hanoi, Vietnam for $350 and sold it in Johur Bahru, Malaysia, at the end of the expedition for $50 to a local motorcycle collector.

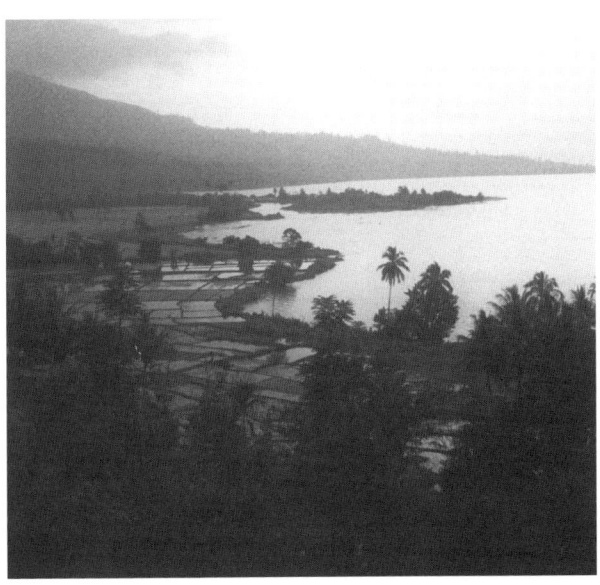

End of the road: approaching Danau Ranu, Sumatra, in April 2004, six months after leaving Hanoi.

The Light Elsewhere, 2013

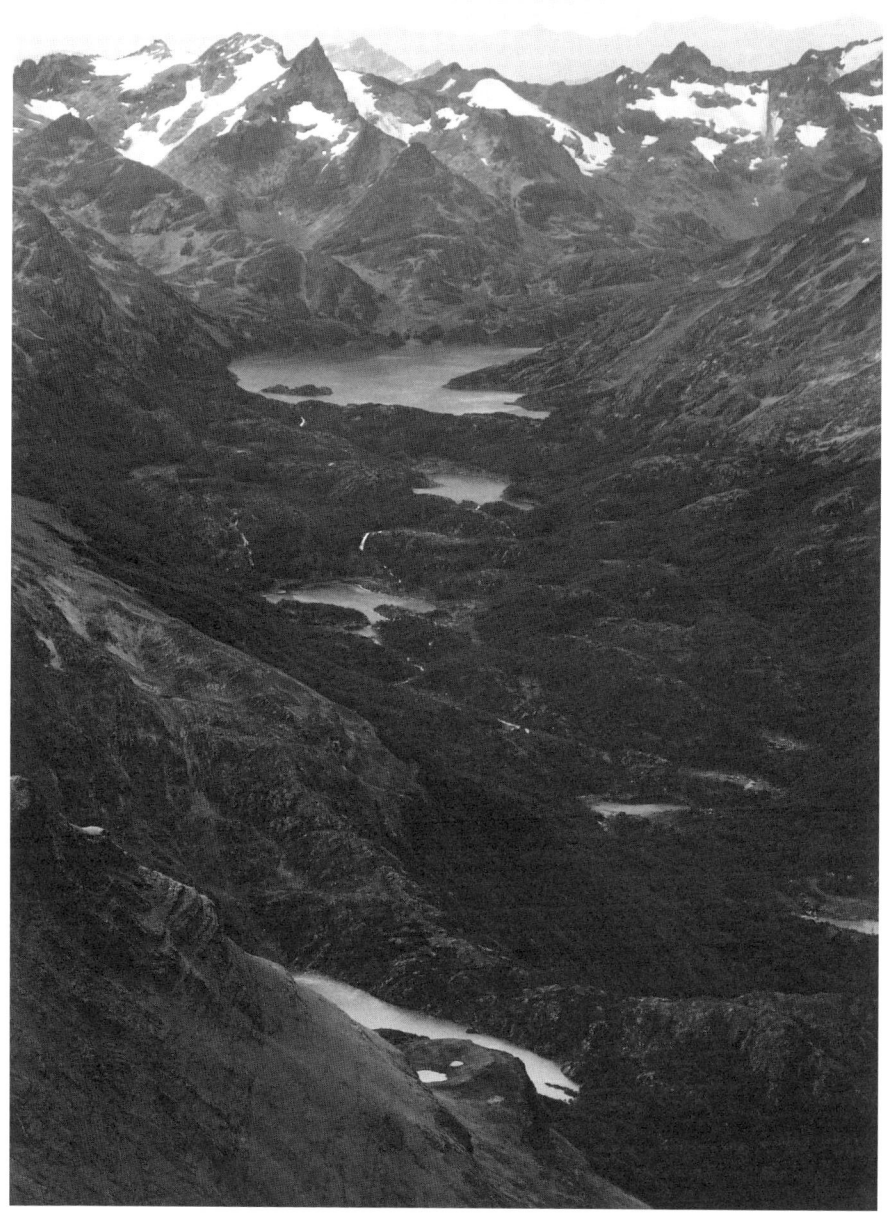

Looking west across the wild interior of southern Tierra del Fuego from the summit of Cerro Tonelli, Ushuaia.

A Short Walk in Tierra del Fuego

The lure of small mountains in big places

Just before dawn, I'm woken by a gust of wind that sounds like a fighter jet with engine failure. In southern Patagonia, the wind is a constant companion, but this was something different altogether. Shearing off the icecap with awesome ferocity, this particular gust audibly impacted the world. For fifteen seconds, the entire superstructure of the Torre Glacier in the Chalten Massif appeared to shudder violently at its foundations. As the gust subsided, the primeval boom of a chunk of calving ice echoed around the mountains, sending a series of micro-tsunamis across the silty brown water of Laguna Torre.

I unzip my lightweight tent and peer out into the half-light. Squinting through the gap in the pines just above my lonely camp, which is perched in a solitary, sheltered enclave on the edge of a ravine high above the moraine, fast-moving strands of altostratus are making serpentine shapes in the monochrome air. I glance up the ravine. On the limit of the tree-line, which is perhaps a thousand feet above me, the stunted pines bend double under the force of the wind.

Somewhere up there in the darkness of the morning, Cerro Torre and Cerro Standhart are lost in cloud. The decision is not so much reached as presented: I'm not going climbing today. In these mountains, the weather is the expedition leader.

I learnt from that abortive solo foray into the Chalten Massif that the weather in southern Patagonia is very much not to be messed with. It's the meteorological equivalent, perhaps, of a 120 kilogram black belt in ninjitsu. What to do next must be determined not by your ambitions, necessarily, but by what the local weather forecast suggests is achievable. Back at base camp, which is a converted shipping container in a windswept garden at the top end of El Chalten, I plot my next move. The 2677 metre peak of Huemul lies just to the southwest. It's a kind of outrigger of the main Chalten Massif. A classic multi-day hike leads up to the high pass, Paso del Viento, connecting the valley of Laguna Toro and the Viedma glacier flowing out of the impregnable wilderness of the Southern Patagonian Icesheet. Many do this hike at a leisurely pace over four or five days. I pack food for just two nights, choosing to go fast and light.

I take the first, easy section up through the pine trees and across the pampas at a slow run, and in a few hours arrive at the first camp by the river that flows out of Laguna Toro. A cold wind is blowing down off the glacier, and straight after dinner I get an early night: tomorrow is going to be a long day.

Waking at first light, I pack up quickly. After a light breakfast of dehydrated scrambled egg and some coffee, I set off. I weave my way around the shore of the small laguna above the camp and reach the river crossing as the sun is rising. The river here plunges through a narrow gorge, about ten metres wide, that's strung with a 10mm stainless steel cable. Getting across is easy but dramatic: you just don your climbing harness, clip yourself in, and simply pull yourself across the raging torrent beneath. Once on the other side, you're truly in the high country of Patagonia.

The glacier is vast and brooding, spilling off the side of not just one but three big peaks. It feels as if the mountains above are literally overflowing with ice, which in a way they are. I make steady progress up the edge of the terminal moraine, stopping a few times to check I'm in the right position. The route-finding is tricky in places, with no trail at all, just moraine debris and scree. After a couple of hours I reach the top of the moraine and pick up a faint, zig-zag path that leads up to the summit of Paso del Viento. What lies on the other side is something truly awesome: the Southern Patagonian Icesheet spreads across the surface of the world in every direction west, north and south like an immense frozen blanket, lined with gently arcing moraine-spines like dinosaur scales. I pause for a while on the summit of the pass, taking it all in. A local guide and his Swiss client come up the other side of the pass, having crossed a section of the icesheet; they're the only people I'll see all day. I'm glad for their brief company in this desolate, extraordinary place. We exchange greetings and swap notes about the routes we've taken. All too soon, we say our goodbyes and I'm heading southwest into the vast, shining wilderness of Patagonia. The afternoon slowly slides into a radiant austral summer evening as I traverse the western slopes of Huemul. Atop a short slope, the landscape changes, and the glacial desolation is transformed into verdant pasture broken with stunted, windbent trees.

The huge, blue-green expanse of Lago Viedma stretches out beyond the mouth of the glacier. I pick my way down a steep, thousand-metre descent to reach a pristine shingle beach at the head of the lake. Pitching my tent a few metres from the water, I reflect on an incredible day's hike: the best forty-odd kilometres I can remember hiking, in fact, anywhere in the world. As the sun falls below the horizon, I fall asleep quickly after dinner to the sound of huge chunks of ice calving off the glacier into the lake. My dreams that night are filled with ice and fire.

The next morning it's an easy hike out to the trailhead under the hot austral sun. Six hours later I'm back in El Chalten, looking at the long range forecast again with moral assistance in the form of a strong coffee in a half-pint glass. It's not good at all. A succession of fronts are pushing in from the Pacific for the next five days, picking up energy as they cross the icecap before slamming into the Chalten Massif. On the spur of the moment, I check the forecast for Tierra del Fuego, eight hundred miles south and the southernmost point of the Americas. It's looking much better: a seventy-two hour window of reasonably high pressure. Sometimes, particularly when on a solo mission, the best ally in the world is spontaneous improvisation. The decision is quickly made, and I book a flight to Ushuaia.

The dark and jagged sweep of the Cordon Martial guards the northern edge of the Beagle Channel, the stretch of open water that separates Tierra del Fuego from the much smaller Isla Navarino to the south. The Channel takes its name from the ship captained by Robert Fitzroy that took Charles Darwin on his celebrated voyages south in the mid-nineteenth century, and provides the last safe passage between the Atlantic and the Pacific oceans before Cape Horn.

I leave the still-sleeping streets of Ushuaia by taxi at dawn, bound for Valle de Andorra, the last outpost of civilisation before the wilderness of the Tierra del Fuego national park rises up to the west. I jump out at the chosen point and watch the battered old Merc disappear back down the dirt road in a cloud of dust. I glance around. I'm alone at the base of a wide valley that extends westwards towards a great cirque of steep and craggy peaks, the tallest still freaked with traces of snow from the recent storms. Early sunlight sparkles from the numerous icefields that remain on the upper slopes throughout the year.

Following a sinuous trail through the dense Patagonian forest, I cross and re-cross several rivers that flow down this valley from the small lagunas in the glacial basin to the west. By mid-morning, after navigating a mile-long flooded glade of pines straight out of Lord of the Rings, I've travelled more than fifteen kilometres, and strike up the steepening slope to the south, gradually breaking free of the dense and waterlogged foliage and into the barren terrain of higher ground.

The trapezoidal volcanic bulk of Cerro Tonelli, one of the highest peaks of the Cordon Martial, rises above the col of Paso de la Oveja that defines the southwestern head of this valley. These mountains are not unlike the famous Cuillin Ridge on the Isle of Skye in northwest Scotland: like the Cuillin, they fully make up for in drama what they lack in height. Despite that fact Cerro Tonelli stands at only just over four thousand feet, it somehow has the look and feel of a much bigger mountain.

Pausing for a brief rest on the lonely plinth of Paso de la Oveja, I check my watch. It's 2pm. Pressure is still stable, the wind light, the horizon clear, and the signs generally good. All the summits to the north and west remain free of cloud. I decide to go for it. The climb up to the summit of Cerro Tonelli is mainly easy scrambling, with just a few short sections of moderate rock climbing across deteriorating rock terraces and boulder-filled gullies steepening with height. In the final few hundred feet it becomes steeper still, culminating in a seventy-degree chimney stuffed with tottering blocks. It's easy for a competent rock climber, but intimidating at the same time. At the apex, I press eject and land just below the crest of the summit ridge.

Right in front of me extends one of the most astonishing panoramas I've seen. The crystal air sparks with snow and ice and light. The shining mountains of the world's most southerly inhabited place extend in every direction except south; that way, the vivid blue of the Beagle Channel merges into darker blue of the Southern Ocean not far beyond. In the far distance, I can just make out an isolated island: Cape Horn. This place is truly the edge of the world. Being up here reminds me of every single reason I started to climb as a teenager on the gritstone edges of northern England, of all the reasons I continue to climb today. For a few intoxicating minutes, I'm swept away by the wild that surrounds me. After a while, though, threatening cloud begins to obscure the upper icefields of Monte Sarmiento forty kilometres to the west, Tierra del Fuego's highest and most inaccessible summit. In the moments before I strike down from the summit of Cerro Tonelli, the white air shifts to grey, and the austral sky grows wilder with the approach of yet another Patagonian storm.

For the sake of speed, I descend the mountain by a different route than the one by which I ascended; a fifty-degree scree run results in a rapid altitude loss of five hundred feet, landing me on the narrow plateau that separates the twin summits of Cerro Tonelli and its southerly companion, Cerro Martial. By the time I reach the plateau, the tops of both mountains are shrouded in cloud. The wind shifts in a weird game of fitful, powerful gusts followed by eerie lulls. It's time to get out of here.

In a few short minutes, I descend a steep gully that's one of the most efficient and also one of the most thrilling ways off any mountain I've climbed. It's a perfectly smooth, deepening chute, steep and marbled with treacherously fine scree. All this means that once speed is built up it's almost impossible to stop, so I half-scramble, half-slide down the gully, which promptly spits me out at the top of a gargantuan cone of rubble: this is the mountain's natural waste disposal system. By the time I'm jogging down the spectacular Canandon de la Oveja, the valley that extends from the shores of the Beagle Channel up to the pass, the storm has swallowed the mountains entirely, and cloud base hovers a hundred feet overhead as I continue my descent. By the time I reached Ushuaia in the early evening, a chill breeze was blowing. The entire Cordon Martial was lost in a great bank of low cloud. White horses had began to rise on the Beagle Channel. I'd been given a precious gift in the short window of stable weather I'd used to climb Cerro Tonelli, a small mountain in a very big place.

For reasons that remain hard to articulate, there was something unusually thrilling about this short walk in Tierra del Fuego. It was the combination of the long approach through that enchanted forest, the steep climb to the summit, the approaching storm, and the fast descent: the quintessential mountain day.

Climbing, at heart, is a journey into the hinterland of the mind as much as an exploration of a physical terrain. If that's true, then Cerro Tonelli was one of those rare quests where the hinterland is clarified all at once, like a rising gust of wind filling a sail.

Climb magazine, 2017

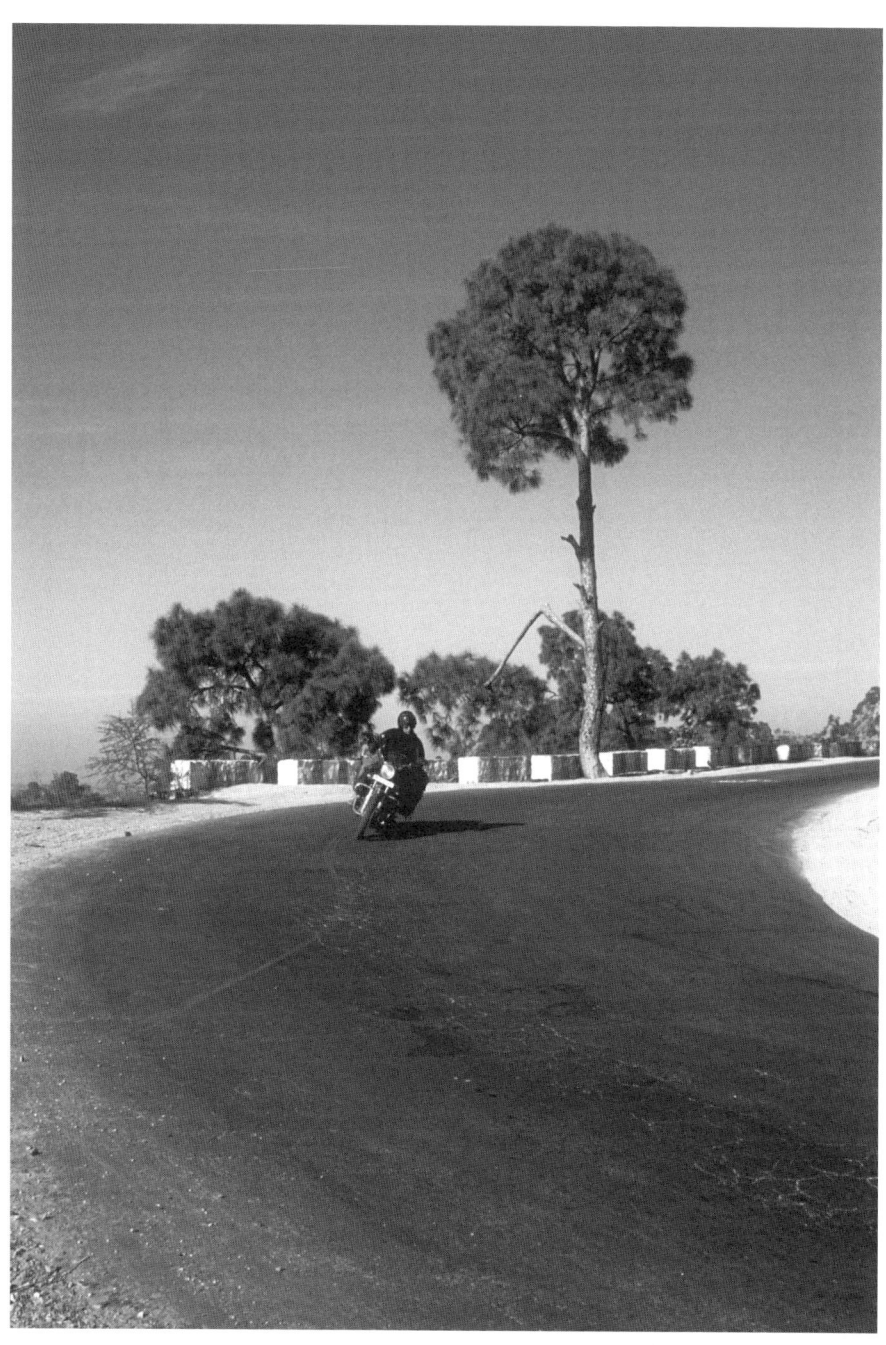

On the road with the Enfield Bullet 500 near Shimla, Himachal Pradesh, India, in December 2005.

Shakti and Dust

Three months around India on a Royal Enfield

"Namaste…? Hello, Mr. Singh?"

As I peer through the darkness of Lalli Singh's basement workshop in a sidestreet of Karol Bagh, I break into a gentle sweat induced by the inhalation of stagnant petrol fumes. My hopes of finding our Enfield Bullet 500 serviced and ready to hit the road are diminishing with every breath.

"Hello mister" one of Mr. Singh's affable young mechanics pipes up.

"I am most very sorry sir, but we have problem. Bike not finished." The boss, it transpires, is away, and the Bullet 500 won't be ready until the day after tomorrow.

Delhi is the kind of city in which a great deal of time and energy can be expended simply in order to get out of it. It's huge, noisy, and extremely congested. After a four day epic of multiple rickshaw journeys between Connaught Place and Karol Bagh, the paperwork was over, the Bullet was finally ready, and the prospect of escape from the metropolis was imminent. My girlfriend at the time, Sarah, was a fellow enthusiast for overland travel. We planned to ride around as much of the Indian Subcontinent as we could between December and March. That was as detailed as the expedition plan got, and it was better for it.

The phrase 'baptism of fire' was very much the moniker for the first day of our three month roadtrip around India. Just a few hours outside Delhi – between the bustling plains towns of Ghaziabad and Haridwar – Sarah and I found ourselves squatting with an alarming urgency in a large field of sugar cane, attempting to camouflage ourselves from curious locals, as the notorious bacteria of the capital began to take serious effect.

After numerous impromptu stops later, our journey gained a new level of comedy as I steered the loaded Bullet at walking pace through a swarm of cyclists and vegetable sellers amid the traffic of Muzzafaragnar, whilst Sarah made impressive arcs of projectile vomit over my shoulder. Amazingly, she even took serious measures to avoid spattering the town's numerous cows with vomit. Cows are sacred to Hindus, and we just escaped the disdain of the locals by a combination of her good aim and a few well-timed swerves on my part.

After a few days in Rishikesh recovering from what was an unpleasant but certainly not uncommon introduction to the Subcontinent, and after giving ourselves an even stronger dose of antibiotics, the magnetic pull of the mountain roads of Himachal Pradesh proved too strong to resist.

The air grew steadily colder as we wove through the foothills of the Himalaya, heading westwards toward that troubled frontier country of Kashmir. We emerged from the Jawahar Tunnel into the sharp December sunlight of the Kashmir Valley after surviving over three miles of terrifying icy darkness filled with diesel smoke. Passing various avalanche warning signs, I realised we were in the precarious position of travelling on a motorcycle in the approaching clasp of the Himalayan winter – and with only one way out should the first snows arrive early.

The weather was fair, but the ever-present thought of a cold front coming down from the Karakoram was enough to make our stay in Kashmir a short one. In good weather, the renowned Highway 1 from Jammu to Srinagar is one of India's most dangerous roads. In snow, it would be much worse, and riding a bike on a road like that would be a game of Russian roulette. We needed to devise an escape plan.

Salman Rushdie has described this region as "a paradise not so much lost as ruined" for good reason; the division of Kashmir in 1947 into two territories in two different countries during the partition of India is widely seen as a mistake that has created an intractable geopolitical problem in this part of the world.

At around 3 a.m. on the morning of December 12th 2005, the shockwave of an earthquake originating in the Hindu Kush measuring 6.7 on the Richter scale hit

Srinagar and generated a series of minor tsunamis on the city's lakes. Woken by this unusual nocturnal intrusion, the seismic wave subsided as our rickety houseboat pitched violently for the last time, and I meditated on the local tale claiming that earthquakes are caused by the writhing of a subterranean serpent. Not being a naturally superstitious person, I am however a keen advocate of respecting local custom, and did not wish to wait and see if the apocryphal snake's powers extended from plate tectonics to the generation of blizzards.

We left Srinagar the following morning, which was later reported as the coldest December morning in Kashmir for five years, wearing literally every single item of clothing we had. Back on the other side of Jawahar Tunnel, and despite almost being forced back through the gargantuan fume-filled pipe by a clamorous posse of Indian soldiers demanding our passport details, visions of imminent frostbite subsided with the first few gusts of wind that came in from the south.

With that warm air blowing up from the Punjab came hallucinations of a distant country; the granite plains of Karnataka far to the south, and of boulders watching over the ruins of lost empires among the lush green of the paddy fields. With that in mind, we made haste out of the icy claws of the coming Himalayan winter, and south towards the sands of Rajasthan's Thar desert.

A couple of weeks later, having covered 4000 kilometres since leaving Delhi, we rode the Bullet into the ivory sand at Mandrem on the Malabar coast on December 30th. The Arabian Sea glinted with preternatural zeal that evening. Seeing the ocean for the first time after a long, hard journey in mountain country is always a powerful experience. We spent a blissful new year on Goa's Arambol beach, celebrating our survival of the world's most dangerous roads with an appropriately poisonous-looking bottle of Maharastran rum. At last, it was time to escape India's potholed highways and homicidal driving for a while.

One of the most striking things about arriving in Hampi, a bustling market town in the state of Karnataka, is the sheer quantity of granite that surrounds the place. By moonlight, the landscape is transformed into a lunar surface, ghostly and highly-charged. The complex ruins of the Vijayanagar capital have been well known by travellers for several decades now, since the first intrepid hippies followed the lead of adventurous Victorian historians.

Even today, Hampi remains an extraordinary place. At the beginning of the sixteenth century, what now lies in ruins was the spiritual and economic foundations of city of some 300,000 people, and the centre of the most powerful Hindu empire in early modern Indian history. The marauding Mughals then destroyed the place and murdered most of its citizens in a very short time. In 1565, the Vijayanagar forces were defeated in the Battle of Talikota, and much of the city of Vijayanagar was sacked. As such, these ruins have a particularly spooky quality.

Rather more recently, the extraordinary granite boulders which provided the raw material for the Vijayanagars' temples and palaces have become one of the focal points for rock climbing in southern Asia. This was our principle reason for a stop-off here.

There are countless boulder problems at Hampi, on both sides of the Tungabhadra river. Many of them follow strong structural features over perfect granite, and just a few hours walking into the hills to the north and west of the river reveals a boulderer's shangri-la: an apparently endless vista of boulder-strewn hills stretching to the horizon, harbouring virtually limitless climbing potential, and much of it still unexplored.

The phrase 'adventure bouldering' is nowhere more appropriate than here, and exploring the wilder parts of the Hampi area is not for the careless or the faint-hearted. This granite moonscape is the home of black bears, leopards, king cobras, and a small population of one of the deadliest snakes on earth – the elusive banded krait. Such creatures are very rarely seen among the established climbing areas; but they are here. A glimpse of an occasional basking crocodile in the marshy shallows of the Tungabhadra is in reality the closest you are likely to get to a dangerous animal here. We spent a glorious week climbing at Hampi in the quiet of mid-January, which is also the coolest time in southern India and therefore the best season for bouldering.

A hundred miles west north west of Hampi, the small town of Badami nestles under a sandstone escarpment that's a complete contrast to the climbing at its better-known granite neighbour. For those prepared to stay in an authentic, noisy and dusty south Indian town, the rewards in terms of exploring a relatively untouched climbing area of exceptional quality are huge. If you're also up for dining on the local dhosa and thali, the staples of south and central Indian cuisine, then even better.

The rock at Badami is of a similar quality and age to Australian sandstone, and reveals constant geological surprises including glistening veins of quartz and monumental synclines. The escarpment that looms over the town's whitewashed stone houses extends for several kilometres to the north and east, forming a topographically complex massif that is relatively straightforward to navigate through compared to the confounding, impenetrable boulder-fields of Hampi. Although rarely exceeding forty metres in height, these cliffs and their satellite boulders take on a stature greater than their actual scale, particularly in the rich, dust-red evening light.

Lots of single-pitch climbs have now been established in the deep canyons closest to the town, and on the west-facing wall above the 5th century Agastyatirtha Tank (a kind of man-made lake). There are a series of remarkable medieval temples cut out from the north-facing crag next to the Tank. These date from the time of the Chalukyan empire, which covered most of the central part of south India's Deccan plateau between the 4th and the 8th century AD, of which Badami was the capital from 540 to 757 AD. The visible presence of Indian history pervades the air around these cliffs on the high Deccan plain. Watching woodsmoke rising from the *havelis* on the edge of town in the evening, surrounded by the ruins of the Chalunkyas vanished past, it is easy to forget the twenty-first century for a few minutes. Well, at least until the silence is broken once again by the reverberation of a giant pressure horn from a state department bus or a brightly-coloured truck. This is the country of endless noise and colour, after all.

Climbing at Badami is an experience shot through with ghostly associations of an ancient India; here you can encounter a very different environment to the modern India of megacities. Yet by a strange process of continuity unique to this nation of unprecedented change, the past is never entirely separated from the present in India, but connected to it by the cultural pathways of one of the world's oldest civilised societies. Among the red sandstone cliffs at Badami from which the Chalunkyas carved their temples, that ancient India becomes suddenly alive, just as it does above the rice fields along the banks of the Tungabhadra river at Hampi. Rock climbing in Karnataka is an experience not to be missed if travelling through the beautiful south of India with some time to spare.

After our long-awaited break from the road to explore Badami and Hampi, we covered a lot more ground: we went south from Goa to the Nilgiri Hills of south Karnataka, the Cardamom Hills of Kerala, then down to Kanyakumari in Tamil Nadu, the most southerly point of the subcontinent where the landmass of this vast nation finally meets the Indian Ocean. From here, we continued south to Sri Lanka.

We headed back north through southern India in mid-February as the pre-monsoon heat was already beginning to build, thickening the morning air like swathes of smoke. Sarah had to fly back to England in a week to go back to work. I had a loose plan to ride back to Delhi over the following month, completing a huge loop around the country. Whether or not I'd actually complete the loop, though, was still to be decided.

One afternoon the following week just after Sarah had flown home, in the middle of a hot afternoon in the arid heart of India with the monsoon rains still months away, I turned the Bullet around. I would head back to Hampi, I decided, and sell the bike there. After riding the entire length of the subcontinent, from Himachal Pradesh and Kashmir all the way down to the southernmost point of India, I'd had enough of Indian roads. The standards of driving in India are truly terrible, and always keeping a watchful eye on the relationship of risk and probability, I decided the road trip was finally over. As I manoeuvered the Bullet through the chaotic traffic of another noisy, dirty town somewhere on the high Deccan, I remembered my near miss with that truck in Java three years before during my motorcycle expedition around East Asia. I'd covered enough distance, I felt, on the dusty roads of India to last a lifetime.

I made good progress on that final day of the road trip from the small village in rural Karnataka I'd left at dawn all the way to the border of Andra Pradesh. Soon afterwards, I hit the main National Highway that links Bangalore with Hyderabad, where I'd have to fork south and then east again back to Karnataka. Here, I stopped at a little roadside dhobi-stall for chai, dahl, and a cigarette, and paused for fifteen minutes before riding on.

Just as I departed, I glimpsed the biggest snake I have ever seen in an immense, overgrown well. Curling into the darkness like a scene from the *Ramayana*, the Sanskrit epic from ancient India, the enormous king cobra quickly disappeared into the murk and mire of the well. What a place into which to descend! In India, the hooded cobra (*naja naja*) is replete with surrounding mythologies; the Hindu god Shiva is often painted with a protective cobra around his neck. Snakes make me a touch superstitious, and just after I saw the cobra I passed by a Hindu funeral procession in a dusty village on the plains. The mourners' songs meandered out into a hot, empty sky. It felt like that huge snake, vanishing as quickly as he came, had come to tell me in no uncertain terms that I'd reached the end of my journey.

I rode west into a blue spring evening on the high Deccan plateau, streaked with shafts of gold-edged sunlight, as thick clouds of insects descended on the cooling rice fields. After a while, the sun went down over the Deccan, and I rode on into the night. I had to continually clear the insects from my helmet's visor. As I neared Hampi, frogs croaked through the darkness from the Tungabhadra river as a huge full moon rose over the plains, breaking the drone of cicadas. Above me, towers of pale granite were shadowlit against a night sky pinned with luminous stars. After a while, I stopped by a lone coconut tree and cut the engine.

Silence – that most elusive thing of all in the world's busiest country – overcame me. At the end of the very last day of my journey around the subcontinent, I finally understood what so many generations of monks and mystics had discovered before me. India is not just another place on the map, but another way of seeing.

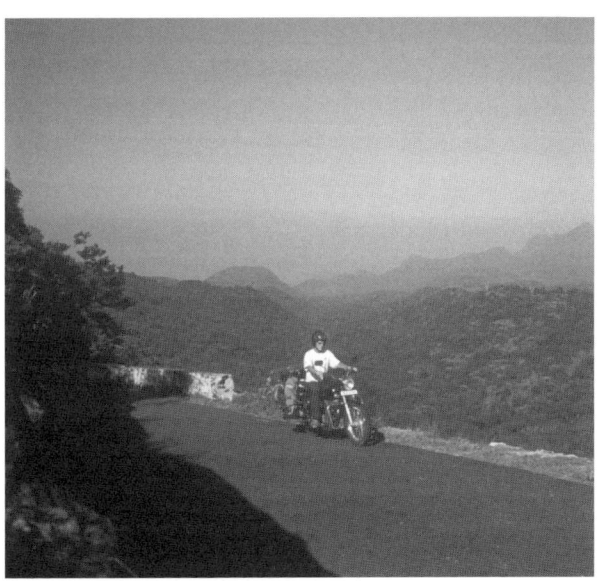

In the Cardamom Hills, southern Western Ghats, Kerala, in March 2006 and 6000km after leaving Delhi.

Climber magazine, 2007

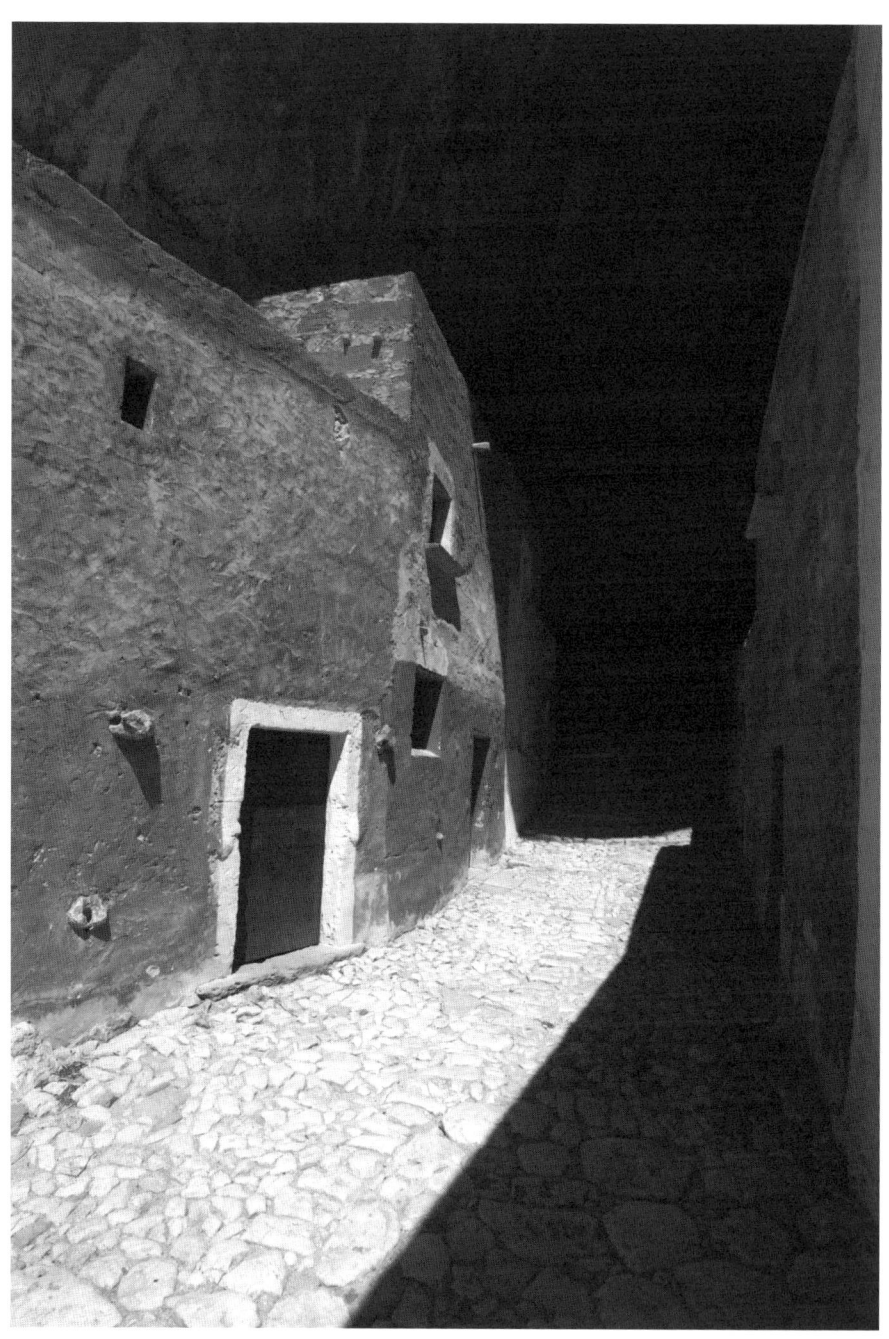
Cobblestone street in a traditional village in San Vito Lo Capo, western Sicily.

Shadows over Sicily

Reflections on a hinterland of blood

Pillars of concrete and protruding girders are staked across the hillside at intervals, like the forlorn relics of a supersized vampire-slaying frenzy. The November sun, falling towards evening, casts weird shadows across the ruinous, deserted construction project. Vagrant goats traverse the deteriorating slopes between the half-finished houses, grazing half-heartedly amongst piles of cement debris. The wind carries the soft clang of their bells across to where I'm standing, beneath the blood-orange escarpment of Bausso Rosso on the opposite slope, surveying the dereliction with amazement.

It seemed the other side of the valley had been recently used for testing a prototype demolition technology, or possibly a new type of short range missile that had proved unusually effective. The scene was more like something from the ruins of Sarajevo in the late 1990s than to the idyllic, rustic corner of the Mediterranean I'd imagined.

Walking away from the cliff at the end of the day's climbing, a black Maserati with blacked-out windows screeched around the corner and blasted past us before disappearing in a cloud of dust. The scene could have been straight from *The Godfather*, but in Sicily, as I would come to learn, the past is always close behind.

It's my first climbing trip to the island in the autumn of 2007 with my girlfriend of the time, Sarah. The wry note in Maurizio Oviglia's seminal 2006 climbing guidebook refers to "one of Palermo's worst urban disasters" overshadowing the crag. This further fueled my curiosity about Sicily's perplexing and brutally violent past. This ghost-development was, I discovered, a visible reminder of the insidious activities of the Sicilian mafia, or *Cosa Nostra* ('Our Cousins') which until relatively recently was one of the world's largest criminal organisations. During the 1980s and into the early 1990s, under the psychopathic leadership of Salvo 'Shorty' Riina, the Sicilian mafia became so large and so financially successful that the organisation could no longer simply re-invest the money it made from crime back into crime alone. Construction, therefore, was a good option. The half-finished building projects that smudge the white hills and blue coasts of Sicily are a decaying reminder of the wealth and influence of the so-called Men of Honour who once controlled much of the life and business of this island.

The more I found out about just how far the mafia's shadow fell across Sicily, the more I began to understand that the story of its evolution from a protection system for fruit growers in the nineteenth century into one of the world's largest criminal organisations was inseparable from Sicily's modern history. Despite its numerous imitations across the world from Japan's Yakuza to Britain's Yardies, there has probably never been an organised crime network quite as effective and as financially successful as Cosa Nostra in the 1980s at the height of Salvo Riina's power as *capo di tutti di capi* (boss of bosses).

Then again, as I would come to understand, Sicily simply isn't like anywhere else in the world. I probably watched The *Godfather* trilogy at an impressionable age, but over those November days of sun and rain, travelling and climbing on the island, Sicily completely got under my skin. We played high-stakes dodgem with phalanxes of revving Fiats in downtown Palermo, climbed soaring open grooves and big, bold face routes at Capo Gallo, high above the white rooftops of Mondello, and refueled with pistachio ice cream and expresso from the gelateria in the town square which appeared to be owned by an enormous, magisterial Persian cat. After climbing at Bausso Rosso one afternoon we got lost trying to find the disused tonnario (tuna factory) in Scopello, ending up on the deserted beach at San Vito Lo Capo and staring up at a thousand foot wall of steep, immaculate limestone. A heavy surf thundered on the breakwater that evening, and the lighthouse started to flash across the darkening sea. A storm was blowing in, making the palm fronds clatter and invisible doors slam in alleyways. The streets were deserted, and the sense of absence in falling light that inhabits the paintings of Edward Hopper infused the salty air.

The next day, we drove east beneath the white bastions of Enna, past Mount Etna smoking through squalls of snow, and south to Siracusa, once the home of Archimedes. After navigating a one-way system so labyrinthine and unfathomable it must have been designed by one of the great Greek mathematician's deranged disciples, we climbed half a dozen routes on beautifully compact, weathered limestone in the Valley of Pantalica, a serpentine ravine riddled by a vast honeycomb of five thousand year-old rock cut tombs; the exact origins of the people who made them are still, amazingly, somewhat disputed by archeologists.

The next day, rain blew in from the south, leaving tidal traces of red sand picked up over Africa across the Roman cobbles of Ortigia, and huge flocks of starlings circled high overhead, drawing black fractals in the grey sky. We got lost again trying to find Cavadonna, the best cliff in the south of the island, stumbling upon it after some very Italian-style driving on dirt roads through a maze of citrus orchards. All too soon, we had to catch the plane home, but before we got back to Palermo, I'd fallen for Sicily head over heels.

Time passed, and climbing trips to other islands came and went. But something about Sicily stayed with me – something that had as much to do with the place itself as its obviously colossal climbing potential. Six years after that first foray, I returned with two of my oldest friends and regular climbing partners.

Arriving in Trapani late in the evening, Ramon Marin, Gavin Symonds and I were met outside the airport by a man wearing a black bowler hat who bundled us into an unmarked van, and drove off at speed into the night. After a few minutes our driver made a hard turn to the right and sped through orchards of orange trees. The air was thick with orange blossom as we pulled into an unlit compound defended with razor wire and paved with broken concrete. Squinting my eyes in the darkness, I spotted a series of metal troughs at the edge of compound, such as those that might be used to feed a family of rapacious pigs. Around the troughs, large unmarked black plastic bags were piled high, their contents completely unidentifiable. What the hell were we doing here? The Sicilian mafia's preferred method of disposing of the corpses of traitors was by dissolving the bodies in large baths of sulphuric acid. I promptly asked the only member of our team who spoke any Italian to press our driver for an update.

"Ramon, any chance you could you ask this chap exactly why we're here?"

Eventually, after negotiations are concluded, three of us pile into an absurdly undersized Fiat and drive out through the orchards before hitting the expressway west for Trapani. As we head for the dark hills to the west, I'm thinking about Palermo magistrate Giovanni Falcone's last journey. Falcone and his colleague Paolo Borsellino were the two men largely responsible for the demise of Salvo Riina's 'Corleonese' mafia and the end of the most brutal phase of Sicily's modern history.

Together, Falcone and Borsellino painstakingly orchestrated the mass arrest of virtually all the key figures in the Sicilian mafia. The so-called Maxi-Trial began in February 1986 and lasted until December 1987. Both magistrates knew, of course, they were vigilantly marked men. As Falcone was driving west from Palermo on May 23rd, 1992, the Men of Honour were waiting above the Capaci intersection. As his vehicle passed, mafioso Giovanni Busca pressed the detonator of a pipe-bomb so massive it ripped up a substantial part of the motorway's central reservation. A month after Falcone's assassination, another massive car bomb killed Paolo Borsellino as he left his mother's apartment in Palermo after visiting her for Sunday lunch. Despite the predictable murders of Falcone and Borsellino, Salvo Riina's stranglehold on Sicily was already on the decline. The Maxi-Trial resulted in 360 mafiosi convictions, and the beginning of the end of Cosa Nostra itself.

There is only one way to San Vito Lo Capo, and only one way out. The road climbs sharply to the west of Trapani, and the citrus orchards of the plains quickly break into narrow terraces of olive and acacia. Arriving by night in a clearing storm, the four mile sweep of the Gulf of Cofano opens under a crescent moon against the black land; white breakers surge beneath the two Moorish watchtowers that proudly guard the entrance to the bay. A thousand years ago the Moors – like the Phoenicians, Greeks, and Romans before them – crossed the Mediterranean and stopped at Sicily, using the island as a convenient stepping stone for trans-Mediterranean trade.

Things had changed in Sicilian climbing in the five years that had elapsed since my first visit; new crags were being discovered and developed all the time. On the first day, I woke to the patter of rain on the roof and the boom of a heavy swell. The storm, fortunately, had mostly blown out overnight. Later that day, with mist rising from the shore and the crag still encased in humid air, I swung through gigantic portholes and fused drainage channels of rust-coloured stone in the hourglass-shaped cave of Pinetta Grotta. The next day, we walked east along the foreshore from the westerly tip of the peninsula to discover some of the most compact limestone I've encountered anywhere in the world at the Grotta Calamancia. Here, as the sun hits the rim of the cliff, dust began to circulate in the heated updraft. We packed up, walked a few hundred metres to a tiny cove, and dived off a rock-cut platform engraved with the graffiti of generations of Sicilian fisherman.

Later, resting between attempts on the sloping platform underneath *Crown of Aragon* – a pristine thirty metre sweep of intricately-featured overhanging limestone – I thought about the ruined complex of smallholdings at the base of the cliff: their collapsed rafters and tumble-down walls told the tale of another, earlier era in the Cofano Gulf when this area had a much larger population than it does today.

Until the Maxi-Trial of 1986-87, the mafia cast its shadow across every aspect of Sicilian life. So wide was its influence and reach, nobody on the island could escape it. As we wandered the quiet streets of San Vito Lo Capo later that afternoon, in search of the essential Italian post-climbing refreshment – coffee and ice cream – it struck me both how easy and how convenient it can be to forget about the social reality of the places we climb. Around the time I started climbing in the late 1980s, San Vito Lo Capo wasn't just unknown to climbers, but to tourism altogether. Whilst travellers were never specifically targeted in mafia-controlled Sicily, if you were unlucky enough to witness an assassination, you could be pretty sure you would be next on the hit list. Any mention of Sicily carried with it the association of the mafia long after Riina's organisation was dismantled, keeping many visitors away.

We continued exploring the climbing around the San Vito peninsula. Making the exposed abseil into The Lost World, I pirouetted around in the breeze twenty metres away from the rock with the Cofano Gulf glittering a thousand feet below, and had to blink as I stared into the shadows of a vast cave dripping with gigantic stalactites.

When you're out there, spare a thought for the invisible Sicily of the very recent past: the island controlled by the *omerta* [the code of silence], and by the culture of fear, corruption, and brutal retribution that was the modus operandi of this most menacing of criminal organisations. And, when you're passing the Capaci junction near the airport that now bears Falcone's name, pass a thought for those two brave men, Giovanni and his colleague Paolo, and for the shady figures in black suits and dark sunglasses they so relentlessly pursued.

Climb magazine, 2014

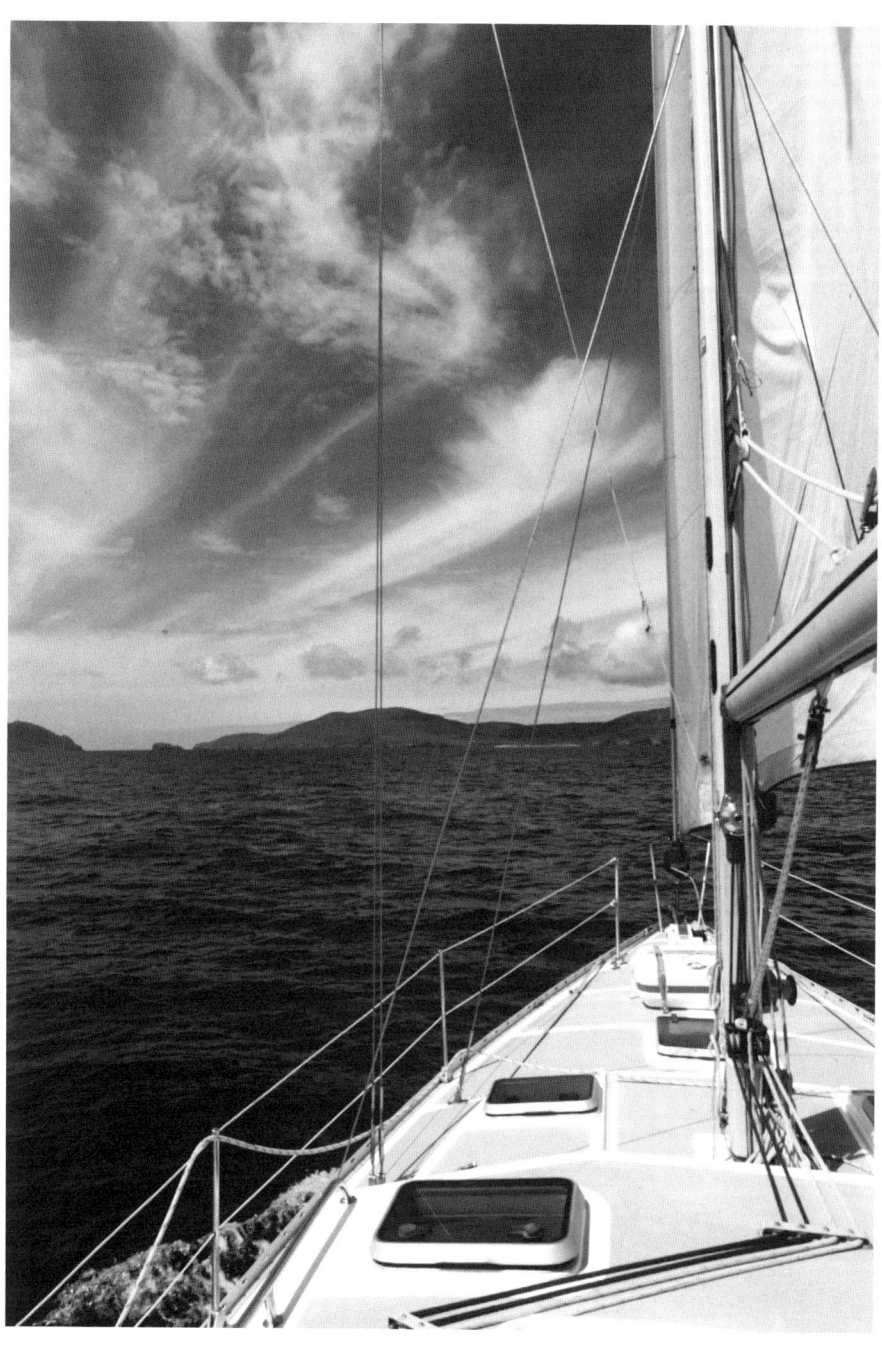
Approaching the east coast of Mingulay, an uninhabited island in the Outer Hebrides, aboard Hecate.

Exploration & the Human Spirit

Meditations on the idea of a journey

The surface of the bay glimmers darkly, like a distant planet. Dawn is breaking. A group of men and women have gathered on the beach as the final supplies are loaded into the twin hulls of the ocean-going canoe pulled up on the sand. None of them have any way of knowing the outcome of their imminent expedition. Once the last provisions have been stowed, the canoe begins to thread its way through the lagoon. It clears the line of surf on the outer reef, picking up speed as it enters the deeper water beyond, propelled now by the wind and the equatorial current. With a following swell behind them, these unknown explorers now pilot their intrepid craft towards the shifting blue line of the horizon. After a while, the island they have departed vanishes astern; the canoe is completely alone on the Pacific.

Ancient Polynesians first discovered and populated many of the Pacific islands in antiquity, using ocean going sailing canoes combined with spectacular skills in natural navigation built up over centuries, and passed down through the generations. The largely undocumented voyages these ancient seafarers made are some of the most extraordinary feats of exploration in the whole of human history. Using rudimentary but highly effective nautical technology, ancient Polynesians discovered some of the most remote inhabitable places on Earth. The Marquesas, for example, were discovered as early as 100 AD: they are 3500 kilometres from Hawaii and about 1500 kilometres from Tahiti, from where their first human inhabitants may well have come. Today, we know quite a lot more about outer space than the ancient

Polynesians knew about the ocean they were exploring. How exactly they made their remarkable voyages still remains an active area of enquiry, both for historians and for modern enthusiasts of the art of natural navigation. Why they made these voyages, though, is a question more relevant to philosophers. Why would you leave Tahiti or Hawaii – places as accommodating to human life as any that can be imagined – for a perilous quest across a vast ocean? Why risk death for the chance of discovery? What's the point of making a dangerous journey at all? Overpopulation and food scarcity were two key reasons in the ancient Pacific. Such questions remain of huge relevance today for those of an adventurous mindset.

NASA's Frank Borman once famously suggested that "exploration is really the essence of the human spirit". If he is correct, then the ancient Polynesians were certainly enacting that essence, just as Borman was as commander of Apollo 8 – the first human-crewed spaceflight to orbit the Moon in 1968. Despite using incomparable technologies, those ancient Pacific voyages have a central principle in common with modern practitioners of exploration like Command Pilot Frank Borman: the acceptance of risk as a necessary part of any exploratory endeavour.

The intriguing, complex links between ancient and modern exploration demand a series of large questions. If exploration is indeed the essence of the human spirit, then is some exposure to risk also a fundamental part of what it means to be human? Perhaps we cannot explore anything without risking something in the process. And if this is true, then does the experience of dealing with risk enable us to properly function as living beings? And what, finally, would happen if we were to live in a world where our exposure to risk was eliminated? These are questions that require urgent attention in the early twenty-first century, where physical risk is being systematically eliminated, via various cultural and technological processes, from Western lifestyles. At the same time, popular interest in adventure sports involving various degrees of risk is currently surging. This dynamic is, in itself, hugely revealing.

First coined by the authors Greg Lukianoff and Jonathan Haidt in their bestselling 2018 book *The Coddling of the American Mind*, the notion of 'safetyism' describes as a belief system which prioritises safety (including emotional safety) as a sort of sacred value. Cultures of safetyism are increasingly obvious as they proliferate in different areas of human activity: from so-called 'safe spaces' for undergraduates to the highly contentious ethical issue of the postponement of death for the institutionalised elderly, and the ever-increasing levels of parental protection of children. It is perhaps the latter that most clearly shows the current cultural dismantling of Frank Borman's beautiful idea about exploration – and therefore risk – being a kind of key force at the heart of the human spirit.

A recent study by the British Children's Play Survey, for example, concluded that primary-school age children in Britain are losing the freedom to play independently. While their parents were allowed to play outside unsupervised by the age of nine on average, today's children are eleven by the time they reach the same milestone, according to the survey. The researchers pointed out that not enough adventurous play could affect children's long-term physical and mental health. Anita Grant, the chairwoman of Play England, said after the study was published that "play outdoors is fundamentally important for children to develop a relationship with the world around them. Adults' protective instincts are not helpful when they restrict exploration and a child's natural instinct to engage with their environment freely." If you're a naturally adventurous person, Grant's point might strike you as obvious.

The following story provides an interesting counterpoint to contemporary narratives promoting the idea of safety as a core moral value. The brilliant film *Following Seas* tells the life story of groundbreaking American sailors Nancy and Bob Griffith, who circumnavigated the world several times between the early 1960s and the late 1970s, were at one point shipwrecked on the uninhabited Pacific atoll, Vahanga Island, for 67 days, and later skippered the first sailing boat to circumnavigate Antarctica. "We all became highly competent individuals", Nancy Griffith recalls, as she remembers how she raised her young family on the ocean waves.

It is a striking line, and a powerful acknowledgement of the moral value of adventure and risk, as well as the obvious hazards of leading a life directed by ocean sailing. *Following Seas* is also a remarkable document of the era before any kind of safetyism existed; at one point, sailing off New Zealand, Nancy reflects on how she once had to dive off the side of her boat to retrieve her toddler (!), who had accidentally fallen overboard when playing unsupervised on deck. "We need to go ashore until these kids have learnt to swim!" she wryly admits in the voiceover. Such nonchalance might seem reckless, but it has the advantage of building competence in children from an early age; it's the opposite of so-called 'helicopter parenting', in other words.

The physical and moral freedom that Nancy and Bob Griffith actively pursued in their sailing careers – and which every adventure-seeker strives for – can be used as a counterpoint to what the American philosopher Matthew Crawford has called "the safety-industrial complex" of the early twenty-first century. In his essay *The Dangers of Safetyism*, Crawford writes of how "safetyism... has been gaining strength for decades and is having a triumphal moment just now. There appears to be a feedback loop wherein the safer we become, the more intolerable any remaining risk appears."

Crawford's insight highlights the profound risks of a culture of safetyism. If left to itself, it promotes the kind of dystopia already evident in the results of the British Children's Play Survey quoted earlier. But Crawford also offers a way out here for independent thinkers. If safetyism requires 'a feedback loop' in which the safer we become, the less we are prepared to take risks, then there's an obvious way to avoid it: let's not prioritise safety over other values. I can think of a few that sound like better options. How about courage, curiosity, kindness, and determination? Being a critic of safetyism doesn't mean that you must expose yourself to high levels of risk; it just means that you don't seek to elevate safety above other moral values.

Returning to those ancient Polynesian voyagers with whom we began, we might answer the questions raised earlier about why it is that human beings are drawn to take risks in the first place. After many weeks – or possibly months – at sea, having endured storms and privations on an unimaginable level, the master navigator (*pwo*) leans over the side of the left hull. His eyes light up as he spots a tiny strand of pale green seaweed in the blue water of the Pacific, then a frigate bird overhead – two signs that land may not be far off. After cross-referencing the frigate bird's flight pattern with the direction of the swell that has brought that strand of seaweed to the canoe, he tells the captain to aim their course a degree to the west. After a few hours, the explorers shout with delight as a smudge appears on the horizon. Green and shining in the sun, coconut trees growing in the sand define the island's perimeter.

They've succeeded in their mission: they have found a new world. At the same time, they've enacted the exploratory urge that's the essence of the human spirit. At some deep psychic level, many of us must still take risks – perhaps great risks – in our lives just as those early explorers did in order to inhabit different spaces, to be elsewhere, and to reimagine our personal journey in the universe.

BASE magazine, 2021

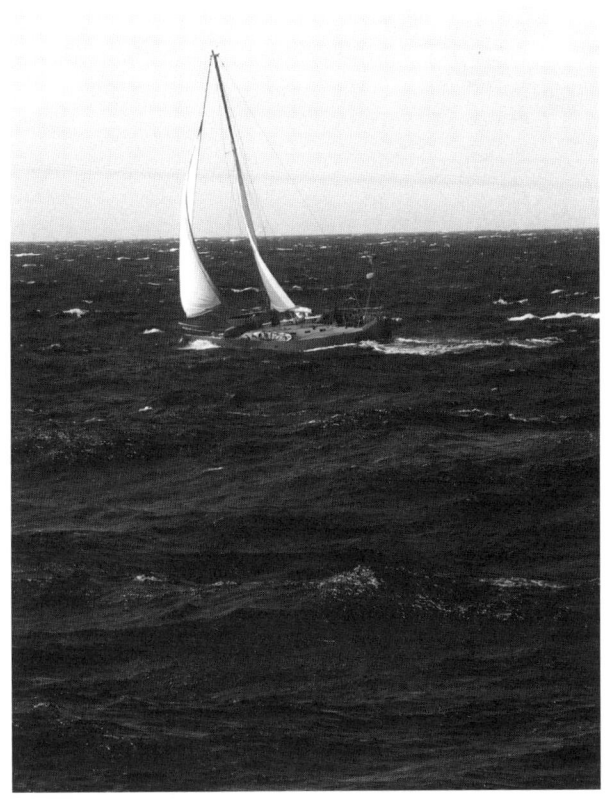

Sailing west from Penzance, Cornwall, in 20 knots of wind and a rising sea.

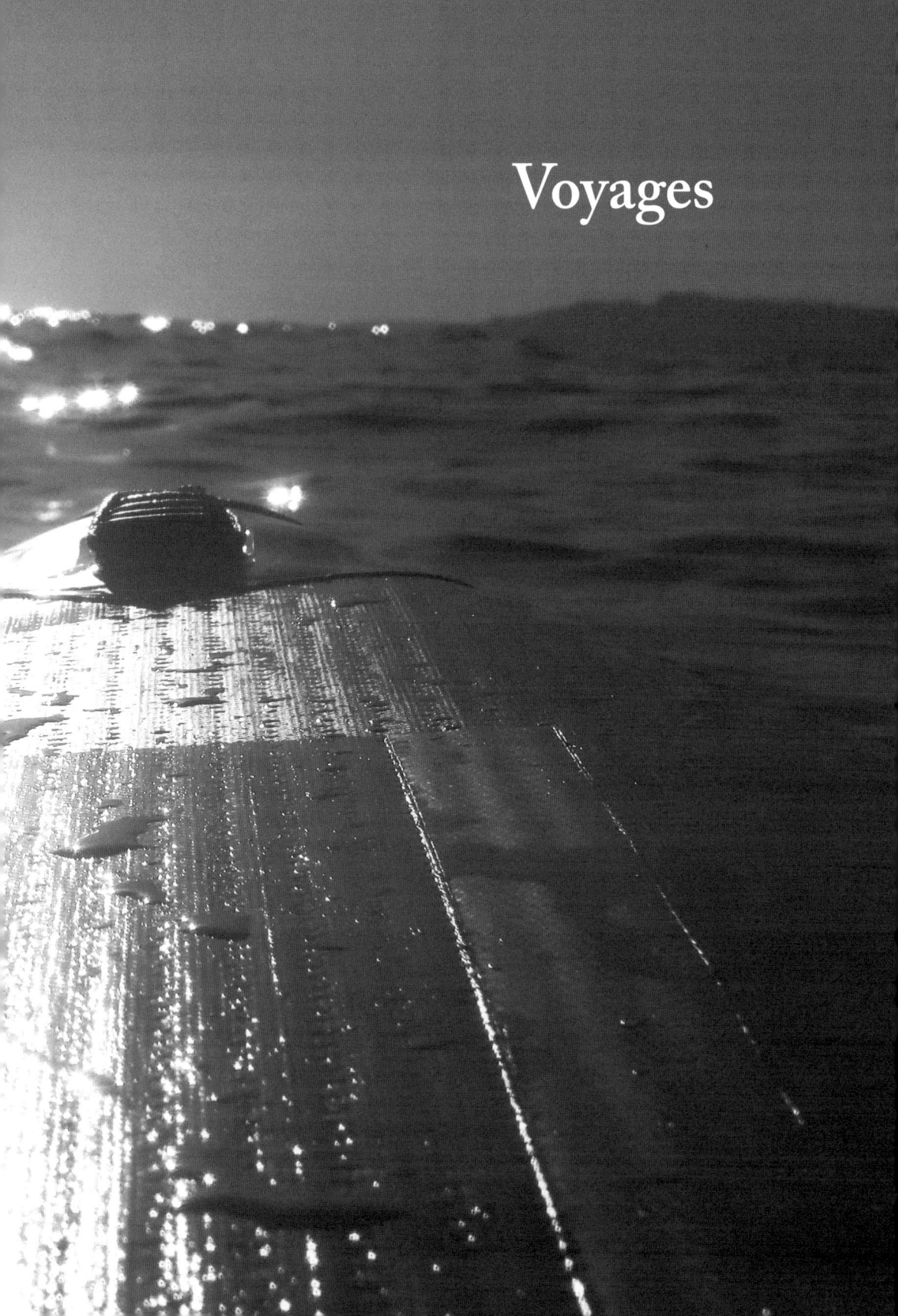
Voyages

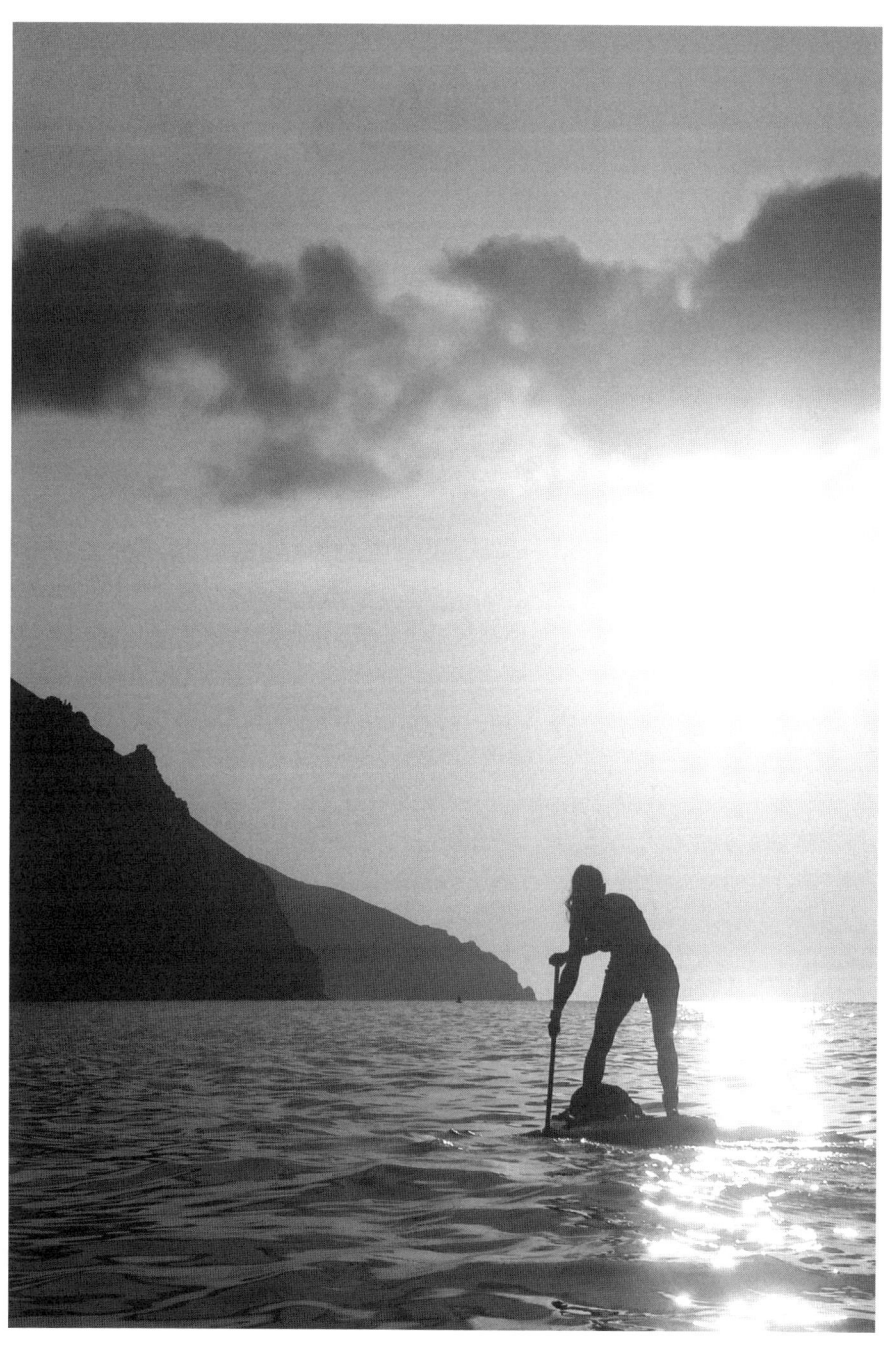
Dorka Fekete heading west on the ebb tide towards Highveer Point at sundown, Exmoor Coast, North Devon.

Days of the Celtic Sun

Southwest England by standup paddleboard

I live near the edge of an island. The sea is an inescapable presence here. It's where you'll end up if you walk out of the door and keep going in a vaguely straight line.

On my local streets, there's often a tang of salt in the southwest wind as it blows up the river from the estuary. At dawn and dusk, wayward gulls wheel across the warehouse rooftops and ride the convection currents through the alleyways. The sea isn't an abstract thing somewhere beyond the horizon: it is the horizon.

At the beginning of the best journeys of your life, you often don't know your final destination; you can't really imagine the end. We're talking about the kind of journey, in fact, when the outcome remains unknown until the very last days. This was one of these journeys.

In 2013, I began experimenting with standup paddleboarding as an alternative to both sea kayaking and whitewater kayaking as a way of exploring coasts and rivers. I was initially attracted to the concept of standing on the water rather than being confined within a kayak's tiny hull. I quickly began to realise that once you've mastered the various techniques, and with the required level of fitness, you can do pretty much anything you can do in a sea kayak on a standup paddleboard.

It was a transformational realisation.

It was this new understanding that directed me towards an unusual adventure: a journey around Southwest England, one of the finest stretches of coast anywhere in the world, by SUP. It is also the coastline I know best, and that's closest to home.

The concept was simple. It would not be a continuous expedition, but rather a series of shorter trips, eventually forming a complete journey around the peninsula. This was going to be a jigsaw-like, jazz improvisation of an expedition composed of separate parts, and completed over several years. This modus operandi was forced by the unique demands of my chosen mode of travel, as well as the necessities of fitting a long distance voyage around work and ordinary life.

When travelling significant stretches of open water using a small craft, good planning and logistics is usually critical. Long trips linking estuaries, headlands, islands and the open ocean become possible with experience. SUPs are considerably more wind-affected than sea kayaks, due to their lack of draught and the fact that your body can act as a sail. Whilst it is possible to paddle upwind, if you want to travel longer distances of more than twenty kilometres, say, in one push, it is a huge advantage to plan your journeys with the wind, the swell, and the tide in your favour. An appealing aspect of this process is the logistical planning itself, which is an art in its own right. Friday nights often must be spent carefully cross-referencing tidal current models, swell charts, and detailed wind forecasts to ensure the weekend is a success.

There can be a considerable level of commitment when launching a paddleboard downwind and with a strong tidal stream running with you. In many cases, after a short period you simply can't go back. You're fully committed to the trip, like the ancient Polynesians were when they set out in their *wakas* [twin-hulled ocean canoes] on their voyages of discovery in the Pacific. I wasn't going to find any undiscovered islands on my journey, of course, but I was about to set forth into a world of wind, salt, silence, and light on a truly epic scale.

It all began in August 2017, when I launched from the small harbour at Porlock Weir on the Exmoor coast on a humid, hazy afternoon, heading west on the powerful ebb current. A couple of hours into the trip, I met my first fellow traveller.

"I think you're bloody mad, mate".

The old fisherman's voice drifted out across the ink-dark water off Foreland Point, the broad headland that extends into the sea off the purple hills over Countisbury. I took his words as a muted compliment; it turned out he'd never seen anyone on a paddleboard off the Exmoor coast before.

It was the final hour of the ebbing tide, and the low chug of the fisherman's inboard engine quickly faded as I headed west. As with most of the larger headlands around the coast of Southwest England, a fierce tide race forms off Foreland Point in choppy conditions, creating big overfalls and standing waves.

Today, though, the wind was light. Green shadows extended across the sea, and it was hard to imagine the violence of the race in full spate. Even so, in half an hour the beginning of the flood tide would be racing up the Bristol Channel from the Atlantic. Only a short distance remained to Lynmouth, where the East and West Lyn rivers meet at the end of their descent from the high country of Exmoor before bubbling into the sea.

The tide forces so much water into the bottleneck of the Bristol Channel that this vast estuary experiences the world's second largest tidal range after Canada's Bay of Fundy. On spring tides, which occur twice a month, the difference between high and low water can exceed *fourteen* metres at Avonmouth. By using this spectacular tidal power to your advantage, you can travel seriously fast along this coast, making relatively long sections possible within the window of favourable current.

In May 2018, after completing various sections of the Dorset and South Devon coast, I pushed off from Lynmouth just before noon on a perfect early summer morning. I intended to use the tide to travel beyond Bull Point and to the very end of the Exmoor Coast at Mortenhoe, where the Bristol Channel meets the Celtic Sea.

A fresh easterly breeze was forecast, and I picked it up as soon as I was away from the shelter of the land, clocking seven knots as I travelled across Woody Bay, catching and riding the following swell in the classic 'Hawaiian downwinding' style. The tide was already ebbing fast, speeding my progress west past the isolated beach at Heddon's Mouth. During World War Two, crews from German U-boats apparently made secret night landings at Heddon's Mouth by dingy to replenish their supplies of fresh water from the stream that flows into the sea. How did this lonely beach look to a German submariner on a dark night in the 1940s?

Piloting a SUP in the open sea for a sustained period requires strength, agility, good balance, and total concentration. The only way to maintain pace and focus is by staying hydrated and by maintaining energy levels; taking regular, short breaks to eat and hydrate tends to be a good strategy.

A few miles west, Great Hangman, one of Britain's highest mainland cliffs, slopes almost a thousand feet down into the sea like a slumbering primeval beast. The wind slackened off here and my pace slowed. An hour later, I'd made it to Lee Bay, and picked up the last of the ebb tide rounding Bull Point's prominent lighthouse. Morte Point was the last obstacle of the day's voyage, a slender prow of rock shelving into the sea in the menacing shape of a crocodile's jaw, and guarding access to Woolacombe Bay – my final destination.

Approaching the point from the east, I picked up a southwesterly groundswell, and my speed dropped back considerably. A tell-tale micro eddy around a lobster pot confirmed my suspicion: the tide had already turned and was running against me. I calculated that if I couldn't make it around the point due to the strength of the early flood current, I could safely land in one of the coves just to the east from where I could pack up the board and reach the coast path. A huge logistical and safety advantage of inflatable SUPs compared to sea kayaks is the ability to land almost anywhere, pack them up, strap them to your back, and walk out to the nearest road. With that plan B in mind, I turned the board nose to the tide, and headed west into the stream.

In local Exmoor lore, it's said that Morte Point is 'the place that God made last and the Devil will take first'. One of the reasons this headland has been the site of so many wrecks is that, unusually, there's no inshore passage of calm water between the point itself and the fierce tide race just beyond it. Even on the calmest days, confused waves break in different directions across the jagged rocks, forming a treacherous cauldron of swirling swell. Compound this with the tide running off the point at four knots, and you've got a seriously challenging environment. Morte Point is definitely not a place to be in the wrong conditions.

The power of the tide became obvious as soon as I was paddling against the main flow as I rounded the point. Staying some way out to sea beyond the offshore rocks to stay clear of the breaking waves, the nose of my board pitched and dived through the overfalls and eddies of the race. The low groundswell coming in from the west amplified the size of the waves, but I cleared the point successfully, and entered calmer water on the westerly side. It wasn't quite over though.

A glance at the rocks two hundred metres to my left revealed my position as almost stationary, despite the fact I was paddling flat-out. The flood current was running hard against me now, hauling me back towards the violence of the race. The prospect of negotiating those overfalls for a second time did not appeal, so checking my speed as I cross-referenced a series of fixed points on the headland, I made a ninety degree turn and executed a rapid 'ferry-glide' across the current to the shore, making final landfall in a sheltered, sandy cove just north of Mortenhoe. In the thirty-three kilometers from Lynmouth, this final section against the tide had been the most challenging – and tiring – by far.

The wild stretch of huge cliffs, inaccessible bays, and reefs between Clovelly and Bude, known as the Culm Coast, is a place of solitude and mystery. It is home to some of the best sea cliff climbing in Devon, and I'd always wanted to see the places I knew so well as a climber from the different perspective of the sea. I set off from Clovelly on a cloudless June morning in 2018, heading west for Hartland Point. I had to deal with a punishing headwind all the way on this section, and finally rounded the point after a three hour tussle, taking the easy inshore passage through the tide race, only to hit a strong eddy current running against me for several hundred metres. This often happens after rounding a big headland and is always something to be wary of. I've noticed it is particularly likely to happen if the headland is a right-angle shape in relation to the coast. St. Aldhelm's Head in Dorset and Dodman Point in Cornwall both produce exactly the same effect from very similar topography.

The following day, between Bude and Hartland Quay, a distance of just over twenty kilometres, I didn't see a single other craft on the water after setting out at first light from Crooklets beach; just a lone hiker silhouetted against the clouds as the sun rose behind the four hundred foot cliffs around Cornakey. Riding the early morning tidal stream half a mile off Lower Sharpnose Point, the best cliff on the Culm Coast for rock climbing, I drank some hot black coffee from my flask as the first rays of the sun lit up the inky contours of the swell. I've climbed most of the routes at Lower Sharpnose, and have wonderful memories of the place. Perhaps, I thought, the great landscapes of the heart are not those that you travel halfway across the world to find; they're sometimes just beyond your own front door.

Further southwest, a little later that summer, the tide was running strongly with me as I cut through the lively race that forms off the point west of Crackington Haven. The drama of the overfalls was overshadowed, though, by the continuous rampart of enormous shale cliffs in the distance stretching southwest to Tintagel as I cleared the point. Some, like Beeny Cliff, are up to 600 feet high.

Passing Beeny, the scene of one of Thomas Hardy's best poems, the coast is a geological marvel of massive, impregnable shale cliffs and offshore stacks. It's truly awe inspiring. The tiny natural harbour of Boscastle, in an otherwise inhospitable coast with not a single other sheltered anchorage, is almost unbelievable when approached from the sea. Between two sheer walls of jet-black stone, a narrow, north-running inlet suddenly appears, only becoming truly visible once you've actually crossed the harbour bar.

As I was packing up the board on the quay at Boscastle, a man and a boy approached me, inquisitive about where I'd come from, and how I'd travelled. I told them briefly about my voyage. The man then exclaimed in a dead-pan tone:

"Well, great to meet you. My name's Robert. I clear out the houses of the deceased." Had I drowned and slipped into a parallel universe? It was a surreal and fitting welcome ashore after paddling one of the wildest sections of the southwest coast alone. Robert and his son even gave me a lift back over to my car in the back of their van; I speculated about how many lifetimes' worth of stuff it had contained as we rattled along the deep Cornish lanes.

A few weeks later, on a day of unreal Mediterranean conditions, I launched through turquoise water and half a metre of beach surf at New Polzeath under azure skies and a blazing sun. Was this Cornwall or Sardinia? Cutting the corner from Pentire Point across to Tintagel took me far out to sea. Rounding this epic headland, famous for its links to Arthurian legend, was the highlight of this part of the journey, with a lively tide race running through deep water directly under the three hundred foot cliffs at the apex of the point.

The equally impressive section between Padstow and St. Ives – passing the great headlands of Trevose, St. Agnes, and finally Godrevy – was completed a little later, over four separate days, taking me to West Penwith and the point where the Celtic Sea meets the English Channel. Rounding St. Agnes Head, under the sheer cliffs of Carn Gowla, was particularly exciting. There was a decent wind-chop, which combined with the clapotis effect [when the waves go in both directions] made the sea pretty lively. I avoided the worst of the tide race here by navigating an almost invisible channel cleaving through the cliff on the seaward apex of the headland that was, at one point, only just wide enough for my board.

If I were forced to choose, the final twenty-eight kilometre section from St. Ives to Sennen Cove – the north coast of West Penwith – is perhaps the finest section of the entire coast of Southwest England. The wildness of the granite and greenstone cliffs, interspersed with occasional tiny bays and offshore rocks between Cape

Cornwall and Zennor Head is simply astounding. I know the crags and zawns of this coastline well as a climber, but to experience the place from the wilderness of the sea takes it to another level. Grey seals, dolphins, basking sharks, and ocean sunfish can often be seen here. What makes this trip really committing, particularly if there's any swell, is the fact that there is really only one safe landing along the entire route, at Boat Cove. You're truly in the Atlantic Ocean here.

Twenty four hours later, I launched from the slipway at Sennen Cove just after low water on another vintage Cornish day. My destination was the small harbour of Mousehole, just beyond Land's End, where the different tidal streams of the Celtic Sea to the north and the English Channel to the south converge, creating lots of complicated inshore and offshore currents: this is the very edge of mainland Britain.

Without much swell to worry about, I navigated close to the cliffs, through granite arches and spooky zawns filled with the boom of the sea. Tidal rapids form along here in the narrow channels between the headlands and the jagged reefs that surround them; perfect for a bit of whitewater action. To say the paddling here is world-class is something of an understatement. About half a mile offshore, somewhere between Land's End and Chair Ladder, the rudder-like dorsal fin of an ocean sunfish flapped lazily for a while against the side of my board before it plunged back into the deep. The *mola mola* is a truly strange fish, feeding mainly on jellyfish at great depths and coming up to the surface to reheat itself in the sun.

Rounding Chair Ladder and passing the coastguard lookout on Hella Point, I'd crossed the official threshold of the Celtic Sea and entered the English Channel. The groundswell dropped back as I turned north, and a fresh westerly tailwind whipped across the deep blue water. Half a mile off Porthcurno, the combined effect of the wind and the last of the flood current carried me east at speed towards Penzance, the first and last sheltered harbour in the English Channel. With the wind and tide in my favour, I reached the shelter of Mousehole quay's solid granite walls in quick time. It had been a truly exhilarating day.

The Lizard peninsula juts out into the Channel to form the most southerly point in Britain. Where the broad shelf of volcanic rock (technically an exposed ophiolite complex) that comprises Lizard Point finally meets the sea, it erupts in a maze of jagged, shallow reefs that criss-cross one another. Add to this the fact that the Point experiences some of the strongest tidal streams anywhere in Cornwall, and you've got yourself a pretty good working definition of a 'ship's graveyard'.

My partner, the Hungarian mathematician and elite waterwoman, Dorka Fekete, and I set off from Coverack for this section on Easter Sunday 2019, heading south for the Lizard on a perfect spring morning. The sea was oily-smooth at first, but by the time we neared Lizard Point, the tell-tale gusts of a brisk westerly were beginning to make their presence felt. We flew past the Coastwatch Station at six knots, and soon passed the Lighthouse looming over Housel Bay on the most southerly point of the Lizard.

Our original plan had been to land at Mullion Cove on the western side of the Lizard peninsula. As we approached the southern tip of Britain – and the place where the Spanish Armada was first sighted in July 1588 – I noticed a menacing looking line of heavy surf on the outer reef to the west of the Point. I then remembered the story of two experienced sea kayakers, Chris Duff and Mick Wibrew, who were both wrecked on Lizard Point in huge seas during their round-Britain paddle. This forced a quick strategy change, and we landed in Polpeor Cove by the ruined lifeboat station on the Point itself. We completed the equally dramatic section from Porthleven in Mount's Bay to the Lizard the following year, in August 2020, landing in exactly the same place as before.

St. Mawes, just east of Falmouth, is like an artist's impression of a Cornish fishing port. It oozes an old fashioned seaside glamour; a hint of what summers used to be like – or what we'd like to imagine they were like – before package holidays to Benidorm. I set out alone from the quay at quarter to eight on an azure July morning in the summer of 2020, heading northeast for Mevagissey, a full twenty eight kilometres northeast.

I turned northeast at St. Anthony's lighthouse on Zone Point, exiting Carrick Roads and entering the open sea, and the brooding prow of Dodman Point appeared dead ahead on the horizon. Dodman, the highest headland in South Cornwall, was often the first land sailors in the past would have sighted when entering the English Channel. It's a stirring thought to think of those men on the creaking decks of sailing ships glimpsing it on the horizon as they returned from months or years at sea.

Pausing for a break close to the offshore island known as Gull Rock, off Nare Head, I took in the scene. The sky was an unbroken, impossible blue. The faint breath of a northeast wind ruffled the sea from time to time, broken only by the distant cry of gulls from the cliffs. I could think of no better place in the world to be on a Sunday morning in early July than out here, half a mile off Cornwall's Roseland Coast. It was like one of those dreamlike days you might recall from a childhood summer: a moment held still in time. Two hours later I landed on the fish dock at Mevagissey in a state of Zen-like rapture, my spirits lifted by the freedom of the sea.

The section of coast between Mevagissey and Fowey is dominated by the huge bight of St. Austell bay. This can be made into a superb trip by launching from Lostwithiel, a few miles inland, at high water; you head down the Fowey river on the outgoing tide, enter the sea, then cross the bay itself. When we did it in the late summer of 2019, a brisk east wind helped our progress west-south-west.

Downwind standup paddling on a following swell is a real delight, once you've mastered the strange art of 'catching the bumps' as the technique is known; essentially it means SUP-surfing a wind-generated wave that isn't fully breaking. Much of the southeast Cornish coast between Plymouth Sound and Fowey was completed this way. Rounding Rams Head, the big, blunt headland west of Plymouth Sound, was a highlight. We completed it as a team of three. Once my friend Gavin Symonds (one of the strongest paddlers in the UK), Dorka, and I got out from the shelter of Plymouth Sound and into the wind-line, we cruised across to Downderry at a steady six knots, with a following swell and a fresh southeasterly powering us along. Who needs the Molokai Channel when you've got the English Channel?

The coast east of Plymouth is dominated by the dramatic sweep of Bigbury Bay. Here, on another day of exciting following seas, Gavin and I sped downwind together from Wembury to Hope Cove. The more spectacular section to the east then leads around Bolt Head, Prawle Point and Start Point, the latter defining the western extremity of the vast 'lee shore' that is Lyme Bay. In the days of sailing ships, this was a trap in strong southwest winds, sometimes driving vessels to their fate on the mighty shingle reef of Dorset's Chesil Beach if they couldn't clear Portland Bill.

Whilst the fast tide races off Bolt and Prawle are fun in their own right, rounding Start Point is particularly exciting. When approaching from the west, the flood tide pulls you straight down towards a pair of huge, flat-topped rocks that lie just offshore beyond the lighthouse. The idea is to keep as direct a course as possible through the narrow gap between the rocks as you rise and dive through the overfalls.

After Start Point, the rest of the South Devon coast is less dramatic and more benign, and the section between Blackpool Sands and Brixham is idyllic. North of Dartmouth, the English Riviera Air Show made things more interesting. As Dorka and I negotiated the offshore Mew Stone rock on this leg, the bassy drone of an Avro Lancaster's V12 Rolls-Royce Merlin engines seemed to make the sea simmer as the plane banked south overhead.

I slowly pieced together the remaining sections of the East Devon and Dorset coast through the summers of 2019 and 2020. Dorset's Jurassic coast between Weymouth and Poole Harbour – and the final section of my journey – is a unique marine landscape. I used to sail here with my dad thirty years ago, and I've spent hundreds of days rock climbing on the sea cliffs of Swanage and Portland, so this place is very special to me. On the section between Lulworth Cove and Kimmeridge, we enjoyed perfect downwind conditions off the jenga-like precipice of Gadd Cliff. Here, the days that my dad and I spent offshore along this coast during my childhood summers came rushing back. Getting lost in the fog, camping on storm beaches, fishing off the back of the boat, refuelling with my grandma's cheese sandwiches; all the good stuff.

The past is, of course, a foreign country, as the writer L.P. Hartley said. The USSR still existed back then, and the internet didn't. But the past is also close behind us, and it shapes who we are today. I'm sure it was those days off the Dorset coast as a kid that instilled my love of the sea that eventually led to this expedition around Southwest England.

I've been lucky enough to have travelled all over the world since those early adventures with my dad. Even amongst expeditions to some very remote places, paddleboarding the coast of Southwest England counts as one of the greatest adventures of my life. It was an ever-evolving psychodrama of things past, passing, and to come: a mission of the heart. There was also, it seemed, an element of poetic justice in the way I entered the expedition's final stages during the summer of 2020 when long distance international travel was effectively put on hold. Some of the best adventures of all, I have begun to understand, are often found just beyond your own backyard. In August 2020, doing some island-hopping in the Isles of Scilly, a pod of bottlenose dolphins circled our boards for a few minutes in St. Helens Gap, just off the north coast of Tean island. Rising for air, vanishing, then appearing again, they moved as silver shadows through the blue water, signalling to us in their own indecipherable language. Where else would you want to be?

On the course of the voyage, I also encountered the hydrographic systems in action in the most direct possible way: the tidal streams, wind currents, and swell cycles that surge perpetually around all the British islands, under the summer sun and through the winter rain. Out there on the water, their results are right there before us. Strange standing waves two miles out. Big eddies in the no-man's-land between cliff and reef. The darkness of the sluicing water before the outer rocks.

The unfathomable power of the sea.

Epilogue

I'm in the middle of Mounts Bay, the great bight between Land's End and Lizard Point in Cornwall. It's a gold and blue September afternoon. My journey around Southwest England is almost done; this is one of the last remaining sections I'd left uncompleted. I'm about half a mile offshore. The old mine buildings of Trewavas on Rinsey Head stand proud and tall against the Atlantic. A lone local fisherman in an open boat is pulling up lobster pots, his blue hull merging at times with the waves; ghostly reminders of the days when tin and fish were Cornwall's great raw materials. In a moment of intense clarity, I inhale the crisp Atlantic air, turning my board west into the sun. The harbour at Porthleven is not far off, but the land behind me blurs into a mirage. Out here, I seek another reality from the one I shall find ashore. I am alone in a world of wind, silence, and light. I am free here; as free as I have ever been. The sky burns the horizon with its brightness, and shards of refracted crystals dance on the dark blue water. Real adventure isn't just about a route across a map, but also a quest for what lies at the heart of your own life. And somewhere on this shadow-sea you might find the unknown world ahead, the voyage to come.

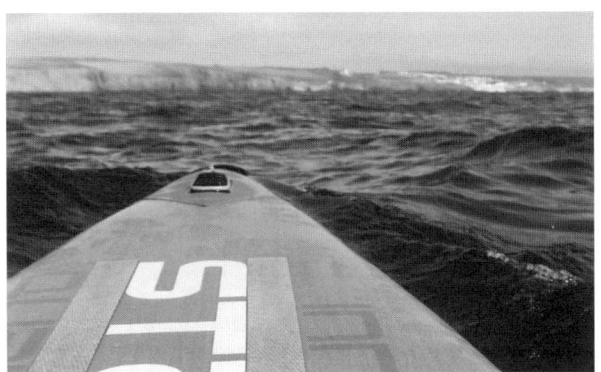

A nautical mile off Porthleven, Cornwall, towards the end of the expedition in late summer 2020.

BASE magazine, 2021

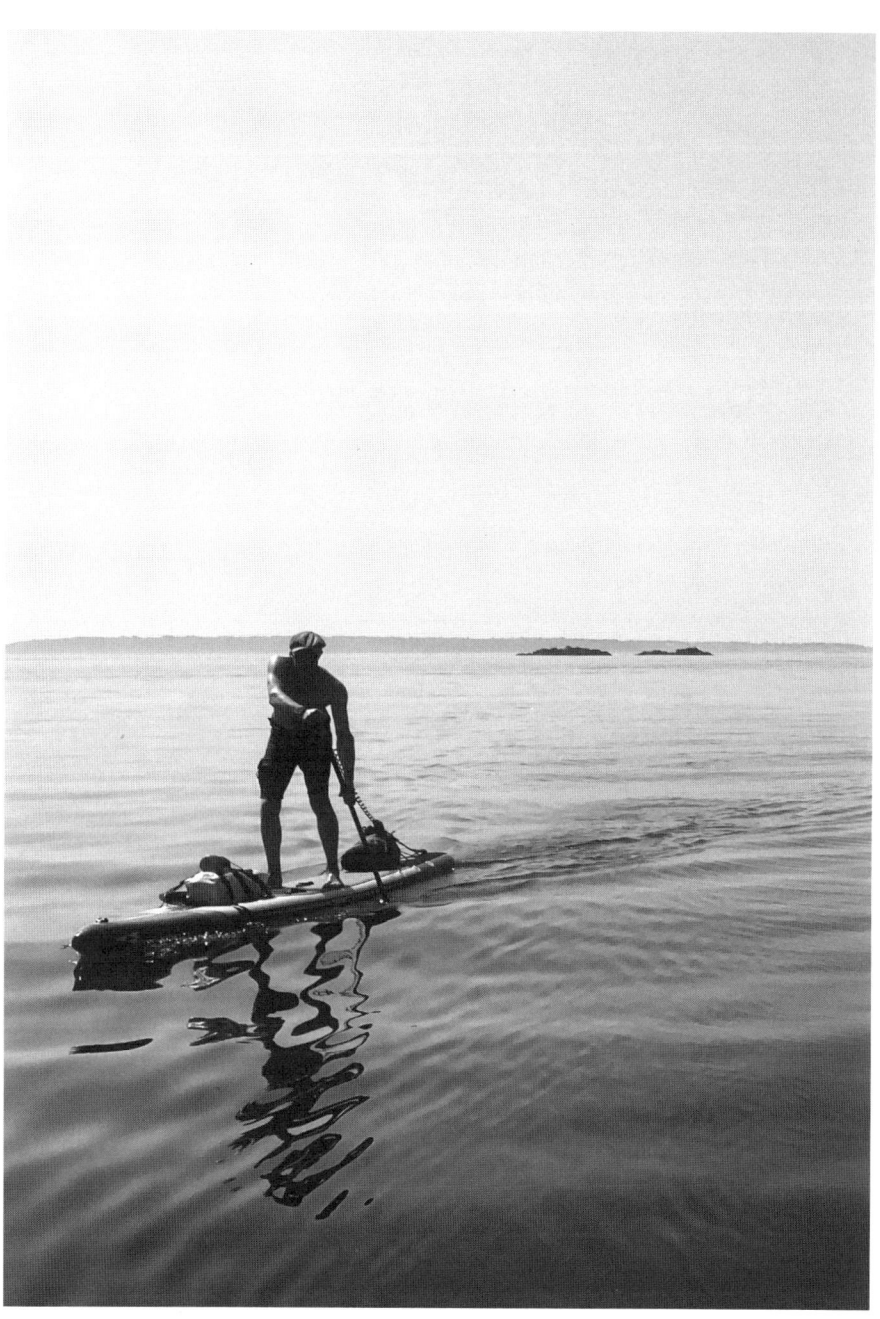

The author at the midway point of the crossing between Guernsey and Sark, in the full force of the tidal stream of the Great Russel. Guernsey is visible in the distance.

Vectors in the Stream

Island-hopping the English Channel's strongest tides

Some of the most memorable things start with a simple, mad idea.

This was one of them.

When you glance at a nautical chart of the English Channel, certain landmarks stand out. The chalky trapezium of the Isle of White at almost dead centre. The huge promontory of Cap de la Hague that marks the northern tip of Normandy, sticking up into the Channel like a proverbial French finger brandished against the marauding English. And the other defining headlands of Beachy Head, Portland Bill, Start Point, and The Lizard, the sea-marks of times past and to come. But something else on this chart – something small but deeply intriguing – pulls the mariner's eye south from those more prominent features and towards the Channel Islands.

Almost at the centre of the archipelago lies the tiny island of Sark; a strange, beautiful, star-shaped dot on the chart. It's arguably the most interesting feature of the Channel Islands as a whole. It comes complete with dramatic cliffs, caves, isolated coves, a tiny but permanent local community, tractors instead of cars for transport, no income tax, and a method of self-government that only parted with the feudal system in 2008 (yes, really). Sark lies just over six nautical miles (13km) east of Guernsey and just over eleven nautical miles (21km) north north-east of Jersey.

For sailors and paddlers, one of the most compelling and also complicating aspects of getting there is the extremely strong tidal current that flows through the channel of the Grand Russel, separating Sark from the smaller islands of Herm and Jethou, and through the narrower channel of the Little Russel, that separates those islands from Guernsey itself. On spring tides, the tidal streams in both the Grand Russel and Little Russel exceed five knots, and around the headlands and various offshore rocks the stream can be as strong as seven knots, and in certain places even more than that: that's the speed of a large river in spate. This makes the crossing to Sark an interesting challenge, to say the least. There is nowhere else in British waters, and possibly in the world, this far offshore that experiences such strong tidal currents.

The crossing from Guernsey to Sark has been done numerous times by experienced sea kayakers, and in good conditions it's an excellent trip by kayak. I could find no evidence of the voyage having been done by anyone on a standup paddleboard, which is of course not to say it wouldn't be possible. A strong paddler in a high performance sea kayak can easily maintain an average speed of six knots in good conditions. Even the fittest standup paddler can't match that. In my experience, working with an average speed of just over three knots is a good rule of thumb when planning long distance SUP trips in the open ocean, based on using a race board or a fast touring board.

The paddling technique and overall strategy required for a trip like this is known as 'ferry gliding', which as it suggests involves making an adjusted course across a strong current in order to arrive, hopefully, at the correct point on the other side. The longer the glide, and the smaller the landing point, the more complicated it is to get it right. Sark is a small island, and it takes almost three hours to paddle there from Guernsey. Since you're transiting across water that's moving faster sideways than you can paddle forwards, you're essentially trying to hit a moving target, so it would be possible to 'miss' the island entirely. Even in good conditions, it wouldn't be easy.

Derek Hainon, the author of the excellent sea kayaking guidebook to the Channel Islands, came up with a useful way of thinking about offshore crossings in the archipelago. He describes these islands as "like rocks in a river". This particular river in question changes direction four times a day, with the six hour North Atlantic tidal cycle. The other thing is that the 'tidal river' between Guernsey and Sark, if you think of it like this, is slightly wider than the mouth of the Amazon – the world's largest river – at Macapá, Brazil, just before it flows out into the Atlantic Ocean. Furthermore, numerous tide races with large overfalls, eddies, and standing waves form in multiple locations around and between the islands. Here were all the ingredients, it seemed, for either a truly memorable quest or an gargantuan epic.

The necessary weather window opened in the middle of June 2022. After carefully cross referencing tide times, swell charts, and the wind forecast, my partner Dorka and I decided the Sark paddle was worth a shot. She's one of the few people I know who's strong and confident enough on a SUP, and on the water in general, to do something like this. She's also one of the only people I know who is mad enough to even contemplate doing so.

The Channel was as calm as a mountain lake as we left Portsmouth on the ferry; these were ideal conditions for what we had in mind. A few hours later, as we approached Alderney, the most northerly and outlying of the Channel Islands, massive swirls and interconnected eddies began to appear in the sea as we entered the Raz Blanchard, or the Alderney Race. The whole thing was on an impossibly large scale: the tide race extended around us for dozens of square miles. The Raz Blanchard forms in the channel between Alderney and the tip of Normandy's Cotentin Peninsula, and the tide stream can reach twelve knots through here at certain times of the year. That's 22 kmph, by the way. The French coast was around thirteen nautical miles south, Alderney was eight nautical miles west, yet even way out here the tide was rolling along like a freight train. On this particular afternoon there was almost no wind, yet still the sea was alive with tidal energy in a way I had not previously seen anywhere else, not even on my extended paddleboard expedition around Southwest England between 2017 and 2021.

There are very few places on Earth where the power of the moon's gravitational effect on the ocean is as evident as in the sea lanes of the Channel Islands. There are only a few other places with a bigger tidal range, like Canada's Bay of Fundy and in the Bristol Channel. But those are inshore waters, not the open ocean. To experience this kind of tidal power in an offshore, deep water, oceanic environment you have to come to the Channel Islands, and in particular to exactly where we would be heading tomorrow.

The sea is a canvas of deep blue, unbroken by whitecaps, as we make the last preparations and checks on the tiny beach just north of Bordeaux Quay. There's a faint breath of wind from the northeast, but nothing of note; the forecast says it'll drop back to less than four knots after three o'clock, when we'll be in the heart of the

Grand Russel. Let's hope it's right. The islands of Herm and Jethou define the eastern horizon, with the southern tip of Sark just visible to the south. Through the anticyclonic haze of high noon, it looks really quite far away.

Is this a good idea? Probably not. Are we going to try it? Damn right we are.

We go through the final, routine safety checks: VHF radio, check. Flares, check. Tow lines, check. Rafting cords, check. GPS, check. Fin bolts, check. It seems surreal that we are actually about to launch on this preposterous mission, but the weather and sea conditions are as good as they get, and here we are.

We push the boards off the shingle stern-first, spin around, and head straight out into the main channel of the Little Russel. The south-flowing tide stream has only just started, and we instantly pick it up as we contour a big rock that lies a hundred metres offshore. The Little Russel is full of rocks, some small, and some almost tiny islands, and they make useful reference points for our drift. We keep the dominant landmark of the Bréhon Tower well to our south; this early nineteenth century fort was repurposed by the Nazis as a gun emplacement during the wartime occupation of the Channel Islands. We make good progress across the central section of the Little Russel, wisely avoiding the sailing time of the Condor fast ferry that accelerates to 35 knots hereabouts as it heads north.

The white sand dunes and dark green swatches of marram grass that define the northern point of Herm resolve into view, and soon we're gliding through aquamarine and turquoise water as the light refracts off the sandy bottom. It's more reminiscent of the Caribbean than Britain. A solitary puffin lands on the water just to starboard, confirming we're in temperate rather than tropical waters after all.

We set foot very briefly on Herm as we portage the boards across a narrow sand bar to avoid paddling around the large reef that extends seaward from the northern tip of the island. A great sluice of dark blue water now rushes in to meet us: the south flowing tide stream on the east side of Herm is now running at full bore. The dramatic, dark red cliffs of Sark's west coast now shimmer on the horizon. We waste no time here, as we want to have time for a break on Sark to wait for slack water for the return crossing. After a very quick break to hydrate and inhale an energy bar, we push out into the stream of the Grand Russel.

Heading away from Herm and an assembled flotilla of yachts and powerboats basking in the sun and shelter of the northeast side, we create an invisible marker point around ten degrees north of the northern tip of Sark, and use that as our sighting line for the tidal vector we want to take across this six kilometre-wide channel through which the water constantly surges. The tide is now flowing considerably faster southwest than we can paddle east, so we need to make for a point that's at least forty-five degrees upstream of our intended landing point on Sark's Grande Grève beach if we are to stand a chance of actually arriving there. This is the most difficult and complicated tidal vector I've yet tried to do on a paddleboard.

A large rock known as the Noir Pute commands the lonely seascape on the northern approaches to the Grand Russel. We keep it to our port side as we make our course out into the heart of the channel. From here, with our corrected course to the north, the tide should pull us down on to the coast of Brecqhou, a small island that lies just off the west coast of Sark, from where we might be able to nip around into the bay where we intend to land. The operative word in that sentence is might, however. The grand plan sort of works, but as with so many things at sea, it doesn't exactly pan out as intended.

As we approach Brecqhou, the large castle completed by the Barclay brothers is prominent on its seaward coast, and provides a valuable reference point. We're closing in, but we're still around half a nautical mile (circa 1km) offshore. As the first big eddies appear that indicate we are entering the faster tide stream created by the point of Brecqhou itself, the castle appears to be moving left at an alarming rate, the same way a platform bench might look as your train departs from the station. It's blindingly obvious that for this final leg to clear this point, we're going to have to paddle seriously hard to get across the tide stream and enter the sheltered water on the south side of Brecqhou. If we don't manage to clear the point, on the other hand, we won't be going to Sark at all. We'll just have to head back out to sea, going southwest with the tide towards the distant coast of Brittany (!) somewhere beyond the horizon, and wait for the stream to change direction in around two hours before heading back to Guernsey. That was the Plan B. At sea, you always need a Plan B. Neither of us were particularly enthusiastic about that one, needless to say.

Dorka is slightly to the north of my position, focused and going flat-out. We're both paddling at full race-pace now, with high cadence strokes and maximum concentration. We only look up occasionally to check our course. Under our boards, the tide rushes and swirls, eddies and counter-eddies. It feels as if we're approaching the far bank of the greatest river on Earth. In a way, we are. It's hard to tell precisely the speed of the tide, but it very much feels as if we're being pulled sideways at six or seven knots – which far faster than our forward speed of around four knots. There's virtually no wind, yet the sea is alive with the pull and torque of the tide. It's an awe-inspiring place to be standing on a 14 foot race board, just centimetres above the sea.

Finally, after what seems like a long time but was really just fifteen minutes, we clear the small rock known as 'Givaude' that lies a few hundred metres southwest of Brecqhou, and we're finally in calmer water sheltered from the main tidal flow.

Grande Grève beach now appears dead ahead: we've made it to Sark. It's almost low water, and the exposed sand glistens darkly in the afternoon sun. A few yachts are anchored in the bay. It's a tranquil scene, so far removed from what we experienced in order to get here. I hear the nose of my board slew into the sand: touchdown. Dorka lands just after me. There are fist bumps and laughter. We did it – or at least half of it.

We carry the boards up the beach, away from the soon-incoming tide. We need calories and fluid, in that order. Salami, nuts and energy bars are inhaled with half a litre of water and the last of the coffee. We hike up the beach, and scramble up the ultra-steep, falling-down concrete steps that lead up to La Coupée. It's Sark's most famous feature, and arguably Britain's strangest and most spectacular road. It is essentially a track along a knife edge ridge that separates Sark from Little Sark, the peninsula on the southern point of the island. There are cliffs on both sides going straight down to the sea. It's too narrow for a car other than possibly a vintage Mini, and there aren't any of them on Sark in any case. It is perfectly designed, however, for a small sit-on-top tractor, the defacto mode of motorised transport on this beguiling little island. On cue, just such a tractor chugs down the hill and across the ridge – or is it a bridge? – as we arrive.

The scene from up here is serene. The large island of Jersey sleeps on the horizon far to the south. Tidal eddies dance around the visible coast of Sark to the east and west, and the route we've just taken to get here stretches out beyond, first to Herm, and then the distant coast of Guernsey glimmering in the late afternoon sunlight. It would be great to spend the rest of this perfect summer's day exploring Sark at a leisurely pace, but we don't have time: the tide is soon going to turn, and we want to get back to Guernsey before sunset.

It's just before five o'clock. We make final safety checks on the boards and to our own personal kit once more. We notice the beach slowly getting smaller: the tide has already turned. In an hour, the last of the south-flowing stream between here and Guernsey will wane, change direction, and start to flow north again. We're going to try to use this briefest of lulls in the tidal cycle to our advantage. You can't really call it slack water here, as this time around high or low tide is described. The sea in the Channel Islands never really stops moving, it just slows down a bit before doing a 180 degree turn and winding back up to the full five knots-plus. Guernsey, here we come.

We push out and paddle west into the sun. The wind has died, and the sea shines darkly in the liquid light. It's utterly idyllic. Soon we're out beyond the point of Brecqhou; it's nothing like it was an hour and a half before, when we came this way through that great river of swirling, eddying salt water. The stream has now fully slackened off. We make a reckoning point just to port of the tiny island of Jethou, which lies immediately south of Herm, to give us a course back across the Grand Russel. The advantage of going across at this moment is that the tide will first pull us slightly south, then back north as the stream changes direction. This makes the vectoring angles we'll need less extreme than on the way across to Sark.

We glide back across the Grand Russel as the midsummer afternoon slips slowly, imperceptibly, into evening. And what an evening it is. One of wildest sea channels in the British Isles is, for now, a filmic scene of total calm. A powerboat heading east for Sark passes well north of us at thirty knots before coming slowly off throttle and circling around. We think perhaps they've seen puffins, but actually they've just seen us. They slowly approach. We give them the thumbs up, and they wave us on. The middle of the Grand Russel is not, I suppose, the most likely place to see a couple of standup paddlers on fully-loaded race boards.

As we approach Jethou, the north-flowing stream gets going. A large rock topped with a brilliant white obelisk called La Grande Fauconniere lies just southeast of Jethou, and makes a useful reference point as we close in. At first, we think we can clear Jethou to the south, but the way the obelisk starts moving as we near the islands proves otherwise. We're being pulled north now by the power of the early flood stream, so we're going to have to use the Percée Passage, the spectacular channel between Jethou and Herm, to get back into the Little Russel and return to Guernsey. It was my intended route, as it's a stunning sea lane that divides the two islands.

We eddy-hop the tide and come close inshore to the outer rocks, and soon we're entering calmer water on the north side of Jethou. The island is privately owned so we don't land, but have a quick break for some snacks and fluid in the shallow channel between Jethou and Crevichon, another islet directly north. Now we're back in the Little Russel proper; it's just a short ferry glide of a few kilometres across to Bordeaux

Quay on Guernsey and our launch point. This time, we keep the Brehon Tower well to port. We're on a true diagonal vector now, letting the tide do at least some of the work of taking us back to base.

The afternoon has slipped into a perfect midsummer evening. A few boats are returning to St. Peter Port or heading out for the night. We get the final tidal vector almost exactly right – third time lucky – and approach the big offshore rock beyond Bordeaux Quay just upstream of the main flow. The tidal stream is going at full chat now, and we come into the final turn of the voyage flying alongside that offshore rock at a solid seven knots. A powerful eddy-line catches my fin as I turn around the rock, and I have to brace hard not to fall, but suddenly I'm paddling out of the tide for the final time and towards the beach. Dorka is just ahead of me. She smiles a wide, bright smile as we come into the bay. A cocker spaniel runs down the beach towards us, barking excitedly as he stares out to sea. He wants a ball from his owner, of course. But secretly I imagine that he somehow knows where we have been today, too.

We did exactly what we set out to do, and give or take a few interesting moments – particularly the tidal vectoring on the approach to Sark – it all went to plan. Completing a quest such as this was the result of a lot of paddling in tidal and offshore waters, and combining that experience with a perfect window of sea and weather conditions. Part of the joy of adventures at sea is in seizing the moment, and being in the right place, at the right time, with the right team. For the voyage out to Sark and back, we certainly captured that magic moment of correct ingredients.

The next morning I woke early to the sound of the wind in the treetops. Heavy weather. I made a strong coffee and drove down to the harbour. The sea was grey, and the Little Russel was a surging mass of wind-against-tide. Big overfalls heaved around the rocks; spray was whipping from the whitecaps. There wasn't a boat in sight, not even a trawler. I could just make out the bulk of Herm through the mist. And beyond it – nothing. I listened to the patter of the drizzle on the windscreen, squinting out to sea, looking for something I knew was there. Sark – that strange, beautiful, star-shaped dot on the chart – had vanished to windward.

www.base-mag.com, 2022

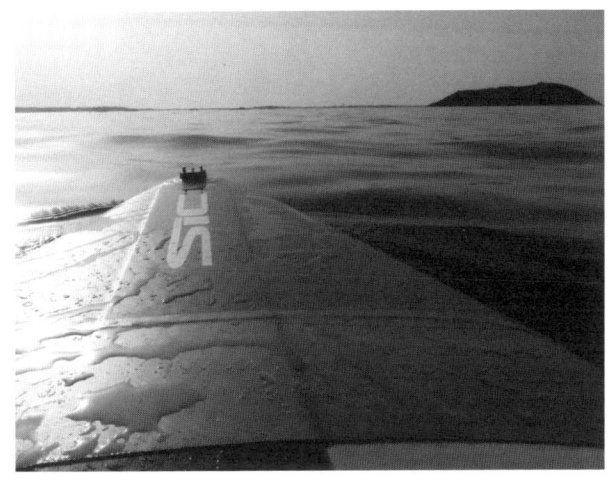

Looking towards Jethou (right) and the distant coast of Guernsey from the heart of the Great Russel channel.

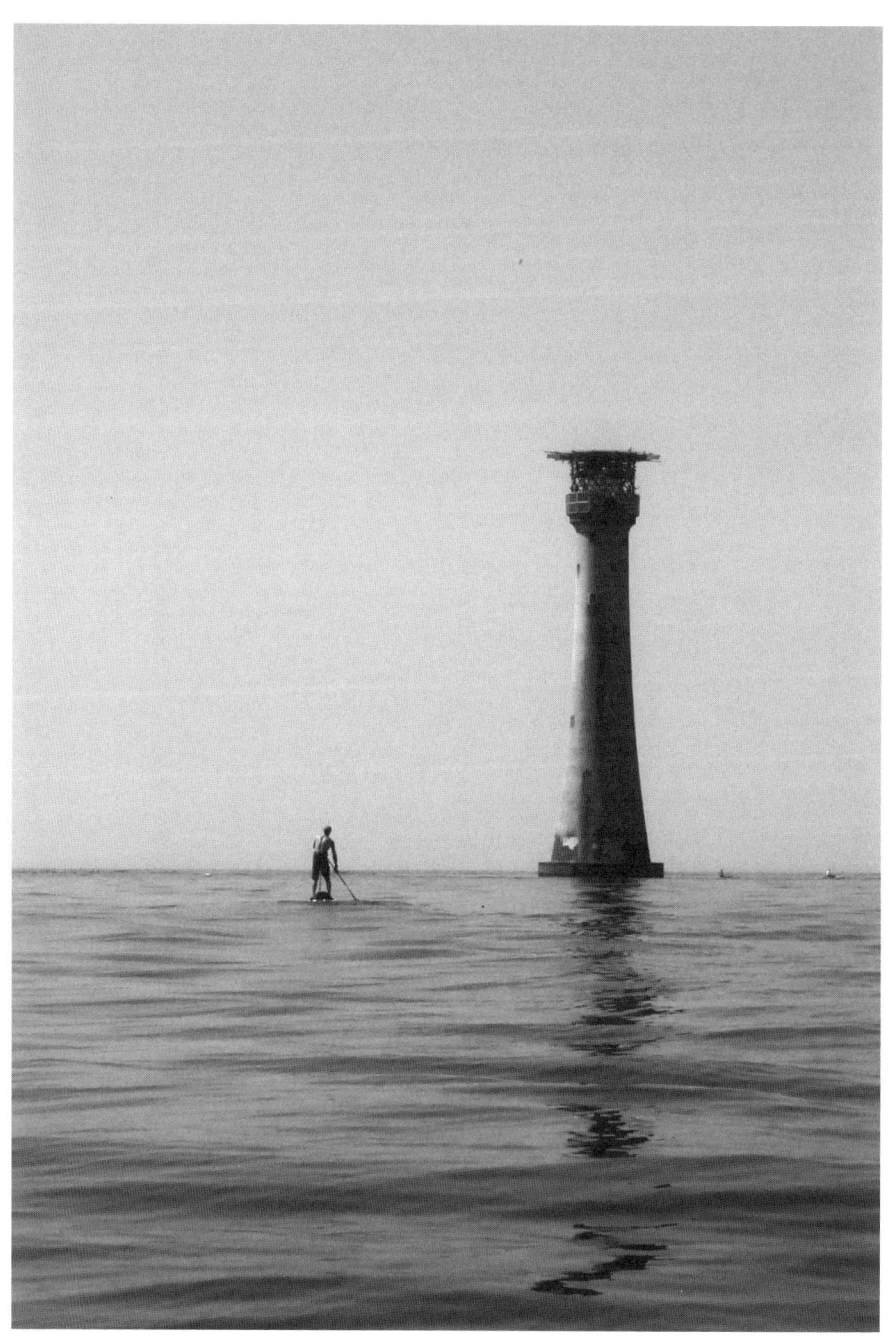

Approaching the Eddystone Lighthouse, the world's first open-ocean lighthouse, twelve miles off Plymouth.

Rock Steady 'Round Eddy

A voyage through time to the world's first open-ocean lighthouse

The English Channel glimmers silver and gold in the early morning sun. The faintest breath of a northeast wind ruffles the water. Nothing but the ghost of a long-lost Atlantic groundswell swirls around the stratified rocks of Polhawn Cove. These are ideal conditions for what we have in mind.

Ten miles out, a solitary column of interlocking granite blocks stands tall in the heart of the sea, commanding the scene off Plymouth Sound like a fossilised version of the Saturn V launch vehicle. It's an object as incongruous as it is weirdly compelling. Rising from a submerged reef of Precambrian gneiss hundreds of millions of years old, it stands resilient against the sea and the weather that perpetually works against it, warning ships of the danger that lies beneath.

It has not always been this way. The Eddystone Reef is a major hazard to shipping in the English Channel, and countless ships have gone down on these rocks. Christoper Jones, captain of the *Mayflower* that took the Pilgrims from Plymouth to North America, remarked in 1620 of the "twenty-three rust-ragged stones around which the sea constantly eddies. If any vessel makes too far to the south, she will be caught in the prevailing strong current and swept to her doom on these evil rocks". Considering Jones's warning, and the growing importance of shipping in seventeenth century England, it's perhaps not surprising that the Eddystone Reef was the site of the world's first ever open-ocean lighthouse, which was completed in 1698 under the direction of the maverick inventor, engineer, and merchant Henry Winstanley. A definitive man of the early Enlightenment, prior to the construction of the first Eddystone Lighthouse, Winstanley had designed a so-called mathematical water theatre called 'Winstanley's Waterworks' in Piccadilly. This remarkable contraption combined fireworks, fountains, and automated mechanisms of all kinds, including a rotating barrel serving hot and cold drinks.

After losing two of his own ships to the Eddystone Reef, and having been told by the Admiralty – with a typically British no-can-do approach – that the rocks were "too treacherous to mark", Winstanley was determined to do something about it. So he began building a lighthouse on the reef himself. Construction began in 1696, and two years later an extraordinary octagonal wooden pagoda stood on the Eddystone. Its upper chamber enacted the purpose for which it was built, and was lit by 60 large candles at a time and a huge hanging lamp. A fire risk assessment, I think it's fair to say, was not conducted. The lighthouse was damaged by storms in the first winter of its existence, then rebuilt in 1699 as a dodecagonal structure with a timber frame and stone-clad exterior. It survived for another four years, during which time no ships, remarkably, were wrecked on the Eddystone Reef. The lighthouse worked.

Winstanley was working on the lighthouse himself when the extratropical cyclone known as The Great Storm of 1703 hit on November 27th. By the morning of November 28th, 1703, no trace of his beloved lighthouse, nor Henry Winstanley himself, remained on the Eddystone Reef. In a strange twist of fate, Winstanley actually got to experience his own cherished ambition. He reportedly had tremendous faith in the lighthouse, and allegedly suggested that he "wished to be inside it during the greatest storm there ever was." It's some way to go, that's for sure.

Modern meteorological analysis of the records of the Great Storm of 1703 have suggested it was likely to have been a Category 2 hurricane, meaning wind speeds of 96 – 110 mph. What tremendous wildness Winstanley must have experienced that November night on the Eddystone in 1703 as he confronted the demise of his lighthouse – and of himself.

After researching the history of the Eddystone Reef prior to our voyage out, the extraordinary character of Henry Winstanley began to loom large. I felt a little like Martin Sheen's Captain Willard in *Apocalypse Now* as he unfolds the classified file on

Colonel Walter E. Kurtz: "the more I found out about this man" Willard says, "the more I wanted to confront him". Since Winstanley died over three centuries ago, I wouldn't be able to confront him physically. But I could at least encounter the site of his final creation in the most direct possible way, from the sea itself.

It was this desire, combined with a window of perfect conditions, that led a team of five of us to head directly out to sea in July 2021 on a ten degree bearing south-south-west from Rame Head, with the Eddystone's modern Douglass Lighthouse just a needle on the horizon. There are not many better places to be than a mile offshore on the British coast on a deep blue midsummer morning. Spirits in our strong, experienced group were high as a light tailwind sped us along, ruffling the surface of the sea from time to time.

A light container vessel with a black and red hull steaming south behind us crackled up on the radio; "Falmouth Coastguard. Number one, number one, over". Was the captain talking of his ship, or of himself? Who knows. But for our team, there was now only one objective: that slowly-growing granite pillar at 12 o'clock, a lone object defining the divide between the wine-dark sea and the sky.

A few hundred metres to starboard, a gannet was diving for fish. An area on the surface just ahead darkened with microscopic bubbles as a shoal of mackerel passed. The bounty of the sea was all around us as the land behind blurred into a mirage.

After three hours, the needle on the horizon had begun to resemble an actual lighthouse. As we approached, tell-tale eddies formed around us, a sure sign that the tide was running through shallower water. A kilometre before the reef itself, we hit an area of overfalls. The strange sensation of being in a tide race ten miles offshore wasn't lost on any of us. It was around high water when we reached the lighthouse, and the reef was entirely covered. The water under us was aquamarine in places as the sun refracted off areas of quartz and patches of sand in the gneiss bedrock. At one point, I caught the silver flash of a bass or grey mullet under the board. The Eddystone Reef is an unexpected, magical marine environment to find this far offshore in a temperate ocean.

The lighthouse itself, of course, dominates the scene. Fifty metres of interlocking granite blocks rising from the waves, impervious to its hostile surroundings and utterly resilient to anything the Atlantic might throw at it. It's a masterpiece of modern engineering. I gazed up at it as we enjoyed a well-earned lunch break drifting on the myriad tidal eddies around the reef, and couldn't help but feel even more admiration for Henry Winstanley and his outlandish wooden pagoda. Today, in near-perfect conditions, it was hard enough just to imagine it standing here on the reef, a great creaking tower with its grand octagonal prism adorned with candles surrounding the central lantern. Conjuring up what it must have been like at the moment it was consumed by the sea that November night in 1703 required an even bolder leap of the imagination. The occasional percussive rattle of an impact driver echoed from high up in the tower. Even on a Saturday, maintenance workers were here, engaged in the constant task of shoring up the structure. Building and maintaining a lighthouse in a place like this is challenging enough in the twenty-first century, let alone over three hundred years ago.

In mountaineering, when you reach the summit you're only halfway there; exactly the same could be said of paddling trips to open-ocean lighthouses. A light south-westerly had been forecast for the afternoon, which would have been a perfectly-angled tailwind to speed us back to England. As is so often the case in high pressure weather systems, the wind forecast was unreliable: a Force 3 south-easterly, in fact, picked up not long after we'd left the Eddystone, causing a bit of chop. We still managed a few short downwind runs before zig-zagging back into line with the tidal vector we were taking to return to Rame Head.

All too soon, the south Cornish coast resolved back again into view; the land had shifted from the mirage it was far offshore back to earth and stone. The small chapel atop Rame Head defined our landfall-marker as we cleared the headland and entered the sheltered water of Polhawn Cove, receiving some puzzled glances from a few anchored yachts as we came in. Clearly their skippers didn't expect a flotilla of loaded SUPs complete with radios and offshore safety equipment arriving here from far out at sea.

As the greatest wilderness environment on Earth, the sea is to be treated with the utmost respect. A big offshore trip such as this should only be attempted with the necessary skills, experience, equipment, and during a spell of settled weather. Every navigator must factor the unknown into their planning, though. As such, there's something in the remarkable story of Henry Winstanley and his original lighthouse on the Eddystone Reef that all mariners – indeed all explorers – can learn from.

In his noble attempts to protect ships from harm, Winstanley created something that had never been seen before in human history – a large and complex structure on an open-ocean reef. Yet he also fell under the fatal spell of his own invention; he could not distinguish its catastrophic shortcomings from the brilliance and ingenuity of the overall concept. With hindsight, plus an understanding of modern stress engineering and materials science, it was inevitable that a lighthouse made largely of wood would not survive very long on a reef in the English Channel.

We know that now, of course, but Winstanley didn't. He built something bold and beautiful out there on the Eddystone reef over three centuries ago that, for a short time, actually worked as an operational lighthouse. In this sense, like all true explorers, he was right out there at the edge of knowledge. The story of Henry Winstanley's lighthouse, then, is not just the tale of the rise and fall of a wooden tower on a reef in the English Channel. It's a story of the expeditionary imagination itself.

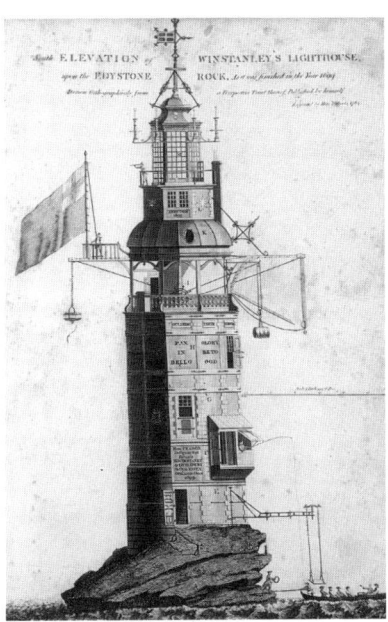

Diagram of Henry Winstanley's original Eddystone Lighthouse, built between 1696-1699, destroyed in 1703.

www.base-mag.com, 2021

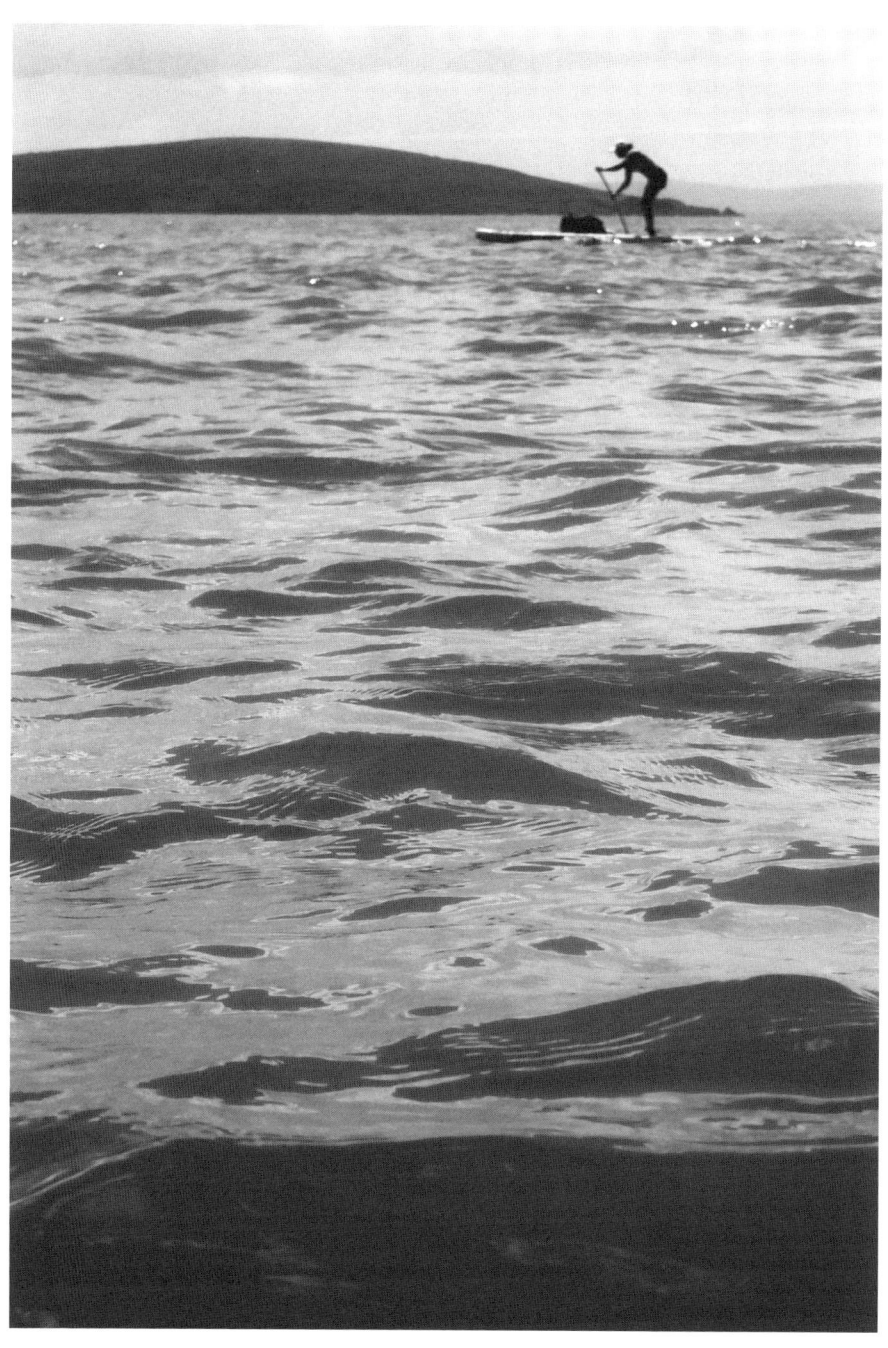
Running with the powerful Bristol Channel tide back towards England from Steep Holm and Flat Holm.

Tidelands

Exploring the unexpected wilderness of the Bristol Channel

In the distance, the bridge stretches across the vast expanse of muddy water like a curving silver blade. The enormous concrete supports of the suspended section silently guard the tidal cauldron surging beneath them: this is the Shoots Channel. High overhead, afternoon light glints against the steel retaining rails of one of the UK's major roads, held in improbable equilibrium above the five kilometre-wide estuary. At certain times of the year, the difference between high and low water under this bridge can be over fourteen metres, the height of an average three story building.

The powerful tidal current is now accelerating towards the bridge as if the water itself were attached to an invisible elastic band. I check my watch: 10.5 knots. This is not a normal speed at which to be moving on a standup paddleboard. I am paddling more to maintain a position in the dead centre of the two huge central supports than to propel myself forward. In front of each of the bridge supports, enormous standing waves rise as current meets concrete, forming multiple whirlpools at either side; think of the bow-wave of a supertanker, and you get the idea. They are clearly things to be avoided. As the bridge rears up directly ahead, I can feel my heart rate rise as the character of the water under my board begins to change; from a smooth, rolling current with intermittent eddies, it begins to shift, first to a series of overfalls, then to something different altogether. The crests and bumps of a massive tidal rapid dance in the near-distance. Spanning almost the entire width of the Shoots Channel, the rapid advances like a monstrous aquatic apparition, a kind of hydrological chimera.

The water suddenly turns dark beneath the board; I've crossed into the bridge's shadow. A hundred metres above, drivers on the Second Severn Crossing are listening to The Archers Omnibus and thinking about purchases at Cribbs Causeway retail park. Below them, I am in a rather different environment. Alone on the powerful stream between Wales and England, I'm being pulled towards some of the wildest tidal water I've ever seen. It's not so much frightening as fascinating; with the skills and kit to deal with what's coming up, I'm not in any danger here. The huge rapid now evolves as I enter it. Upstream of the bridge, multiple chains of small standing waves pile on top of one another for hundreds of metres as the tide flows over the shallow shoals surrounding the limestone reef of Black Rock. On either side of the rock, the standing waves morph into crazed colonies of miniature whirlpools, spinning off from one another like the isobaric map of the central Atlantic during peak hurricane season. This is the sluicing, surging heart of the Bristol Channel; even on a perfectly calm day, it's an extremely intimidating environment. The serious nature of the place is compounded by the extreme infrequency of safe landings. From here, there are just two places I can feasibly land in the next twenty miles: Beachley and Chepstow.

My strategy was to now enter the river Wye on the flooding tide, the mouth of which lies a short distance beyond the bridge. After what seems like a disproportionately long time, but on later reference was actually just fifteen minutes, I clear the worst water of the Shoots Channel. With the Welsh coast now just a few hundred metres to port, the hull of the board suddenly falls silent as the parallel mudbanks of the Wye estuary open up dead ahead. I'm in calmer water at last, and glide the final two kilometres upriver to land on the floating pontoon at Chepstow, which leads rather auspiciously into the garden of The Boat Inn.

Making the crossing from England to Wales via the Shoots Channel by kayak or SUP compresses a big adventure into a relatively short journey that can be done, potentially, on an evening after work. It also brings you headlong into the liminal, transitory, ever-changing and uniquely beautiful environment of Britain's tidal estuaries, of which the Severn is the largest.

As Mark Rainsley points out in his superb *Southwest England Sea Kayaking* guidebook, this trip is "characterised by mudflats and motorway bridges… an outwardly unappealing prospect, [but] it's actually one of the region's most challenging paddles."

There's a deeper point about the relationship between the environment and the exploratory imagination here. Sometimes it's in the most unlikely of arenas that the most compelling adventures can be found: the Bristol Channel is certainly one such arena. Fringed by the industrial outskirts of two major cities, this place remains a haven for wildlife as large container vessels approach or depart Portbury Docks, one of the UK's major commercial shipping hubs. Despite – and in some ways because of – its rudimentary, industrial, occasionally post-apocalyptic aesthetic, the Bristol Channel is a place where you can explore the British coastline on its own terms.

I live just a short distance from the Channel's murky waters. Sometimes, in the full force of a winter gale, I might go for a run along the coast to watch the wind work against the powerful tide, creating standing waves metres high. At such times, the power and wildness of nature is concentrated here. This place is essentially the diametric opposite of a highly curated Instagram feed. It's not obviously photogenic. There's usually not many folk about. And if you screw up, you'll drown.

Perhaps the finest trip of all in Britain's biggest tidal compressor is the voyage out to the two tiny islands of Steep Holm and Flat Holm, an unlikely pair of desert islands salubriously positioned between the fleshpots of Penarth and Weston Super-Mare. These islands are in fact fascinating places, appearing to float improbably on the sea from certain angles. Steep Holm, the higher and more dramatic of the two islands, has the character of a kind of derelict, abandoned Atlantis. It's essential to do this trip on a day of light winds and fair weather, due to the considerable and complex tidal vectoring required, and the risk of missing the islands completely if blown off course.

Whilst Steep Holm is now uninhabited, the relics of human activity are everywhere; derelict Victorian fortifications, abandoned cannons, a spooky WW2 searchlight post, and even a ruined tavern above the island's only landing beach. It's a bit like Mad Max with seagulls.

The best strategy for the trip is to arrive at Steep Holm just before low tide, and then make the short four kilometre crossing from Steep Holm to Flat Holm at slack water, which lasts for less than thirty minutes around here. Timed right, you'll arrive on the latter just as the tide is beginning to flood. We took a quick break here on a shingle beach that's only visible for around an hour either side of low water under the striking lighthouse on the southern tip, which is rather imaginatively called 'Lighthouse Point'.

Suitably refreshed, we headed back for England and our launching spot on Sand Point, the headland north of Weston Super-Mare. The return journey is slightly more complicated than the outward leg, as it involves a ferry-glide of roughly ten kilometres across the Bristol Channel's main tidal stream. After a while, though, we hit the swirls and bumps of the tidal rapid that flows off Sand Point, and arrived back at our launch spot on the beach to the north after a memorable seven hour quest into the deepest wilderness of the Severn Estuary.

Gavin Symonds, Dorka Fekete and I completed the trip out to Steep Holm and Flat Holm in early summer 2020, as the UK was emerging from the strange isolation of the first Covid lockdown. International travel was all but impossible; the government's revised guidance was to 'stay local'. If our island-hopping trip between England and Wales was an imaginative interpretation of that advice, so be it. Social distancing, after all, is quite straightforward in the middle of Britain's largest tidal estuary on a standup paddleboard.

I pieced together various sections of the English and Welsh coasts of the Bristol Channel during various trips, spread over several years, and I've now travelled the entire length of the Severn Estuary and the Bristol Channel by SUP, all the way from the upper tidal limit at Gloucester to the Exmoor Coast, where the estuary meets the Celtic Sea. The more time I've spent on these waters, the more I've discovered about them, and the more I've become intrigued by them.

Unexpected stuff always happens out here: the late September light refracting in white and gold over the Severn bridges as we approached from the north on the powerful ebb current; finding that magical landing-spot on an expanse of blue-grey sand the size of five football pitches in the middle of the Channel downstream of Lydney; the hard graft of battling a sudden crosswind turning the massive bend at Frampton-on-Severn; approaching the rapid upstream of Newnham surrounded by floating tree-trunks like prone primeval monsters; the strangeness of the tide flowing in both directions at the same time under the Severn Bridge at Beachley; and the eerie, endless slap and surge of that powerful current under the hull.

The vast expanse of this great estuary, like its counter-directional tides, is both a gateway to the past and a portal into the future. It's a shapeshifting, visionary space of extraordinary power and beauty. Some of the most interesting landscapes on Earth are actually those that have been significantly modified by humans, just like this one. The Nile Valley in Egypt is another example of this idea.

With the constant ebb and flood of the tide, the ghosts of the Bristol Channel's nautical history move across the same water that runs past large industrial plants, monumental bridges, nuclear power stations, wind turbines, and various other infrastructural components of an advanced twenty-first century economy.

Because the notion of adventure itself is a human concept, it is actually perfectly well adapted to spaces that have themselves been shaped and changed by humans over time – just like the Severn Estuary. True adventure doesn't require a wilderness completely devoid of human life, like the high mountains or the polar regions, to enact itself to the full.

In more than one sense, this estuary is a living representation of the adventurous mindset of human beings: the long voyages to unknown lands that once departed from here; the immigrants who first arrived here; and the adventures in technology, transport and energy that are now enacted here. By exploring this unique wilderness on its own terms, strange secrets will appear and vanish in its mud and silt: the lost channels of vanished rivers, abandoned jetties, shifting shoals, and the ghostly shadow of a medieval harbour overlooked by cranes assembling one of the world's biggest nuclear reactors.

And then the tide roars back in from the Celtic Sea to the west, covering the littoral land completely, making it seem for a few brief hours as if none of it had ever existed at all.

BASE magazine, 2021

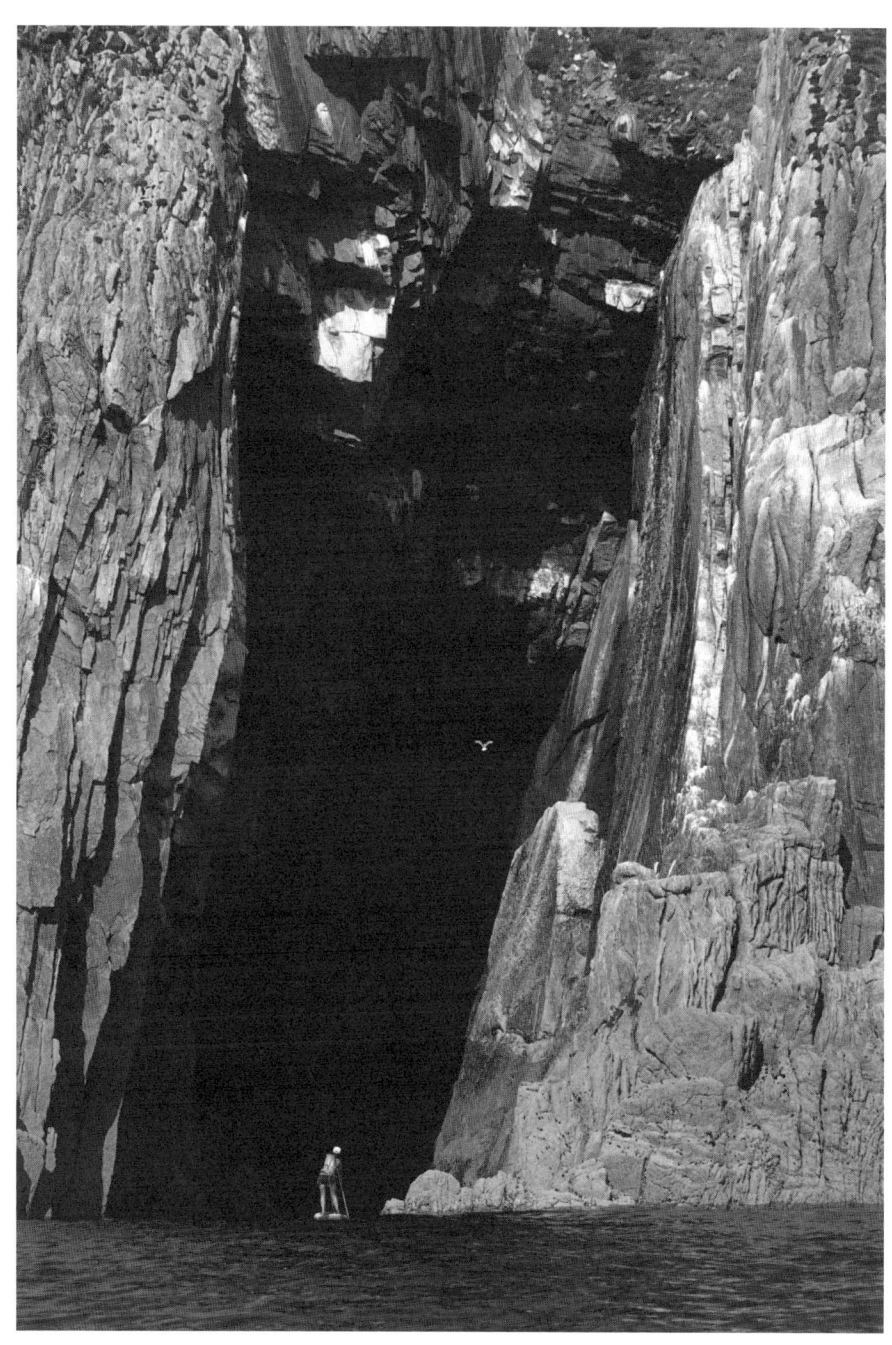

Dorka Fekete exploring the west coast of Lundy after the voyage out from North Devon in August 2022.

West by Northwest

The crossing to Lundy

We left at the crack of dawn. It's always the best time of day to set off on a proper mission; that heady mixture of coffee and bleary-eyed adrenaline fires you up every time. And today we needed to catch the early morning ebb tide to reach our destination: a small island almost twenty miles off the North Devon coast, marking the intersection of the Bristol Channel and the Celtic Sea.

The road northwest was empty and silent. Crows and buzzards circled over deadstill treetops under a blood-orange sky. There was barely a breath of wind: a big high pressure system had hovered stubbornly over Britain for the past few weeks, creating textbook-perfect open ocean paddling conditions.

The small cove at Lee Bay was mirror-smooth when we arrived. A few gulls floated lazily, like specs of dust on molten glass. Far out in the bay, the tell-tale lean on a lobster pot buoy confirmed what we already knew. The ebb tide had just started to flow, and would speed our progress westwards into this perfect summer's morning and the heart of the sea.

The idea of getting to Lundy on a standup paddleboard completely unsupported had been much discussed within the small team of enthusiasts I paddle with regularly. We knew that the island had been reached before on a paddleboard, by long distance paddler Fiona Quinn, on her Lands End to John O'Groats trip. Fiona – probably very wisely – used boat support for the sea-based section of her trip. As far as we knew, whilst the trip gets done regularly by experienced sea kayakers, nobody had paddled to Lundy unsupported by SUP before. In really good conditions, it would be very tempting to try.

Lundy is a remarkable little island. It was England's first statutory Marine Nature Reserve, and has its own endemic species of cabbage. It also sports an extremely colourful history involving plenty of piracy, Ottoman occupation, and a crackpot one-man monarchist movement by a chap called Martin Coles Harman in the 1920s, which featured a new coinage system organised in denominations of 'Puffin'; he was fined five pounds and fifteen guineas by the House of Lords, but he kept his spurious claim to being King of Lundy.

It's a full thirty-five kilometres, or just under twenty miles, from Lee Bay to the landing beach on Lundy's southeast point. It's a decent distance, and makes this a fully committing offshore trip; once you're being carried westwards out to sea in the tidal stream off Bull Point, which marks the western end of the Exmoor Coast, you're definitely not going back to England until the tide changes direction. At the same time, the strong tidal current certainly helps your progress west.

The paddle out to Lundy is a famous trip by sea kayak, and in good conditions it's a fairly straightforward paddle for the experienced. SUPs, however, are not only considerably slower than sea kayaks but they're also a great deal more affected by the wind. The additional leeway (the technical term for the degree to which the wind blows a vessel sideways through its forward progress) is several orders of magnitude higher on a SUP. We realised that a key factor in pulling this voyage off would be to take advantage of a day of very light wind, ideally less than five knots, and with minimal swell.

Another complicating factor is that the tidal stream doesn't flow exactly westwards towards Lundy from Lee Bay, but rather west-south-west. For this reason, you need to use a northerly tidal vector (a kind of correcting angle of course) on the initial ten kilometres to put yourself in the right position for the tide to 'drop' you down towards Lundy. If you plot the correct route on a chart, it looks like a gentle arc rather than a straight line. Adding any wind into this equation – other than a perfectly direct tailwind – would not be ideal. Too much wind from either the north or south might create a scenario where you could 'miss' the island completely. And that would probably not make for a good day out. Our friends Gavin and Jonny had made their voyage successfully two days before, and in slightly windier conditions than we found today. The plan was for my partner, Dorka, and I to rendevouz with them on Lundy's landing beach just before their return trip on the afternoon flood tide.

This particular morning, however, offered precisely those perfect conditions we'd sought; minimal swell and nothing but the faintest breath of wind from the east-northeast. After twenty minutes of rapid-fire sorting of boards and gear, we pushed off from Lee Bay's tiny cobblestone slipway just after eight o'clock, and without another boat in sight. Heading out into the tide to paddle to an island that's twenty miles offshore is a thrilling feeling, that's for sure.

As we passed the prominent lighthouse on Bull Point, which marks the western end of the Exmoor Coast, we caught the full force of the ebb tide running west. The addition of a lightweight sea kayak compass screwed to the camera mount on the bow of my SIC Maui board using an improvised GoPro attachment made the tidal vectoring for the next ten kilometres a whole lot easier. I just set the needle for ten degrees west-north-west and away we went, carried by the ebb stream into the heart of the Celtic Sea.

Lundy itself shimmered on the horizon dead ahead. It's a strange and wonderful thing, this little island: a three mile long slab of granite sliced off from mainland Britain and miraculously dropped in the middle of the Celtic Sea between North Devon and Pembrokeshire. A small pod of porpoises breached the water not far to the north, rising and falling in that effortless, arcing motion, coming in towards us for a while before heading further out to sea at speed.

After we'd left Bull Point well behind us to the east, the sun's heat was building; the wind that had ruffled the sea earlier had slackened right off. I checked our progress on my GPS: 18.2 kilometres. We were pretty much right at the mid-point of the crossing. A pair of gannets were diving for fish a hundred metres to starboard. A few oystercatchers floated on the sea not far to the north, breaking the silence of the sea with their distinctive trilling call as they took off. Lundy, meanwhile, was getting bigger; we'd easily be there in a couple of hours by my reckoning. The weather was perfect, but the wildness of the situation wasn't lost on either of us. Ten miles offshore in the middle of a strong tidal stream, we were in quite the location to be standing on a pair of fourteen foot race boards, and without any other vessels in sight.

The island itself, meanwhile, was expanding to fill the westward horizon; we were not far off now, approaching noon. A small cruise ship approached us heading east, having spent the night in the lee of Lundy's east coast. The slow line of its wake rose ahead, forming a sudden, momentary groundswell on an otherwise calm day.

We came in to the landing beach on exactly the right course, well to the north of Rat Island, a tiny islet off Lundy's southeast coast which creates a powerful back-eddy in the main tidal stream. This eddy actually pulls you south and east away from Lundy and back towards England if you get caught in it, as Gavin and Jonny experienced two days before, so it's best avoided. The boys were waiting for us on the beach as we came in, and there were smiles all round and even a couple of cold beers for the two of us on arrival as a bonus. We'd done it in four hours twenty minutes, which is a reasonable time even in a sea kayak. The perfect sea state and weather conditions were a big factor in the swift crossing. It also goes to show, though, that with a favourable sea and light winds, you really can do pretty much anything you can do in a sea kayak with a paddleboard.

The four of us rested on the beach for forty minutes at low water, waiting for the tide to turn for Gavin and Jonny's return trip to England. Dorka and I were then intending to paddle around the island itself that afternoon, adding another fifteen kilometres to the trip and completing what we thought might just about be the best paddling day trip in Britain. All too soon, we were leaving the beach as a full team of four, and heading out together into the tidal stream once more. Just off the point of Rat Island we said our goodbyes, wishing the boys a good trip back as we turned west under Lundy's South Light. The drama of the cliff-encrusted and cave-studded west coast now awaited us.

As soon as we turned the corner of Lundy's southwest point, the first members of what would be a very large host of grey seals appeared, following us as we eddy-hopped the tide with that beguiling, dog-like curiosity. The stream was running against us for this first section of the west coast. We cruised upstream using the back-eddies around the offshore rocks and the micro channels away from the main flow as our route; further out it would have been harder going against the full force of the tidal current.

The further north you go up the west coast of Lundy, the wilder and more dramatic the scenery becomes. I've climbed extensively on Lundy's cliffs and know the coast well, but to see it close-in from the sea creates an even more spectacular impression than observing the coast from the cliff top paths. Once Battery Point is passed – with its distinct Napoleonic micro-fort and gun emplacement atop a striking rock archway – things really start to get interesting. Our host of seals followed us with gusto into the wide bay that encases The Devil's Slide, Lundy's most famous cliff; a smooth slab of red and gold granite that falls from the cliff top straight into the sea for four hundred feet. Here, it was so calm that we actually swam right off the boards, without needing to land, in the narrow inlet just north of the Devil's Slide. Half a dozen seals were in the sea with us, their silver shadows darting and turning underwater. Rising and breaching the surface with that distinct, strong exhalation, they stared back at us with their wide eyes and spiky whiskers. What do they think of us when they observe us in this way, these wild creatures with such strangely human qualities? Who do they think we are?

It is no wonder that the myth of the Selkie – the half-seal half-human spirit – emerged in Celtic myth and folk tradition. For me, the seals' brief, curious, free-spirited companionship is always a gift; we take part in their existence when we meet them, as they do in ours. With no natural predators in English waters, the seals are truly the masters of their dominion here, and it is for this reason they greet us with such intrigue and camaraderie. There are very few wild animals with whom it is possible to have this sort of exchange in their native environment.

Heading north from the Devil's Slide, we picked up the full force of the early ebb tide running around North Light; a ferocious race forms here, and even today, in perfectly calm conditions, we could see the crests and breaking waves of a big tidal rapid just north of the point, only a hundred metres from the rocks. Just imagine what it would be like here in a fifty-knot winter gale.

All too soon, though, we were clearing North Light and entering the completely sheltered water of the east coast, and leaving the drama of Lundy's western cliffs behind. We were now approaching the forty-five kilometre mark, and both starting to feel it in our shoulders and legs. But despite the fatigue, the final section down the east coast felt like a victory lap; it's an easy cruise back to base from here, really, and the seals followed us most of the way. We made landfall back on the shingle beach just before six o'clock; we'd been paddling for over eight hours and on the water for over nine. And what a day it had been; just over fifty kilometres of standup paddling through some of Britain's most dramatic coastal scenery, including a long offshore crossing and a circumnavigation of a truly spectacular island. You'd be pretty hard pushed to find a better day of open water paddling in Britain than this one.

We'd planned to paddle back to England the next day, but the forecast changed and the high pressure started to break with an ominous thundery disturbance; huge clouds started to build over Exmoor and Hartland Point to the south the next morning, the wind started shifting and veering, and the visibility dropped right off. Being caught in a thunderstorm on a standup paddleboard ten miles offshore is not, by any stretch of the imagination, a very good idea. It definitely wasn't the best weather for the return leg, so we decided to catch the ferry back the following day. To complete one-off adventures like paddling out to Lundy and around the island in a day, it's all about choosing the right conditions, using the right kit, and never believing it's going to be okay if doesn't look okay. On this occasion, the sea was calm but the weather wasn't good, so it was the right decision to abandon the return trip.

In the end we didn't need to catch the ferry; we met up with a friendly yacht crew and hitch-hiked a lift back to England later that night on a small cruiser, *Cutty Wren*, on her way back to Weston-super-Mare. We departed our bivouac in the fisherman's cave by the slipway at 2 a.m. to paddle out into the bay and meet skipper Mark, Andrew and Alan, who welcomed us aboard with much-needed coffee as we made our getaway from the island on the early morning tide.

Coming into Ilfracombe at dawn the next morning, the sky was a vast, tungsten grey sheen; the high pressure had finally broken, and heavy weather was coming in. By the time we'd had a much-needed greasy spoon breakfast and returned to Lee Bay, the sea looked decidedly hostile as a strengthening breeze worked against the tide, turning the water off Bull Point into a frenzy of whitecaps. It couldn't have been more different than when we set off just two days before on that perfect summer's morning.

A standup paddleboard is an extreme type of craft to use for open ocean, offshore trips. It's also one of the most rewarding ways to travel across wild stretches of water. It can be both the very best and also the very worst vessel for open water travel, and these two things can sometimes become true on the same voyage. On that particular August day we paddled out to Lundy, though, there was nowhere else in the world I would have rather been than following the compass bearing west by northwest, bound for that shining strip of stone in the middle of the Celtic Sea.

www.base-mag.com, 2023

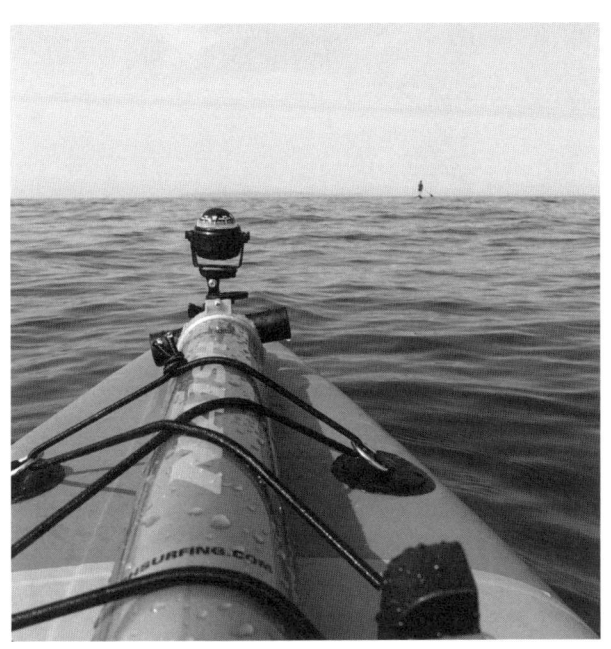

Seal's eye view of Lundy from the midway point of the crossing, with Dorka Fekete setting the pace ahead.

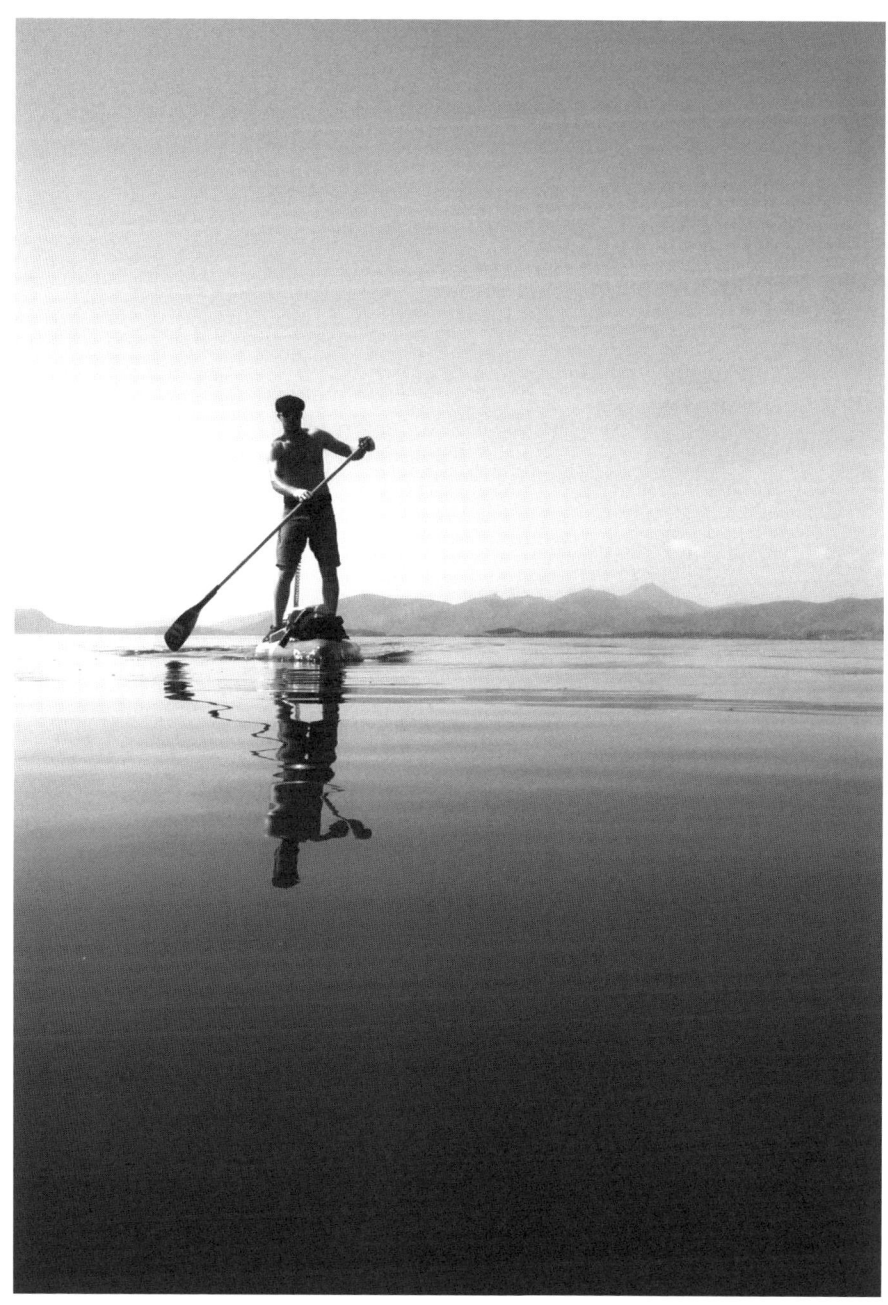

The author halfway between Skye and Harris during the crossing. The prominent mountain in the distance is An Cliseam, the highest point in the Outer Hebrides.

Stream of the Blue Men

Paddling from the Inner to the Outer Hebrides

The Northwest Highlands of Scotland are one of the very best places in the world to explore by sea. A yacht, sea kayak, or even a standup paddleboard are the ideal form of exploratory transport in this most spectacular of regions. In 2021, I made a short trip from Elgol on the southern part of the Isle of Skye across Loch Scavaig and back around the island of Soay, right below the awesome Black Cuillin, which was more than enough to get fired up for a more substantial aquatic quest in the area.

In good conditions, experienced sea kayakers often paddle across the Minch, the great sea lane that separates the Inner and Outer Hebrides. Its name in Gaelic, Sruth nam Fear Gorm, means 'Stream of the Blue Men'. In Scottish folklore, these mythical beings were reported to swim out from the Shiant islands to capsize passing ships, but could be thwarted by skilled pilots.

My partner, Dorka Fekete, suggested after our trip out to Lundy in 2022 that we should see if it was possible to do this crossing, too, by standup paddleboard. She's one of the few people I know capable of doing a trip like this by SUP, and also one of the only people I know who's intrepid enough to suggest it. It's about the same distance as the Lundy trip but considerably more difficult. This is because the tide doesn't flow with you, as with the Lundy crossing, but rather across the course you must take. You therefore need to use tidal vectors all the way, adjusting your course to take into account the constant lateral drift from the tide.

There are various ways of doing it, but we settled on the classic crossing from Camas Mor on the north coast of Skye to Tarbert on the Isle of Harris. This has the major advantage of being able to return directly on the CalMac ferry to Skye. Towards the end of a prolonged period of high pressure in the early summer of 2023, the sea conditions were as good as they get. We camped at Camas Mor after the long drive north from the south of England, and woke to an azure morning that was much more akin to the Aegean than the north of Scotland. The next couple of hours were preoccupied with sorting gear and getting the boards ready, along with the crucial preparation of a hearty breakfast. By eleven o'clock, we were finally ready, and pushed out into the Minch at low water from the shingle beach.

The islands of Fladda-chùain glimmered to the northwest, and in the distance we could just make out the dramatic profiles of Harris and Lewis against the western sky. In the anticyclonic haze, they looked very far away. It can be intimidating setting off on an offshore crossing like this for a destination you can barely see, but that also defines the beauty and full commitment of offshore paddling.

We passed the prominent rocks that lie about four kilometres offshore well to our south; we would follow a north-westerly tidal vector for the first section of the crossing, because our destination is due north-northwest but the flooding tide will be dragging us northeast for the next few hours. The stench of guano is overwhelming downwind of the rocks, which are home to a vast collection of seabirds. We see razorbills, kittiwakes, cormorants, and the first of a huge number of puffins floating on the water beside us. Not long after, a submerged bow-wave creates a strange swelling in the sea about a kilometre to the west, and the distinctive fin of a Vanguard-class nuclear submarine appears. From their base in Faslane on the Firth of Clyde, the Royal Navy use the Minch extensively for submarine training exercises, and the arrival of a Chinook helicopter above the vessel at the same time confirms that's exactly what's going on here today.

Time passes and we continue on north-westwards. The submarine and helicopter disappear, and a pair of dolphins breach the water to the north. We are surrounded by an unbelievable number of puffins, far more than I have ever previously seen at sea. By 2 p.m. we're pretty much right in the middle of the Minch, more than ten miles off the Isle of Skye and a similar distance from the east coast of Harris. On a day like this, it's really quite the location to be on a paddleboard.

Our cruising speed starts to decrease slightly as we approach the Isle of Scalpay, a small island which lies directly off the east coast of Harris. We're still following the northwesterly tidal vector we've been using all the way across, which has put us in exactly the right position on our approach to the Outer Hebrides. When we take a short break, Dorka notices on her Garmin GPS that our drift angle has changed: the shape of Scalpay, as I had suspected it might, is deflecting the tide hereabouts. The stream, therefore, is against us, and the final few kilometres on the approach to Scalpay are by far the hardest part of the trip, with the Shiant islands looming large in the distance. Would the Blue Men of the Minch swim out to scuttle us?

I ask Dorka what precise direction she thinks the stream is flowing in. Directly due northeast, she says. The tide is now taking us away from the Outer Hebrides and towards, er, the Shetland Islands, which are 400 kilometres away across the North Atlantic Ocean. Dorka has a PhD in theoretical probability, so I tell her she should set the final tidal vector for our approach, as getting it wrong is not an option. "Let's go 290 degrees west-northwest, that should do it" she replies. And that way we go, paddling into the falling sun.

The bright red-and-white striped Eilean Glas lighthouse on Scalpay's eastern point stays to our north, although the tide is still trying to drag us eastwards and out to sea on the final approach. We decide to make landfall in a small cove on the south side of Scalpay to have a proper rest. As we pull the boards up on the barnacle-encrusted rocks, I realise we've just crossed the Stream of the Blue Men and made it to the Outer Hebrides. It's a great feeling. A pair of Arctic terns swoop in to greet us, their distinctive hooked wings sketching shapes in the evening air.

After a much-needed break, we're back on the water and weaving through the sheltered channels and maze of tiny islets and islands off the southwest coast of Scalpay. Finally, our destination appears in the gold light of evening, and we land on the floating pontoon of the Tarbert marina just before eight o'clock after almost 40 kilometres of paddling. Heading up the hill to the Harris Hotel for a well-earned pint, I reflect on what's perhaps been the best offshore trip I've yet done on a paddleboard. In the midsummer evening light we can just make out the shape of Skye in the distance, like a long mirage across that tide-shifting sea.

www.monographmedia.com, 2023

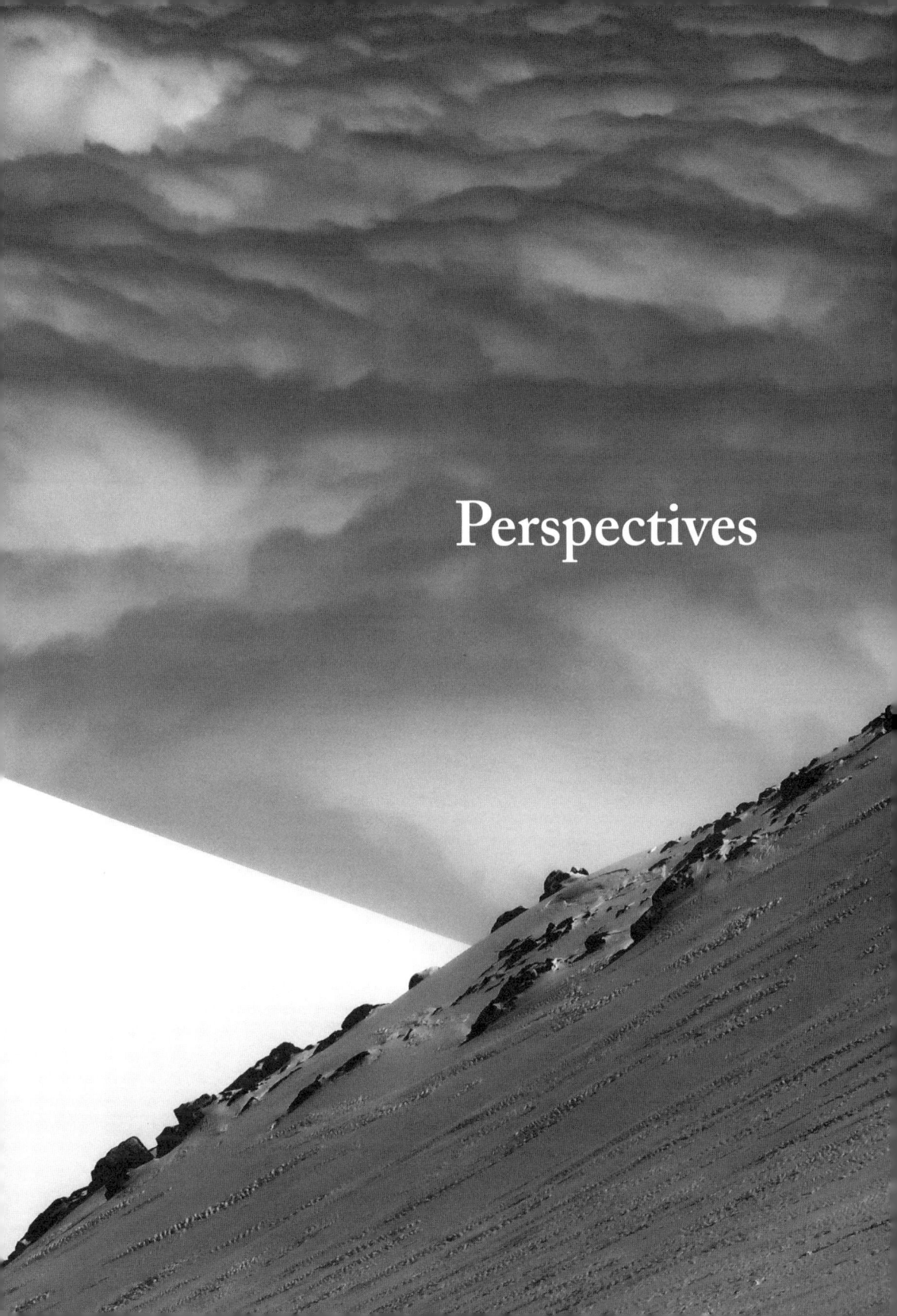

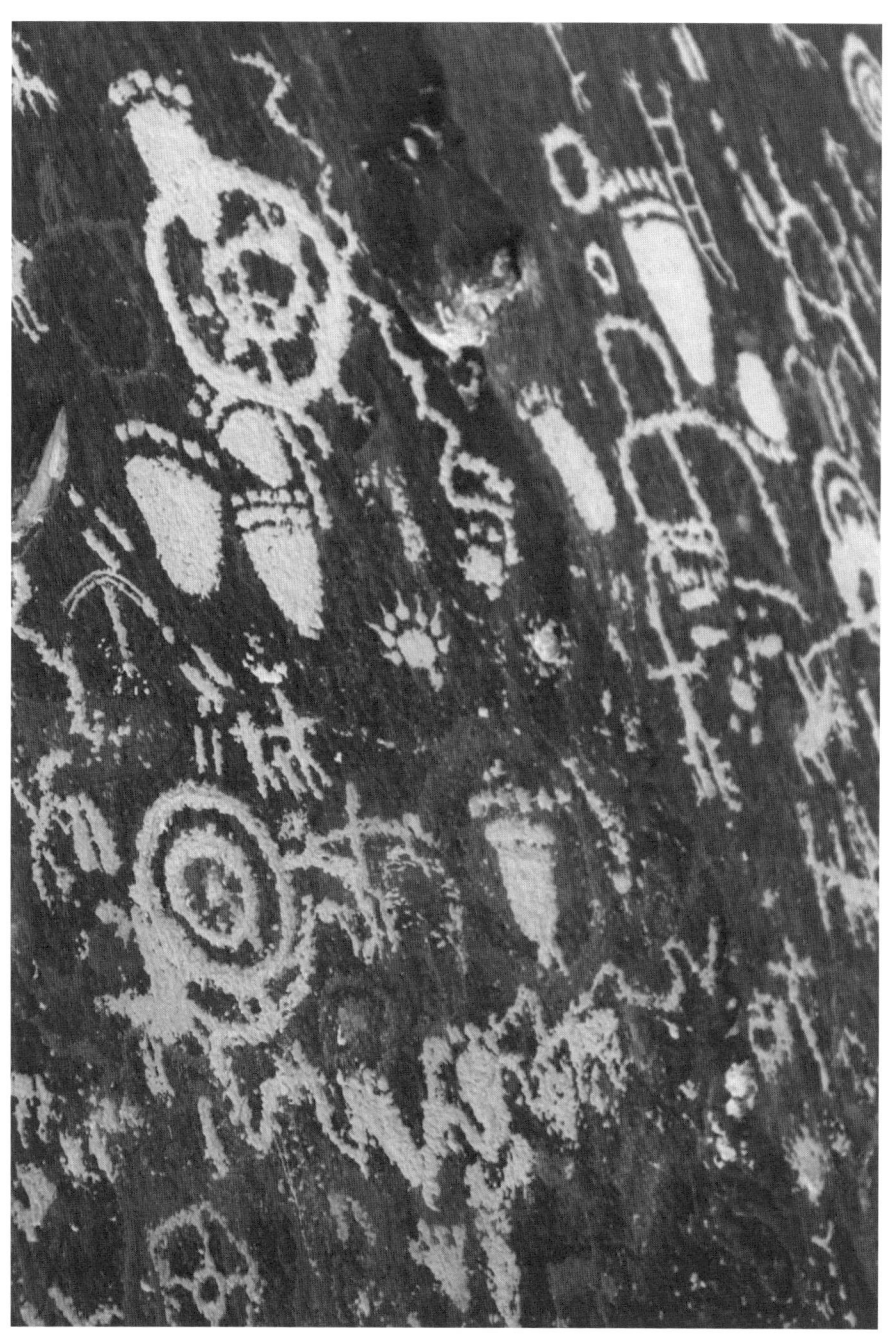

Petroglyphs on Newspaper Rock, Canyonlands, Utah.

The Painter in the Cave

Rock art, climbing, & journeys of imagination

In the evenings, sometimes, I think of the past.

Passing the sign for Lascaux, I lifted my right foot off the accelerator, drawn to the inside lane by instinct. I was driving north through southwest France on my way home to England after a long climbing trip in Spain. I filtered off the autoroute and followed winding roads through the autumn landscape of the Dordogne, all reds and greens crossed with silver blades of reflected light where floodwater striped the fields.

After a while, I approached the cave entrance. There were only a few people around, and the last colour was beginning to drain from the sky. I stopped and cut the engine. The exhaust ticked as it cooled, like a slowing clock.

Almost directly beneath my feet, in the Vézère limestone, lay one of the oldest and most elaborate collections of Paleolithic paintings ever discovered. In the summer of 1940, when Hitler's army occupied France, French teenager Marcel Ravidat found a hole in the ground whilst out walking his dog. Later, with three of his friends, Ravidat slid down the entrance and landed on the dark floor of a cavern. Venturing deeper, they found the walls covered in 2,000 images of animals, human figures and abstract symbols made by the people who lived in this region more than 17,000 years ago. Nobody had entered the cave since they left and Ravidat arrived.

His discovery was like Howard Carter finding Tutankhamun's tomb: a total one-off.

Who were they, those distant artists with their brushes? What did they think about their world and what they experienced there? And, given the harsh conditions of their existence, why did they create these elaborate, detailed paintings on the fire-lit walls of this deep cave? Modern scholars have suggested that the cave painters lived in a complex civilisation with an extensive mythology that grew from their encounters with a teeming natural world (Gregory Curtis, *The Cave Painters*, 2008). In the Hall of the Bulls, one of the centerpieces of Lascaux, astonishingly accurate representations of aurochsen, the wild ancestors of cattle, dance and circle across the walls. Tossing their heads, the painted animals cavort in a rising vortex toward the narrowing of the cave ceiling. To me, these painted creatures seem like messengers of lost stories that are profoundly part of us, in ways we cannot completely understand. And as a climber, I'm drawn by the connection I feel between what we do on rocks and what the Paleolithic artists painted on the very same stones.

Standing near the cave entrance, I thought about the scarcely comprehensible gap between me and the Stone Age people who lived here when these paintings were created. The English philosopher John Gray has written of how "the rarest type of questioning intelligence [is] one that does not aim for solutions to problems but shows that looking for solutions is often the problem." As the chill of evening began to fall, I remembered Gray's observation, and I wondered what kind of sublime magic took place in the minds of the first people to apply the first dash of paint on the cave walls. Had they ever seen a painting before? How did their companions respond to these paintings when they encountered them? Did the paintings change the societies that produced them? It occurred to me that the mysterious process that draws people to climbing might be in some respects quite similar to the process that led those Paleolithic artists to paint the walls of Lascaux, like moths fluttering toward a source of light, searching for something they urgently desire without necessarily knowing the reasons why.

Long before humans began to inscribe the earth with the physical narrative of our presence, in the form of dwellings, boundaries and roads, we were making things that would, much later, be identified as art. This process took place across the entire Paleolithic world, from the limestone caverns of Europe to the desert escarpments of Central Asia. Using the stone walls of underground chambers as their canvas, our ancestors painted elaborate images that look like hunting excursions, dance rituals or star maps.

Although the specific reasons they put their makeshift brushes to stone canvases remain disputed in the archeological community, they seem to have been using the rocks to reflect important aspects of their existence, to make powerful representations of what they understood and believed in, and to comprehend better their relationship with the world. To a climber, this might sound strikingly familiar.

Like that moth clattering against the shadows, intoxicated by the radiance beyond, many climbers are also seekers of infinite worlds, perhaps the same ones that Stone Age painters apprehended in the visionary dark in which they worked.

At the base of the climb *Pink Flamingo,* a soaring thin-fingers crack at Indian Creek, Utah, a small image is carved into the cliff. It resembles a stalk-like bird, giving the route its name. Many years back, on a windswept April morning, my partner and I walked up to the base, keen to try the climb. I'd heard of its identifying flamingo; it was created, perhaps, by a solitary Ute hunter as he rested under the cliff while a squall swept through the canyon. I had no idea of just how close this delicate, almost ethereal engraving was to the dark seam edged with white chalk like a burst winter drainpipe's sidelong ribbons of ice.

Two other climbers were ahead of us. One of them started up the first moves, deftly and deliberately jumping his right foot about ten inches past the petroglyph, leaving it completely untouched and at peace. Later, as a fingerlock momentarily slipped, he fell at the crux. I could see that he and his partner would be engaged on the route for a while. I had a plane to catch in a few hours, and I was eager to climb, so I left them to it: two figures dancing on that indigo wall as dust devils grew like brushfire from the boulders below.

Over the following months, I thought a lot about Indian Creek's pink flamingo and the climb that bears its name. I thought about the two climbers on the route that wild April day, about my own desire to climb the line, and my need to feel an invisible bond with the unknown creator of that tiny, stone-winged bird.

Climbers and mountaineers also mark rock walls with our passage in different ways: some traces are remarkably subtle, like a piton in a niche, scarcely visible until just a few metres away. Some are obvious even to a distant observer with no knowledge of climbing, like the brilliant flashes of reflected light from tiny expansion bolts. Others shimmer with more ephemeral patterns: the white scattergraphs of chalk left on cliffs of colored stone; the translucent, splintered stars of axe placements in flows of blue ice. These are the fleeting ideograms of the vertical world, each with its own history, memory and meaning. We strive to make sense and subtext out of these transitory signs in stories, photographs and films.

If climbing can represent something important about the human condition, then the traces we leave on the places we climb are surely its physical artifacts. At their worst, these marks can be degrading effigies of ambition, like the stumps of trees cut down to ease access to routes. At their best, they can be delicate and profound symbols of the great things that humans can make from the viscera of the world. The link, then, between approaching a rock with a view to climbing it and approaching it with a view to creating art on it seems at once obvious and also perplexing. Since the late nineteenth century, initially in Europe and North America, rock climbers have begun revisiting some of the cliffs and remote caves where Stone Age people once lived. Today, a great many climbing areas contain or are extremely close to sites of important prehistoric rock art. This physical proximity can create its own tensions.

The vast open cave of Cova Gran at Santa Linya, Catalonia, is now known for some of the world's hardest sport climbs, but it was once the home of prehistoric humans: remains have been discovered in the cave from both the Paleolithic and Neolithic periods. The southeast Pyrenees was then an important, densely populated place, possibly inhabited by both Neanderthals and Homosapiens. There is no prehistoric rock art in Cova Gran itself, but when the relentless summer heat fires the walls of the cave and bakes the dust white on the ground, archaeologists come here to try to reassemble stories from 20,000 years ago. Then, in the coolness of winter, rock climbers return to piece together their own narratives, configuring their own patterns on this incredibly steep, compact and golden limestone.

From the top of the route *La Mare del Tano* in the autumn of 2012, I looked across the cave floor below. A dozen or so men and women congregated in the dust: some of them were sitting on boulders, resting, contemplating. Others stood talking in small groups. A woman was stretching her limbs. At the far side of the cave, a man was preparing a small fire to warm his hands. It was a scene, I realised, strikingly similar to that which you might have seen in Cova Gran 10,000, even 20,000 years ago. All you'd need to do is to replace the brightly colored duvet jackets, ropes and quickdraws with animal skins, basic wooden utensils and stone tools.

In some ways, modern scholars' inability to find a single, absolute and definitive key to interpreting cave art represents the gap between what we imagine and what's really there, just as climbing ambitions reflect the rift between dreams and reality, and between what's written in guidebooks and the real vertical terrain we encounter. In *The Cave Painters*, Gregory Curtis writes of how "all that the Paleolithic people preserved by word of mouth – all the poems, songs, languages, customs and social order – is lost and cannot be recovered. It is possible that fragments remain in our own ancient myths, but we can never know for sure." So many of the climbing stories that have been told and preserved for a time by word of mouth have since disappeared with the deaths of individuals or the fragmentations of specific cultures or social groups. The absence left by vanished history is an enduring presence, too, in the climbing world.

In February 2014, I wandered around the alpinists' graveyard in Argentière, France. And in the falling light of a winter's day, the reality of this loss grew as stark as the black and white skyline of the Dru 8,000 feet above. Most of the dark granite headstones were covered in drifted snow, both obscuring and accentuating their inscriptions. All those silenced stories that lie buried with the dead, I realised, speak to us in the way that the rock art of the past does, communicating something of an irrecoverable, extraordinary world, bodied forth only in engraved markings on certain, significant stones.

It seems likely that some Paleolithic rock art was produced as part of ritual activities, such as dances, feasts and coming-of-age ceremonies. Images appear that look like a shaman or holy person, sometimes distinguished by their wearing of a fringed garment. In rock art from Altai and Mongolia, figures depicted in similar garments have been associated with flight. Both dancing and flying have clear corollaries in climbing. The act of climbing can be a form of performance art, just as choreographed dance or parkour are (parkour is an urban adventure sport involving running over buildings). One of the greatest French rock climbers of all time, Antoine le Menestrel, was a dancer by profession. In his heyday in the mid-1980s, he moved on rock with the quickness and keen grace of the first flame on dry tinder.

The innovative British climber Johnny Dawes once told me that he thought "some very interesting things could happen in the gap between climbing and parkour in the future." When I mentioned the flying shamans of Mongolian rock art, he looked out of the window into the abstract grey of a Sheffield street, and an enraptured smile appeared on his face as if he were reconfiguring some lost supernatural trick that someone had once understood, sitting under a rock somewhere on the Steppes of Asia a long time ago.

The search for new routes has been one of the defining features of my climbing career. The limestone sea cliffs of Pembrokeshire are just a couple of hours from my home. Some of my first ascents there are among my most memorable experiences in climbing. At the heart of exploratory climbing is a visceral creative engagement with the world; a search for natural openings, hidden sequences and patterns, for intuitive routes through challenging terrain. In this sense, our desire to make first ascents is a representation of the curiosity, ingenuity and ambition that have driven humans to become so successful as a species. In a parallel context, those Paleolithic artists who first raised their brushes of fibre or hair and their blotters of plant matter were representing the scarcely imaginable drama of the beginning of that process, embodied in the light that seems to spark from the animals that dance across the walls.

Climbing can seem like a magical act: a kind of transformative elixir. In more than three decades of the pursuit, I have known metamorphoses of many kinds: transformations in desire and ambition, in capability and commitment, and in my understanding of the vast range of meanings that the act of climbing itself can have.

In the autumn of 2013, halfway through the crux section of *Archimedes Principle*, one of the best climbs in the Grampians, my left foot picked up a fragment of debris and slipped slightly on a smear. The route follows a gunmetal water streak up a huge, smooth cliff of iridescent, blood-orange sandstone hidden in Australia's remote western Victoria Range. It is the only line of weakness up this part of the Eureka Wall, and I was climbing it in a single 160-foot pitch, using a lightweight rope. With a sudden, instinctive urgency, the fingers of my left hand closed on a first-joint crimp. The last gear was twenty-five feet below. Consciously, I knew I was in no real danger: the gear was solid, the ground was far away, and there was nothing to hit. I took a deep breath, shook off the panic and continued climbing, exhaling hard, listening to the garrulous laughter of the kookaburras in the topmost branches of the highest gum trees. I finished the route with lighthearted abandon, lifted by the birdsong as I swung out along the flying break to join a giant arête.

At the end of the day, I followed my climbing partner, Ramon Marin, up the forked-lightning hairline fissure of *Return to Gariwerd* on the Eureka Tower. On the apex of a narrow ridge of crenellated sandstone, we looked out across the Victoria Range: two tiny humans alone in a wildness of tangled eucalyptus and colossal skies.

My eye traced the outer limits of the landscape, its contours falling into vanishing creeks and rising to meet the battlements of interconnected ridges. We might have been the only people, perhaps, in forty square miles. Until the first European climbers came here a few decades ago, the sole travellers through these high ridges of the Grampians were Australian Aboriginal people, who sometimes painted the rock walls of places with special cultural significance. Like a transect through thousands of years of human history, modern climbing tales collide with the much older histories of the people who once lived here. Could climbing bring me closer to what those earlier inhabitants knew, thought and felt about this place?

In the shadow of the Taipan Wall in Australia lies the Gulgurn Manja Shelter (meaning "Hands of Young People"). A narrow trail winds through dense thickets of eucalyptus to the low rise of a small sandstone escarpment that looks out across the vast, shimmering green of the Wimmera. Breaking through the bush, I came to a shallow cave, its roof blackened by centuries of fire smoke. At the back, unmistakably, were the prints of small human hands – those of children or young teenagers – painted in red ochre. On the right, there were delicate representations of emu footprints. Emus were one of the staple sources of protein for Aboriginal people, as well as a symbol of their connections to the environment. One link between those little handprints and the prints of the emus' feet seems striking: images of the food that makes children strong alongside the handprints of the young themselves as they grow from childhood to adulthood. Near Gulgurn Manja lies Ngamadjidj Shelter, where groups of white figures are painted on the low roof of a cave. Almost nothing is known publicly of the meaning of these paintings. Perhaps they represent early encounters between the Jardwadjali people who lived in this region for tens of thousands of years (and some of whose descendants live on the Wimmera today) and the first European settlers who arrived in the nineteenth century, or perhaps they have another meaning that the the Jardwadjali haven't shared with outsiders. We just don't know.

Not until recently were some of the Aboriginal names restored to the rock art in the Grampians, which the Jardwadjali called "Gariwerd." And not until 2005 did the Jardwadjali win recognition of their native title in the Wimmera. In a sense, the arrival of climbers in the region is part of the wider story of the imposition of a Western culture in Australia that has largely usurped Aboriginal ways of life. In another sense, it may be through pursuits like climbing that visitors can start to reconnect with some of the intimacy the original inhabitants had with the surroundings and appreciate the true extent of their environmental knowledge.

I personally would not have seen these extraordinary Aboriginal paintings if it were not for my interest in climbing.

A difficult and complicating aspect of this conundrum is that today, at the time of writing, Parks Victoria has effectively banned climbing on most cliffs in the Grampians, including those described here, even though they are home to some of the very best sandstone climbs in the world. The justification for the restrictions is that climbing might damage the Aboriginal rock art, and also that climbing itself is an intrusion and a form of trespass on traditional Aboriginal lands. Most people in the Australian (and global) climbing community don't agree that these are valid grounds for preventing access to some of the world's best sandstone.

If there is a modern metaphor in the strange proximity of those pallid, spectral figures on the walls of Ngamadjidj and the white lines of climbers' chalk on the glowing escarpment far above it, it is that climbing is indeed a route back into our ancient human past. Through climbing, perhaps, we might learn about the lost stories of people like the Jardwadjali. I don't agree with the idea that climbing, or anything that takes places in areas where indigenous people once lived, is some kind of colonial act. This is a currently fashionable idea, and might have led Parks Victoria to the climbing ban. I personally believe they were profoundly wrong in their decision to effectively close a large part of one of the world's best climbing areas. However there are many crags around the world which have stories of complex access, and I feel that in the case of climbing in the Victoria Range the future may be a rather better place than the present.

I left the cave that day in October 2014 and walked back up to the plateau, the crenellated sandstone glowing green and brown in the austral sun. Had I not been a climber, I would not have come to this quiet and sacred place in the middle of nowhere, just as I might never have turned off the autoroute to visit Lascaux four years before. My interest in climbing – an interest in what people can do on rocks – took me to these caves, revealing secret layers of the past I had not known were there.

Mystery, complexity, beauty, sorrow, enigma: all these, all at once. Is it a wild fantasy to imagine that climbing at its best, like the rock art that early humans painted on the walls of their caves, might offer us not just a phenomenological link to the past, but also a brief escape from the inexorable passage of time?

I scrambled through the bush to a point directly above the cave paintings, and I looked back up toward the incandescent orange slice of Taipan Wall, gleaming and magnificent, where I'd been climbing the day before. Through my camera's powerful telephoto lens, I could pick out the distinct white trails of chalk on some of its signature climbs in the strong afternoon light: *Anaconda, Venom, Snake Flake, Serpentine.* There they all were, a collection of tall, meandering sketches in white brushstrokes on a blood-orange surface. Rock art, perhaps, of a kind.

Nietzsche once said that "art is not an imitation of nature, but its metaphysical substitute, raised up beside it in order to overcome it." As I turned to leave Gulgurn Manja under the bright heat of the Australian sun, the imprints of so many hands on the cliff above and in the cave beneath, visible and invisible, seemed to mingle in the thickening air.

A light wind picked up, blowing whorls of dust around on the flat sandstone beneath my feet. At that moment, I felt the overwhelming absence of what was not here anymore on the Wimmera, of the innumerable stories of people who lived here tens of thousands of years ago. At the same time, I sensed the presence of climbers high above me on Taipan Wall, making shapes and patterns distinct in place and time; artworks that are now transported into the future, too, like a gift.

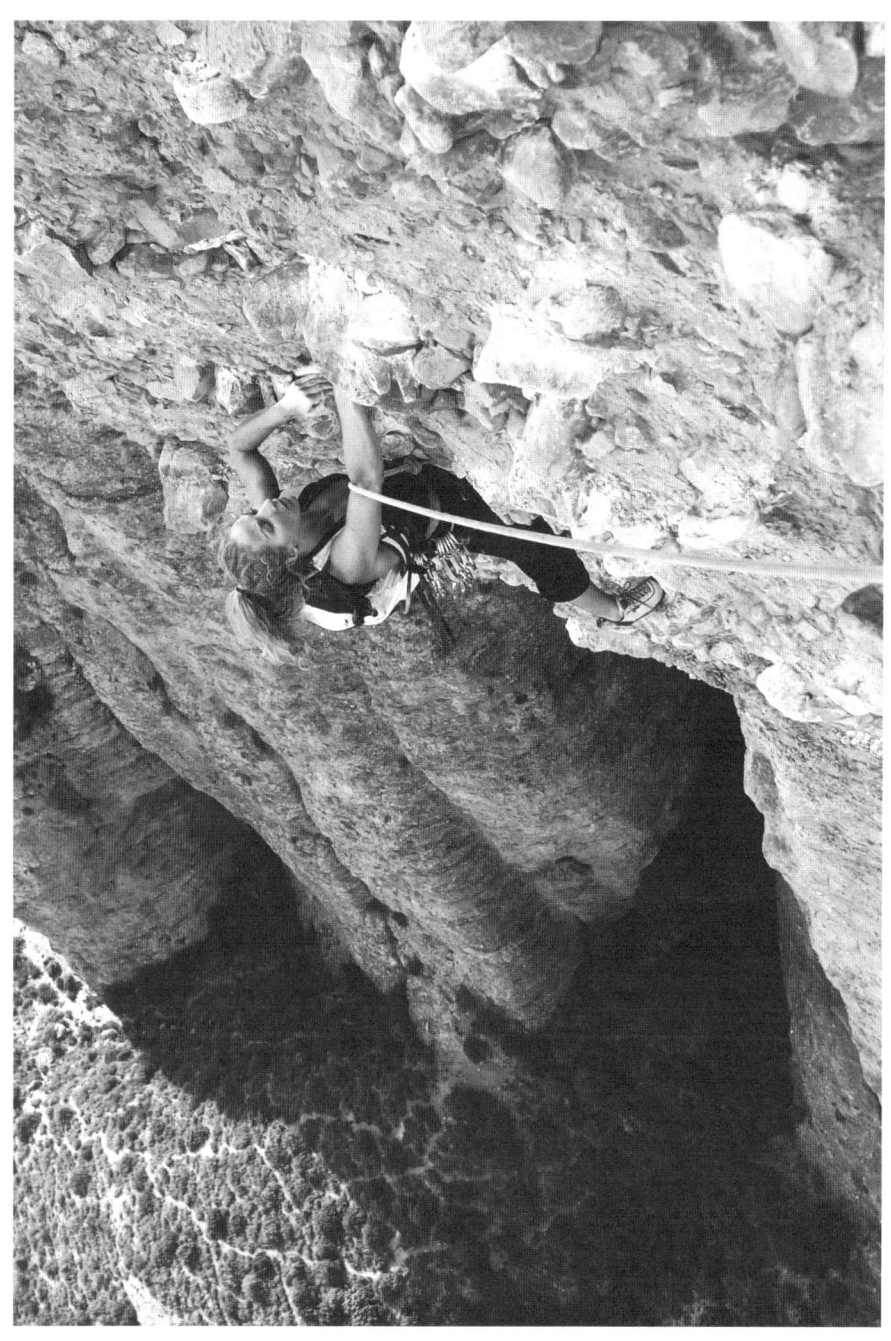

Madeline Cope high on the mega-line of Fiesta de los Biceps (7a, 240m) at Mallos de Riglos, Aragon, Spain.

Rituals of Faith

The spiritual value of adventure in a secular age

A chilly, gusty October evening in northern Spain's Sierra de Guara mountains.

I'm chopping an onion with a blunt clasp knife by the light of my headtorch. A small group of climbers from North America and Australia surrounds me. They're huddled in down jackets, sitting on backpacks, and drinking local tempranillo to take the edge off the cold. Bright eyes flash under dark hoods; stories are shared from rocks around the world. The Yanks, Chris Weidner and Alex Honnold, have just flown in from a rain-sodden trip to Siberia, where they were guests of the Russian Climbing Federation. I'd met the itinerant Aussies a few weeks before; the Americans showed up just two days ago. It's a familiar scene in climbing culture; a group of people from across the world meet up, either as the result of a loose plan or by chance, somewhere way off the beaten track. Their different lives, different nationalities, and different cultures are instantly united by a shared passion and lifestyle.

I've assumed the role of team chef on a lot of climbing trips. Over the years I've probably cooked enough al fresco *pasta arabbiatta* to feed a small battalion, not out of any particular sense of duty, but because I enjoy cooking. You're likely to be hungry after a long day in the vertical, and even basic food will taste great. But there's something else, I think, that makes campfire cooking uniquely rewarding.

The process of making a nutritious meal in a wild place with basic equipment, and the shared, social act of eating it with your companions, underscores something important about the value of adventure. Climbing, like cooking, can be pursued for pure pleasure. But as soon as you start to try hard out there, whether it's on a two hour bouldering session or a two month expedition, climbing involves us in a ritual experience of great value and importance. The evidence for this can be clearly seen in the boulderer carefully brushing holds and chalking up before stepping on the rock; the alpinist calmly filing her axe picks ready for a pre-dawn departure; the sport climber shaking out near the apex of a limestone cave, controlling his breathing in preparation for the crux to come. Through all these seemingly simple acts, there's an undercurrent of ceremony and vigil.

Along with such ritual patterns, it is strikingly obvious how the old rhythm of life – rising with the light and going to sleep not so long after dark – resumes as soon you're living outdoors. At such times, the relationship between climbing and hunting begins to resolve itself: rising, preparing, setting out, pursuing, confronting danger, and, if you're lucky, coming home with something.

Adventure sports are not just about sporting achievement; they're also a link with our past, with our primal existence, and an affirmation of what it means to be human; a modern interpretation of the pursuit of sacred experience that is part of almost all traditional cultures. In Western society today, it is possible to live entirely within constructed spaces, within a virtual, digitalised experience defined by consumption and artifice. But there is no possibility for ritual within this environment. Adventure sports offer a way of creating new rituals in the modern world, using new technology.

Since humans are social animals, a large part of the value of ritual experience comes not just from what the ritual itself is, but also from the fact it is shared with others within a similar belief system. In an invigorating blog about living with a Dallas-based religious cult for a year when she was twenty-one, the writer Mary Wakefield apprehends what might be important about living a shared existence guided by belief: "I'd had a terrific year, I decided. I wasn't so daft as to call myself a Christian, I thought, but I had caught a glimpse of what it's like not to spend your life squirreling away treasure, but to live your faith."

An insight like Mary Wakefield's could have emerged from a month within the nomadic caravanserai of the global climbing community just as easily as it did from the pseudo-religious personality cult in which Wakefield ensconced herself in Texas in the late 1990s. I think I know which I'd rather choose, but that's a personal decision at the end of the day, and I admire Wakefield for doing something I most certainly wouldn't be able to do myself.

I clearly remember a minor revelation I had on waking up just after dawn after a bivouac in the mountains of Oman's Musandam Peninsula in February 2005. I was in my mid-twenties – that most romantic and energetic time of life – and on a two week exploratory climbing trip with a good friend after a long stint of work. As the sun rose over the limestone towers to the east, it occurred to me that I needed to live the climbing life and also an exploratory life for every second I was able to; a life true to my closest-held beliefs. Through the ritual of climbing, therefore, I was able to practice my faith.

The writer Richard Dawkins has argued extensively in his hotly-debated book *The God Delusion* that religious people are in various respects either misled or just plain wrong in their belief in God, or any kind of deity. I don't agree with him. Even committed atheists, like myself and Dawkins, need to find spiritual meaning in the world beyond the prism of our own lives. Dawkins doubtless gets this – maybe quite a lot of it – from his huge fame as a celebrity scientist and writer. I get it more modestly from my passion for adventure sports and exploration in all their myriad forms.

If I have understood anything else about climbing in the years since that desert bivouac in Oman, then it is that the ritual and romance of adventure needs what Barry Lopez calls "a pattern of grace" to sustain itself. This might simply mean believing what you're engaging in has value, both to yourself and to others, and then carrying that belief through everything else you do. Even in a post-religious era, human beings still need faith to sustain themselves. It's possible that in the digital hyper-reality of Western life, we may need even more of it now than ever before.

Faith requires ritual, and climbing and adventure sports just happen to have such informal, spontaneous ceremonies in abundance: waking up on a crystal morning after a desert bivouac, or simply cooking something up over a camp stove for your hungry companions at the end of the day.

Climb magazine, 2016

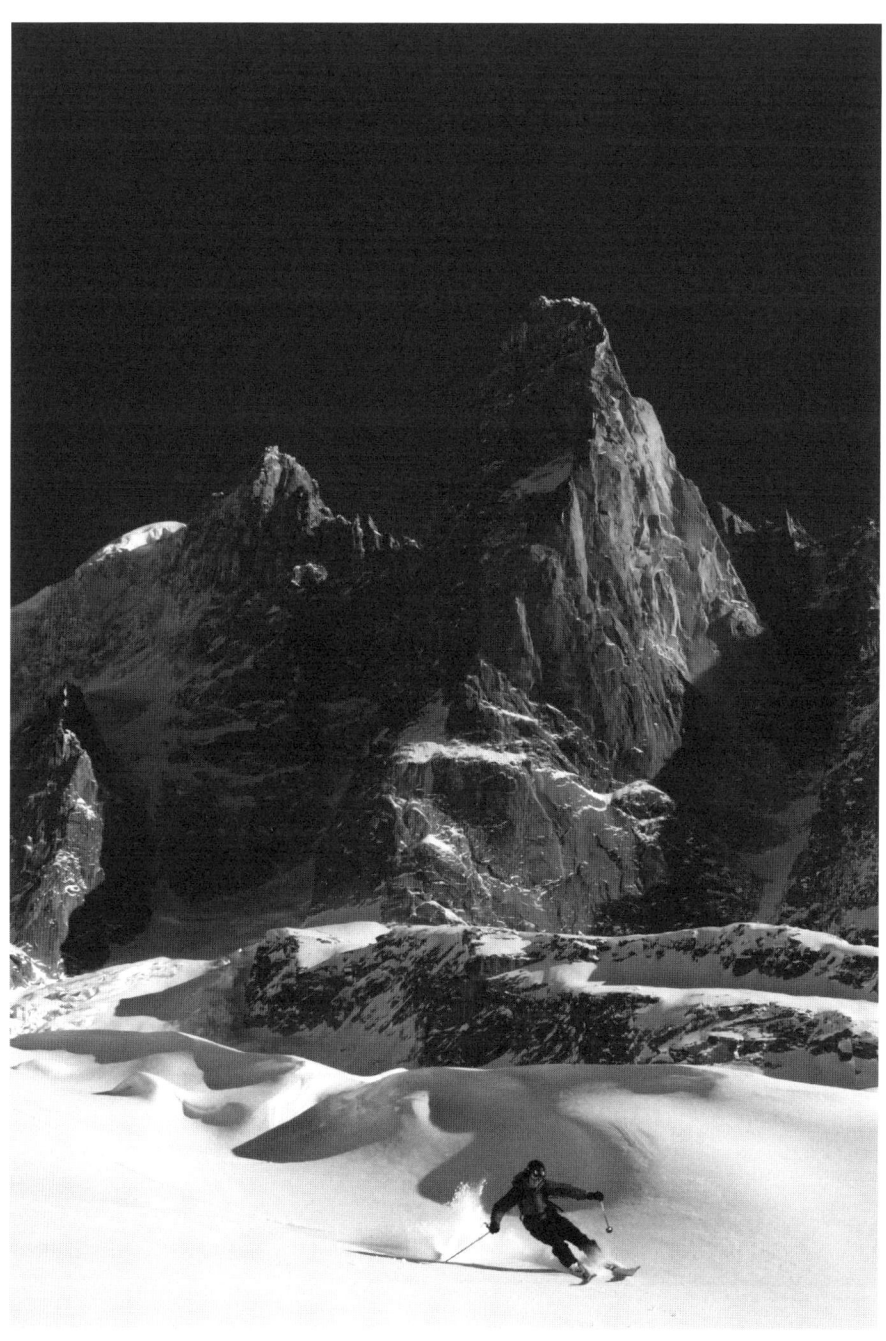
Skiing deep powder on the Pas de Chèvre under the Aiguilles du Dru, Mont Blanc massif, Chamonix, France.

The Hunter in the Mind

On environmental intelligence & risk awareness

Once in a while, when I'm heading out, I think of the hunters.

They're somewhere in Africa a long time ago. They're tracking a big animal, or perhaps several animals. They're working as a small group. Constantly aware of the danger inherent in what they're doing, they never let their attention stray from the task in hand. Their environmental intelligence is highly tuned; they're always looking for signals in the landscape. Tracks in a dry river bed, an area of displaced grass, a shadow moving across the plain; all could signal prey or predator, opportunity or danger. Since their lives involve frequent contact with extreme risk, what keeps them safe above all else is one of the oldest and most powerful of behavioural paradigms: fear.

From an evolutionary perspective, fear is the means by which all animals learn to predict potentially harmful events. And just as it was crucial to the survival and success of Palaeolithic people, it is also vital to the safety and success of any modern adventurer or explorer. Whatever you're doing out there, fear can be one of the sharpest tools in your survival kit. Channeling the mindset of those unknown hunters somewhere in Africa in the earliest days of the human experience can be extremely useful. In simple terms, this means being constantly alert to changes and signs in the environment, and responding to them quickly and effectively. The following story may explain why it's a good idea to remain constantly alert to those changes and signs we perceive in the world around us.

In 2009, I made a serious but fortunately escapable mistake whilst off-piste skiing in a remote part of the Ecrins massif in the western Alps. It had snowed heavily for two days as a cold front pushed southeast across Europe, pulling Arctic air and lots of precipitation with it: a dream combination for any big-mountain skier. My friend Giles and I were making the most of the excellent powder. First, we did a few tree runs, our skis frequently vanishing in a deep layer of fresh snow as light as champagne froth. Once we'd warmed up, I suggested that we should ski the wide bowl just to the north of the forest, as the snow would be even deeper there, away from the shelter of the trees: powder skiing Nirvana!

So off we went without much further discussion, taking a high line along the top of the bowl. We dropped in at a tiny V-shaped couloir right at the apex, and enjoyed almost four hundred metres of descent in some of the deepest powder either of us had ever skied. On reaching the base of the bowl, we noticed through the swirling cloud that there were several big cracks in the surface layer of snow on both sides of the upper slopes. Because of the topography of the mountain, any avalanche triggered high up would wipe out the whole lower part of the bowl. We were very obviously in an extremely dangerous position. Giles, who's a very experienced mountaineer, glanced up at the bowl before he turned back to me and said with characteristic Englishness: "Spot of bother here. We've just skied into a death trap".

I think we both felt the classic fear conditioning response of goose bumps (electrodermal activity) and a raised pulse as we realised just how high the avalanche risk was: the same feeling that our Palaeolithic hunter would have frequently had out there on the African plain. It was obvious that we needed to get the hell out of the bowl, and get out quickly. Without another word we traversed as fast as possible skier's right to reach the relative safety of the forest, where we stopped for a short break. Less than an hour later, as we were skiing safely in the trees, a menacing roar echoed across from the bowl, like the sound of a freight train emerging from a tunnel. A large avalanche had swept down the entire thing, easily big enough to bury a pair of skiers. A rapid response to imminent risk is often key in these situations for a successful outcome to prevail. I haven't made a mistake like that in the mountains since.

We went somewhere we simply shouldn't have gone given the conditions, but we got away with it. Fear was the key, though, to getting us out of that dangerous position – and out quickly. But it's also worth asking why we chose to ski that bowl in those conditions in the first place. I think we both knew deep down it was too risky, yet the lure of fresh snow allowed enthusiasm to overrule pragmatism.

Approaching fear as an instructive tool rather than as an obstacle to be overcome is a solid approach for anyone heading out into a wild environment. In relation to adventure sports, the hackneyed phrase "overcoming your fears" is fundamentally misleading; most of the time, you should actually be paying attention to what frightens you if you're in a potentially dangerous environment rather than trying to overcome it. Most of the time, human beings feel fear for very good reasons. It's worth remembering that a neurological fear response to an actual physical threat like goose bumps (electrodermal activity), a raised heart rate, and increased breathing are completely different things to "overcoming the fear of failure".

It's also worth asking how the fast brain / slow brain theory formulated by the Nobel prize winning psychologist and economist Daniel Kahneman has modified our understanding of the fear response. In his groundbreaking 2011 book *Thinking, Fast and Slow*, Kahneman shows that the human brain has two distinct and sometimes conflicting modes of thought. 'System 1' is fast, based on instinct and emotion, whereas 'System 2' is slower, based around logic, problem solving, and deliberation. Going back to my skiing story, it was System 1 that got us out of the bowl quickly, but System 2 was also very important; it was this slower, more analytical system that informed us of how serious the avalanche risk was, and how exposed we were.

Over the years I've learnt that the deployment of both of Kahneman's neurological systems can be vital for safe practice in dangerous places. Whilst I've pushed deep into the danger zone numerous times, I'm proud to admit I've also backed off on quite a few occasions. The skill of knowing when to turn back is fundamental, I think, to survival on serious adventures.

I've backed off climbs, for example, because I thought the weather might soon change based on observation of the clouds (System 2 in action). But I've also backed off climbs because I suddenly felt nervous about the quality of the rock (System 1 providing instant feedback); and I've turned back whilst paddling on the ocean because the wind shifted from cross-shore to offshore (System 2 suggesting a dangerous situation). The most important thing of all, I think, is to realise that turning back out there doesn't mean failure. In fact, it means quite the opposite: it means living to fight another day.

Returning to the Palaeolithic hunter with whom we began, it's likely that hunter-gatherers would have used Kahneman's System 1 and System 2 thought systems when tracking and hunting dangerous wild animals. They would have needed System 2 to plan and execute the hunt, and they would have needed System 1 if something went wrong and they had to make a quick escape from danger; they would have also used System 1 in the heat of the kill. There is a real sense in which the psychology of modern adventure sports is not that different, in several important ways, from that of the hunting expeditions of prehistoric peoples. The source of the danger itself may be different, but the ways we can mitigate and respond to it – and control it for the best results – are broadly the same as those our distant ancestors would have used.

It isn't just in response to danger, either, that we might use these ancient skills in the context of adventure. The practice of natural navigation, which means finding your way using signs and clues in the environment rather than maps or technology, is another realm in which we can channel a kind of outdoor intelligence that goes back many thousands of years.

Successfully navigating through wilderness terrain, or at sea, without modern technology is an enormously rewarding process; I cannot recommend Tristan Gooley's work on this subject more highly. In his 2016 book *How To Read Water*, Gooley quotes the master Polynesian navigator Chad Kālepa Baybayan (the Polynesians are among the world's best natural navigators) to help explain why natural navigation might be a worthwhile activity: "What it does is sharpen the human mind, intellect, and ability to decipher codes in the environment. For me, it's the most euphoric feeling I've ever had". Based on my experience of offshore paddling, I'd agree. Getting to your destination at sea or in the mountains without maps or a GPS is a hugely rewarding experience.

An intriguing question arises here about motivation. Do we practice adventure sports in order to conjure up our distant past as hunter-gatherers? So next time you're out there, faced with an important decision, keep that Palaeolithic hunter in your mind. By bringing our oldest instincts and skills to bear in a modern context, we might reinterpret them, and discover the earliest foundations of human intelligence anew.

BASE magazine, 2020

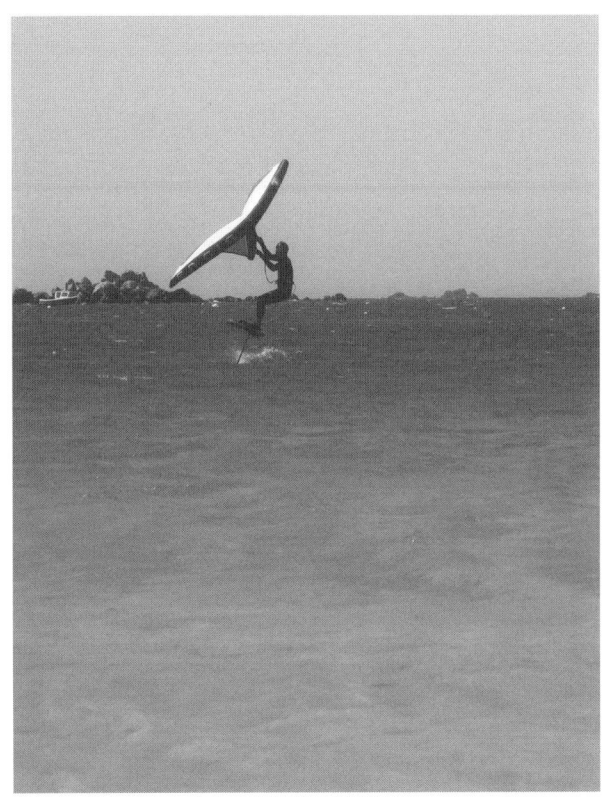

The author wing-foiling off Guernsey, Channel Islands: the skill of 'reading water' is an ancient natural art.

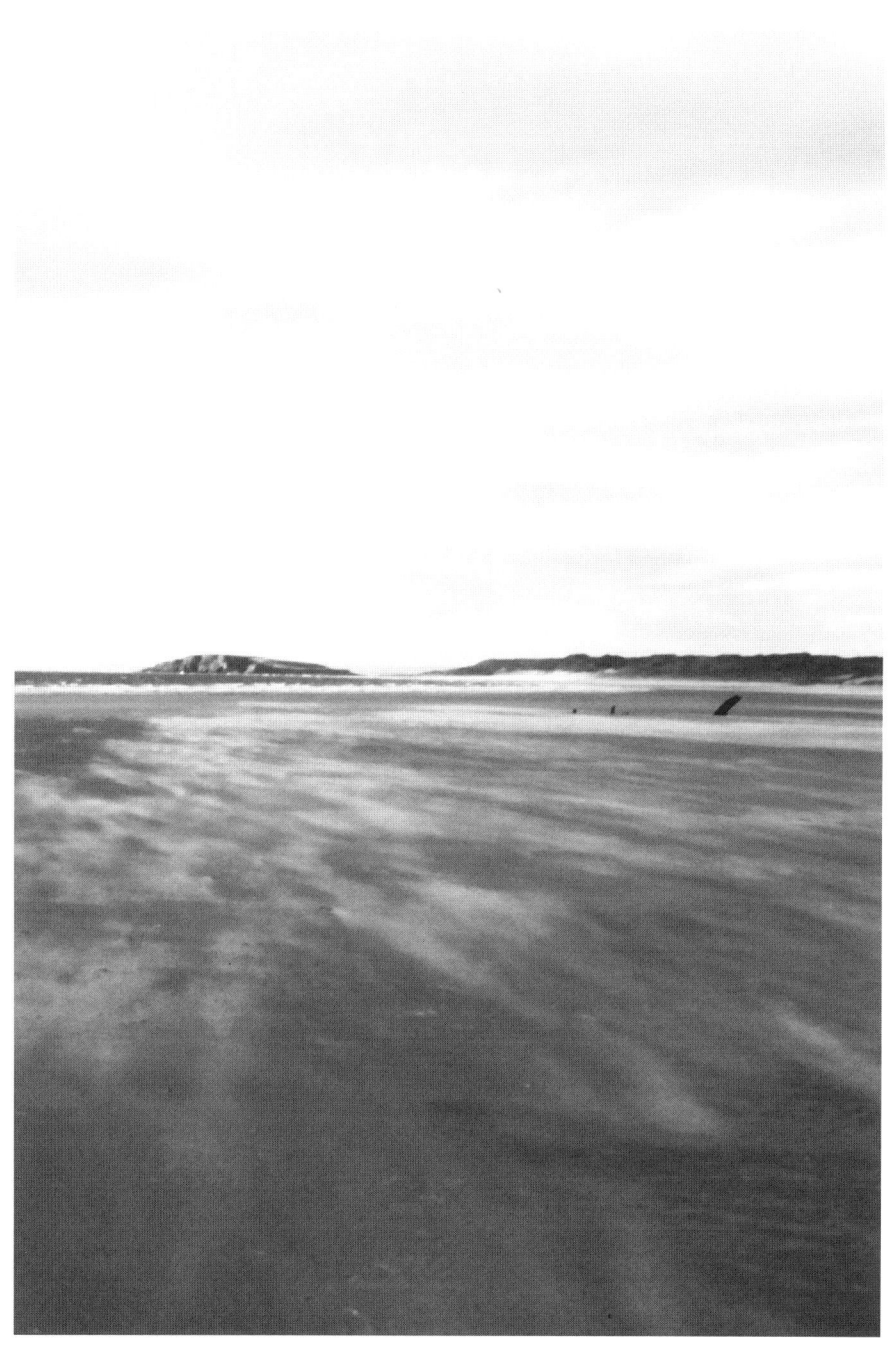
Wind patterns visible on the tide-washed sand of Rhossili beach, South Wales, from Shipwreck Cove.

A Pattern of Grace

In defence of holistic climbing

At full stretch my fingertips catch the undercut hold. I stand on the only good foothold for fifteen metres in half-balance, pressing my left knee against the rock. For ten or fifteen seconds, the friction of denim against rough limestone is the only respite I get from the anaerobic intensity of the twenty-eight move sequence.

Somewhere over my right shoulder, the sun is sinking into the sea to the west beyond Shipwreck Cove, a pint-sized super-crag at Rhossili beach in South Wales. As I hyperventilate on the undercuts, my oxygen-deprived brain registers the penumbra of the surrounding world. Voices of children on the beach. Two women walking a dog. Waves breaking far out on the sand. Herring gulls calling.

Sky, stone, water, light. The climb itself. All this.

Five seconds later and four moves from the finishing jug, I fly into space. Not for the first time and certainly not for the last, I swoop through the air in a widening arc. My partner lowers me off and I land softly on the boulders. Staring back up at the climb I'm trying, the pristine central line of *Helvetia*, I smile. I don't know how many times I might take that same fall from those same moves before I get to the top of the route. I'm enjoying my dialogue with the climb, and curious about what it's telling me about what I need to become the best climber I can be.

A few days later, I did the third ascent of *Helvetia*. The following year, I established some other hard sport routes at Shipwreck Cove, some of them link-ups of existing climbs. The hardest of these additions, a left-leaning line through the steepest part of the cliff, took me about twelve days of attempts spread over two summers. I finally did the first ascent of *Fata Morgana* on a bright, cool afternoon of perfect climbing conditions in the early summer of 2016. I named it after a complex and unusual type of mirage, usually seen by sailors out at sea. By the time I finally climbed the route the goal of actually completing it no longer really mattered. I was simply engrossed in the process of trying to climb it, and once I did it there was both a feeling of sadness that the process had ended as well as a quiet sense of satisfaction.

It's taken me twenty years to understand that the point of trying very hard 'project' climbs has little to do with the act of finally doing them, and everything to do with what you learn in the process of attempting them, the same way the process of practicing a martial art is more important than any actual combat you might engage in as a result of your skill.

Arriving in this state of mind isn't easy. It takes lots of hard knocks; walking away from routes that get away, and from boulders that remove the skin from your fingers as if you'd had an unfortunate accident with a cheese grater. It takes falling off the last hard move on the last day of the trip, and it requires you to support your mates when they're going through it all too. And most of all, it takes doing all this and still having a laugh, still enjoying just being outside and doing something out of the ordinary, and remembering that the ability to go climbing at all is a gift that most are never given. The subtext of all this is that whilst particular, specific climbing goals are undoubtedly important in the short term, in the long term they are far less important than the visceral experience of the vertical world in all its power, beauty and enduring complexity.

In this sense, to live the climbing life to the full it seems necessary to discard the rhetoric of contemporary business culture and a large section of the sports coaching community, and its core fallacy that focusing on clear goals is central to all successful human endeavour.

For people who become obsessed with them, goals are often problems rather than solutions. In the opening of his remarkable collection of short fiction, *Resistance*, Barry Lopez' narrator Owen Daniels writes of how we should "reject the assertion… that humans are goal-seeking animals. [I] believe they are creatures in search of proportion in life, a pattern of grace."

This is a profound and far-reaching idea that is probably more important today than it has ever been, since goals are seemingly more important nowadays for many people than ever before. Lopez's point is really that we should strive to lead a qualitative rather than quantitative life, and to put moral values before goals. It's also a great principle for getting the most from your climbing, too.

In any sport, goals are clearly important. But their significance has been elevated to a degree where it has become difficult to distinguish between goals themselves and the broader quest for excellence in general, and our more complex desire for meaning and completeness in what we do. Once you've achieved a specific goal, you're faced with a tricky question: what's next? The Samurai concept of *bushido* – a code of moral principle which stresses the importance of frugality, loyalty, and the study of martial arts – is the antithesis of this, suggesting instead that we must simply live and act according to the values we wish to uphold in the world.

The central paradox of any goal-oriented culture is that achieving a specific goal doesn't actually mean you've achieved the mastery to which the Samurai aspired. It just means that you've achieved that goal. As Tom Richardson once rather brilliantly pointed out in relation to K2, one of the world's most dangerous mountains, "many people forget that climbing a mountain only means that you've climbed that mountain."

This is the essential problem with goals: they might be useful in the short term, but they don't grow old gracefully in the long term. Some people end up frustrated at not being able to achieve certain things. This is very unwise. A better plan might be to aspire to what Ed Douglas wrote about in relation to Joe Brown, one of Britain's greatest climbers of all time, in a piece for his 80th birthday in 2010:

"He always found new directions to explore and kept that sense of excitement and freshness in his life, many decades after discovering the thrill that climbing first offered him." The goal, in other words, was not a specific one, but maintaining an adventurous approach to life. If you can achieve this in climbing, or in anything else you enjoy doing, then you're going a long way towards living less like a banker and more like a Samurai warrior, which might be a very good thing.

Climb magazine, 2014

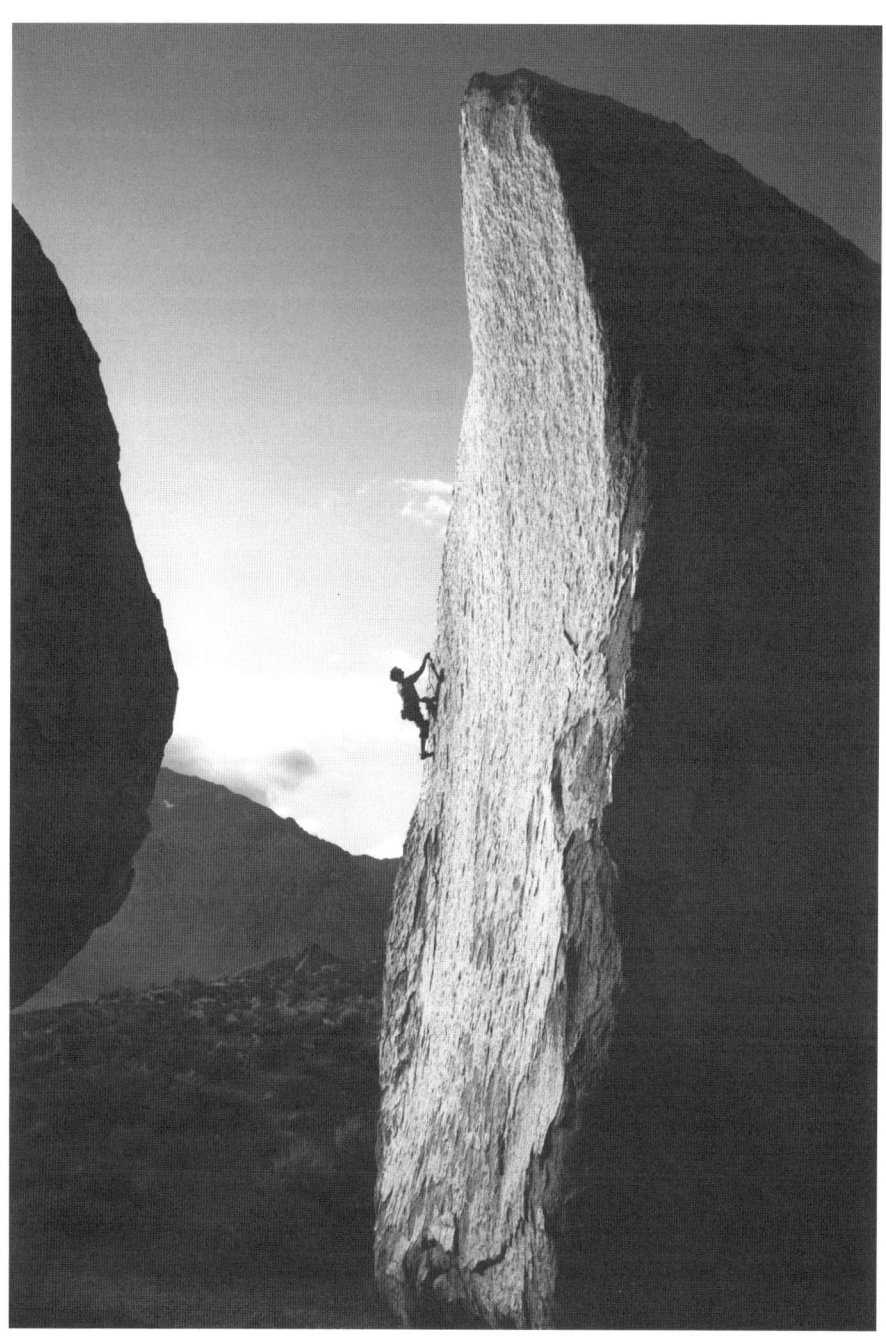

The author soloing on the giant granite monoliths at Bishop, Sierra Nevada, California.

The Path of the Warrior

The link between high-risk adventure and warfare

Extreme adventures such as solo climbing, free diving, or wingsuit proximity flying bring a unique intensity of experience. Very few sporting activities offer this level of consequence to action. Like a warrior on the battlefield, on a solo climb or a wingsuit flight if you make a serious error or just get unlucky, you'll die. It's that simple. But why choose to do such things? And how might we justify – if at all – the very high level of risk such activities often involve?

Like the rampart of a fortified palace on an impossible scale, the West Buttress of Clogwyn Du'r Arddu commands the lonely landscape it surveys from high on the flanks of Snowdon, the tallest mountain in England and Wales. For the adventurous rock climber, it's simply one of the best places in the world to be on a fine day.

The slab rises for over five hundred feet until it merges into the late afternoon sky. I can't see the top. From the tiny ledge at the base where I'm fastening the velcro on my rock shoes, I can't actually see the middle section of the route I'm about to climb; the whole thing just reels skywards. Soon I'm ready, and I set off. I climb slowly and carefully, making every move with attention as I try to find a rhythm between the rock and my own breathing. I quickly gain height. After ten minutes I'm past the second belay stance. A few pitches higher, I arrive at the crux of the whole route, several hundred feet above the ground; a delicate rising traverse above the void between the wall and the inky waters of the mountain lake a long way below.

Here, a precise and perfectly balanced sequence of moves unlocks the upper section of the slab. Now I can see the turreted boiler-plates of Cloggy's upper ramparts, where the huge cliff abruptly stops and the mountain begins again. In what seems like a brief moment, the slab has disappeared into the shadows beneath, and I'm blinking as the sun begins to filter over the line of the ridge a hundred feet above. I check the time; I've been climbing for just over half an hour. Yet it seems like much longer, so intense was my concentration and focus for that short time. The climb was not a difficult one for me, I'd climbed well within my limits, but the experience of being alone and high up on one of the greatest rock faces in Britain was unforgettable: a sense of complete control and being entirely in the present moment.

Through something like free soloing, perhaps, we can achieve the experience of mindfulness sought by the practitioners of Buddhism. I went up to Cloggy that hazy summer afternoon simply to climb a classic line I'd always wanted to do. And I specifically wanted to climb it alone, without ropes, using skill and experience instead of climbing hardware as a safety system. Free soloing will always be the ultimate style of ascent, but it's not to be recommended as such. I've never had a serious accident whilst soloing, but I've definitely had a few near misses. Those close encounters with mortality taught me a great deal about how far, and how fast, it's possible to tread the thin line between courage and stupidity.

As I grew older, I began to understand a lot more about the role probability has in increasing the level of personal risk with every repeated exposure to danger. British climber Tim Emmett once told me he decided to stop wingsuit proximity flying "because I realised there was no way I could continue doing it without losing my life to it". Soloing may not be as dangerous as wingsuit flying, but it still carries big risks. So I began to reduce both the number of climbs I did in this style, and also their difficulty. But I didn't stop soloing entirely. For as long as I can climb, I'll never stop soloing altogether, for a simple reason. Free soloing is an experience that throws you into the heart of your own life, and into the core of what it means to be human.

The good thing about free soloing is that it's not necessary to solo a route that pushes you way outside your comfort zone to have a great experience. The best soloing experiences, in fact, take place on routes well within your limits. That way, you can appreciate the situation without being overwhelmed by it. Being completely terrified is never a positive experience, but a mild sensation of danger definitely is.

Risk means different things to different people. For some, just walking to the top of a mountain like Snowdon might be an adventure of a lifetime. On the other end of the scale, the kind of soloing I've done in my climbing life looks relatively casual in comparison to the exploits of some of world's most accomplished solo climbers, such as the late Swiss super-soloist Ueli Steck, or American climber Alex Honnold, whose free solo ascent of *Freerider* on El Capitan is, in my view, the most impressive sporting achievement in history from the perspective of psychological control.

Solo climbing and mountaineering are activities that channel an extreme intensity of experience. If you get everything right you'll be okay, but if you make a serious mistake – or if something happens that's completely outside your control – you might die. The risks of things like soloing and wingsuit flying are much closer to those for soldiers in a military operation than the risks of conventional sports.

The American artist, author, nightclub owner and bullfighter Barnaby Conrad famously claimed that "there are only three sports: bullfighting, motor racing, and mountaineering; all the rest are merely games" [the phrase is frequently mistakenly credited to Ernest Hemingway]. The point behind Conrad's brilliant, if caddish quip is that any sport requires some degree of exposure to danger – perhaps even a risk of death – to have real meaning. If you're not risking something, he seems to be saying, then you're not experiencing very much either. You might not be a fan of bullfighting for very good reasons, and both motorsport and mountaineering have been made much safer than they once were by far better technology, but the point remains.

It's a powerful message, and underscores the old argument about sport originally being a psychological and physical replacement of warfare in ancient Greece, and a preparation for it if required. Far from being a modern invention, it's likely that sport is as old as human society itself. Some of the 15,300 year-old cave paintings at Lascaux in France, for example, depict scenes of sprinting and wrestling. The fact that the vast majority of organised, competitive sports today involve minimal danger does not undermine the intriguing link between sport and warfare; you only have to watch five minutes of rugby, polo, boxing, or American football to see the point.

According to the academic Vicente Quintero, "the interpretation of the link between war and sport supports essentially two theories. The first is that sport is complementary to war, that it stimulates [martial] attitudes and behaviour, and has an educative function in training personnel for combat. The second theory is that sport represents an alternative to war. This means that sport can be seen both as a 'safety valve', which has the effect of diverting aggressive tendencies away from war-like violence… and [also] a manifestation of these same aggressive and competitive tendencies. Even with the most modern and sophisticated research techniques, there is still no definite and convincing explanation to the sport-war relation." I'd argue that both of the intriguing theories Quintero puts forward here could be true, and quite possibly at the same time. Think of the Christmas Day truce of 1914, when British and German troops put down their guns to enjoy a game of football on the Western Front, only to return to their artillery positions on Boxing Day.

It is not a coincidence that the Olympic Games developed and flourished in Ancient Greece within an increasingly martial political environment, when Sparta was a rising power and Athens was the incumbent power. As the historian Thucydides pointed out, this scenario often leads to war. It's known as "Thucydides' trap", and has correctly predicted many major conflicts like World War One, when Britain was the incumbent power and Germany was the rising power. And it remains of great relevance today to the standoff between the United States and China.

The clash between Athens and Sparta flared up most significantly in The Second Peloponnesian War (431–404 BC), and became the longest and most expensive war in ancient Greek history. The Ancient Greeks left a huge legacy, part of which placed sport at the heart of Western civilisation. In relation to what the Greeks achieved at a civilisational level, it's worth asking if the rising popularity of self-organised adventure sports in recent decades is a kind of cultural response to the risk-aversion evident in so much of Western society: the constant drive to minimise danger, control behaviour, and to mitigate personal responsibility. Adventure sports are a good way out of this moral cul-de-sac.

The reason I love free soloing is that it is everything the risk-averse mentality is not. It can remind us of the mastery we are capable of as we move fast and efficiently in a dangerous environment, making survival-critical decisions, and acting with absolute concentration for every second. Like the Greek facing the Spartan across the plain, here we must use all of our fitness and intelligence to prevail, as we rediscover the old order anew.

www.jottnar.com, 2017

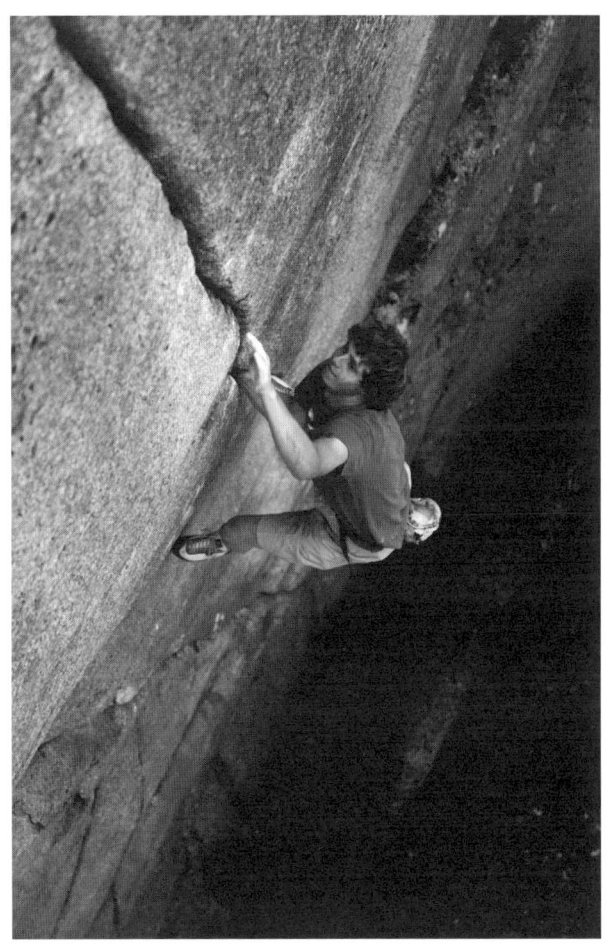

Leading Canadian climber Will Stanhope soloing the testpiece Zap Crack (5.13) at Squamish, BC, Canada.

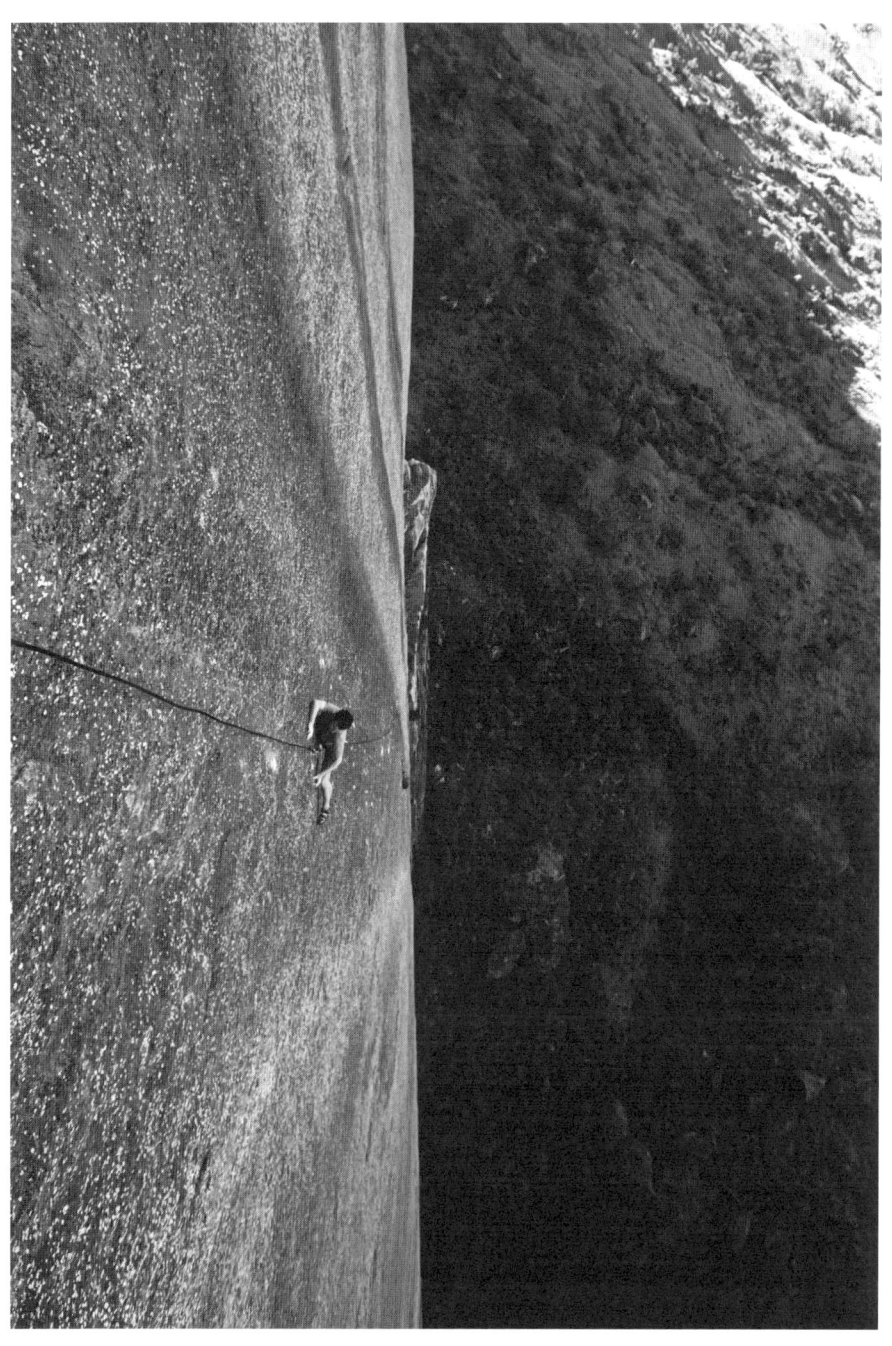

James McHaffie practicing the moves on Tough Enough? (8b+ max, 400m) Karimbony, Madagascar.

A Letter from the Free World

Adventure as a humanist project

"There were riot police everywhere, tear gas, rubber bullets. Protestors stoning the cops. I remember this girl in the crowd who'd just been hit with a truncheon, her blonde hair matted with blood. It was complete chaos."

On a grey and foggy morning in 2008, over coffee on the Boulevard Saint-Michel, I was listening to my dad talking about the civil unrest of May 1968 in Paris, where he was studying at the time. I was in the city for a few hours, on my way to Madagascar and another climbing expedition.

In early January 2015, violence returned to the French capital in hyper-modern form. Just across the Seine from where we'd been talking, two gunmen burst into the offices of Charlie Hebdo on Rue Nicolas-Appert in the 3rd Arrondissement, killing twelve people including its editor, the renowned cartoonist Stéphane 'Charb' Charbonnier, in response to the magazine's mockery of Islam.

The attack was a particularly striking example of the collision between the extremist ideology of radical Islam and the single most important element of any advanced democracy: the existence of free speech within a plural, tolerant society. At the same time, the fact that the attackers also shot and killed a Muslim police officer, Ahmed Merabet, possibly shows the extent of the divide between Islamic extremists and most of the world's Muslim population. In comparison to the creation of the satirical cartoons championed by Charlie Hebdo, the practice of climbing and adventure sports seems like an apolitical act. Yet this is because we perceive these activities from the privileged position of living in a society in which they are possible at all. There is not much of a climbing community in North Korea, South Sudan, or Somalia; basic existence is hard enough for most people in those places. The opportunity to practice adventure sports is a privilege of a prosperous society and therefore a luxury of the developed world. The freedom to go climbing and have adventures, in this sense, is a political and economic freedom as much as a physical one.

The correlation between the right to offend politically or religiously in a free, pluralist society and maintaining the right to take physical risks is a striking one. If you're reading this, you'll probably already know that to live an adventurous life it's essential to entertain risk in some form, at some level. Just as Charlie Hebdo pushed the boundaries of risk in terms of publicly acceptable offence, climbers tend to push the boundaries of acceptable risk in a physical and sometimes moral context.

The British mountaineer Alison Hargreaves, one of the best alpinists of her generation, died high on K2 in 1995. In the aftermath of her death, some pundits in the British press (who knew nothing, incidentally, about climbing) were highly critical of her decision to climb K2 as a mother of two young children. They were of course ignoring the fact that a great many male climbers had, for decades, gone to the Himalaya – where some of them died – as the fathers of young children. The stuffy double standards and residual male chauvinism of 1990s British society were starkly apparent in the absurd stance of the commentariat after Hargreaves' death, which showed in a striking form how climbing can be a sort of moral barometer of the society that participates in it.

However different political cartoons may be from climbing, both depend on one thing for their most basic survival: a society that says 'we think that's a bit crazy, but okay, go for it. You're free and welcome to do that here.' This was the point that the critics of Alison Hargreaves sadly missed. The freedom to participate in sport more widely is significantly dependent on living in an open, tolerant society with a productive economy. The fact the Taliban sought to ban all sport when it controlled Afghanistan in the late 1990s shows the link between sport and a free society: when the latter disappears, the existence of the former is threatened or even ceases to exist.

In Iran's oppressive theocracy, women who participate in self-organised sports like climbing can be arrested and detained for various reasons, including not wearing a headscarf or not being accompanied by a male member of their immediate family. Nasim Eshqi is one of Iran's leading climbers. When we climbed together in Turkey in 2011, she told me about how hard it is to be a climber in Iran if you're a woman: the threat of arrest, she told me, is a constant shadow. More recently, in 2022, widespread protests erupted across Iran following the death of a young Kurdish-Iranian woman, Mahsa Amini, in police custody. Nasim Eshqi recounts the mood the 2022 protests have kindled for her and other women like her: "For a long time, I have been chasing freedom in every corner of the mountains but we women in Iran are not free. Now the same women who have been oppressed for 44 years, since [the Islamic Republic] began, are in the street shouting for freedom, and they want the world to hear them." Eshqi's powerful point here is that the physical freedom to climb and the political concept of a free society are intrinsically linked.

Getting on the plane to Madagascar that night in Paris, I thought about my dad's stories of May '68, and of how fragile that bridge is between order and chaos, between civilisation and anarchy. And of how fortunate I am to live in an era where the former, just about, holds sway over the latter. An act of violence against free speech should strengthen the desire for free speech as a moral value. The Charlie Hebdo attack in 2015 highlighted the vital importance of free speech in an inclusive society, whilst undermining the ideology behind the violence itself. Things like extremist ideology or a police state will never be as powerful as the human desire for freedom.

As a final note, it's worth remembering that the very existence of adventure sports depends on almost everything that radical Islam seeks to destroy: openness, opportunity, meritocracy, and equality between women and men. And, of course, the motivation for adventure itself must surely depend to some extent on the attractive idea that what really matters is what you do in this world, not what you imagine might exist in the next one. If that's true, then the value of adventure, as with satire, depends on how much we're prepared – like Charlie Hebdo – to risk going against the grain of the acceptable and into the realm of the subversive and the radical. Explorers, artists, and free thinkers are the diametric opposite of religious fanatics.

Climb magazine, 2015

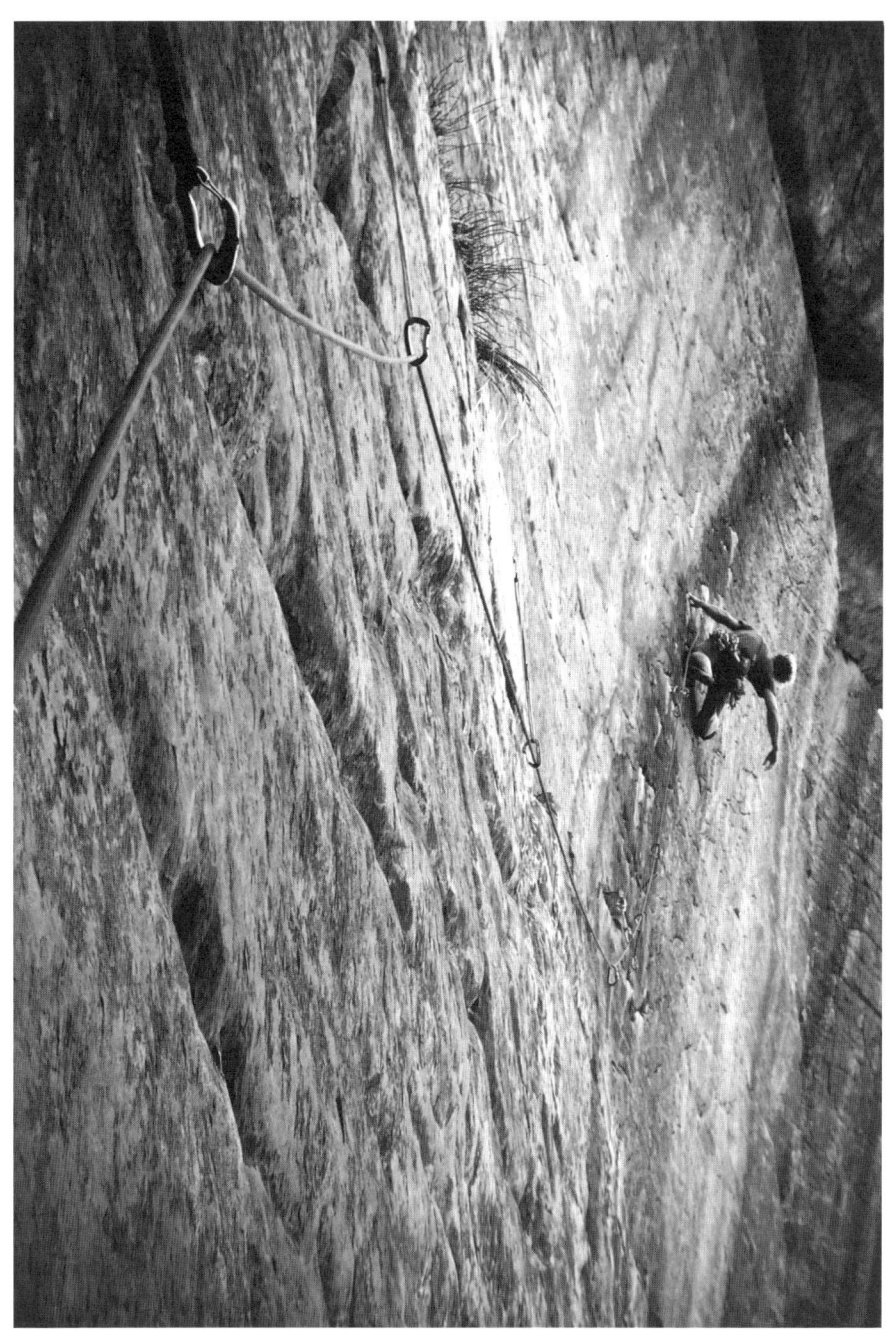

The author high on Archimedes Principle (26/E6), Eureka Wall, Grampians, Australia. Photo: Ramon Marin.

The Pursuit of Sport

A short story about climbing & travelling

The skeletal shadows of charred eucalypts stand tall and silent as we stalk between them, like watchful totems. We're breaking trail through the fire-scorched bush of the Grampians, glancing up at the escarpment above, searching for the imprint of Muline Cave amid the crenelated sandstone of the Victoria Range. My friend Ramon Marin cross-references the guidebook as I squint into the middle distance, my eyes falling on a telling darkening of the horizon at three o'clock: "That's it!"

Seven hours earlier, we'd got off a plane in Melbourne after a twenty-three hour flight from London, jumped in a hire car, and driven five hours into the outback to find this remote place that was supposedly home to one of Australia's best cliffs. We hadn't slept properly for two days, but that didn't matter: finally we'd found it.

The excitement builds quickly as we approach the base of the wall. Somewhere in the thicket overhead, a kookaburra shrieks. After a while, the whack of a pulled rope and the sardonic lilt of a New South Wales accent breaks the air.

"G'day mate, good t' see ya. Welcome to the Gramps." A long-legged figure strode down the trail to meet us. It was my friend Doug McConnell, who I'd last climbed with in Spain three years before. The extraordinary coincidence of seeing him here, in the middle of nowhere in the Australian bush, after travelling for 48 hours straight, was a great way to kickstart what quickly turned out to be one of the very best climbing trips of my life.

As our time down under came to a close, and after climbing more stellar routes than you could shake a bushman's stick at, I began to wonder about that encounter with Doug on our arrival. I wondered through what activity, other than climbing, you could fly to the other side of the world, drive for several hours into the middle of nowhere in a remote mountain range, hike through the untracked bush for half an hour, and then meet a friend completely by chance? A truly exceptional coincidence like that I've just described is possible through climbing, for two key reasons. Firstly, because the places to which climbing directs us are often extremely isolated and few people, if any, ever go there. And secondly, because the international climbing scene is a paradigm of our globalised, ultra-mobile, hyper-connected world; a paradigm only intensified by the fact that climbing is still small in relation to other, more mainstream sports.

Chris Bonington, Britain's first professional climber, once said that "climbing is an incredible vehicle for exploring the world". This is even more true in today's globalised arena than ever before. Personally, I appreciate the places the vertical world has taken me – and the people it has connected me with – as much as the experiences of action on the rocks and in the mountains. Why else, were it not for climbing, would I have travelled as widely as I have?

Another intriguing question arises from all this. If climbing is a perfect excuse for international travel, is it more accurately described as a sport, or as a pursuit? The question has been the subject of intensive debate between different elements of the climbing community over the years. If you concentrate mainly on climbing's athletic aspects, you'll call it a sport. But if you enjoy its adventurous elements but are not so enamoured with its athletic side, you'll call it a pursuit. Both terms thus seem fair.

Yet because of the exponential explosion in the popularity of the most obviously 'sporty' genres of climbing over the past two decades or so – bouldering, sport climbing and in particular indoor climbing – it seems appropriate to re-examine this most curious of questions about the vertical world. The climbing and mountaineering community is full of people, and you might well count yourself among them, who love climbing but who don't feel the need to turn it into an activity with a largely athletic focus. Instead, what this group of people might seek are adventurous journeys to interesting places, and spending quality time outdoors with good friends.

A well-travelled and wise climbing partner of mine once told me that rather than trying to climb certain routes of certain grades, what he really relishes in climbing is simply going to crags all over the place that he has not visited before. His approach strikes me as an invigorating riposte to the increasingly sport-focused atmosphere of modern climbing, in which the athletic paradigm of the climbing gym is so frequently translated directly to outdoor crags. There's nothing ostensibly wrong with this, but it does reduce the possibility of encountering the deeper experiences that the less athletic and more adventurous aspects of climbing might offer.

Could it be the case, then, that the ideal condition to aspire to as a climber is some combination of exploratory traveller and sportsperson; a nomad, as it were, and an athlete at the same time? Through combining these different modes of being, we could perhaps best experience the full richness and possibility that climbing reveals without compromising either.

To anyone who's still wondering what the real answer is to the question I posed earlier, I'd suggest that climbing is very clearly both a sport and a pursuit *at the same time*. There's also no reason an athlete cannot also be an artist; many of the world's greatest sportspeople achieved just this duality. Think of Mohammed Ali, Diego Maradona, or Aryton Senna, and the point becomes clear.

The great thing about the nomad-versus-athlete conundrum is that we can choose how little or how much sport we wish to inject into our own climbing. You can climb for your whole life without even touching a resin hold, a campus board or a pull up bar, and you'll still be just as much of a climber as the one who spends four evenings a week at the climbing wall. As Ben Moon has very wisely said: "There are plenty of people who've done too much training and not enough climbing".

Climb magazine, 2015

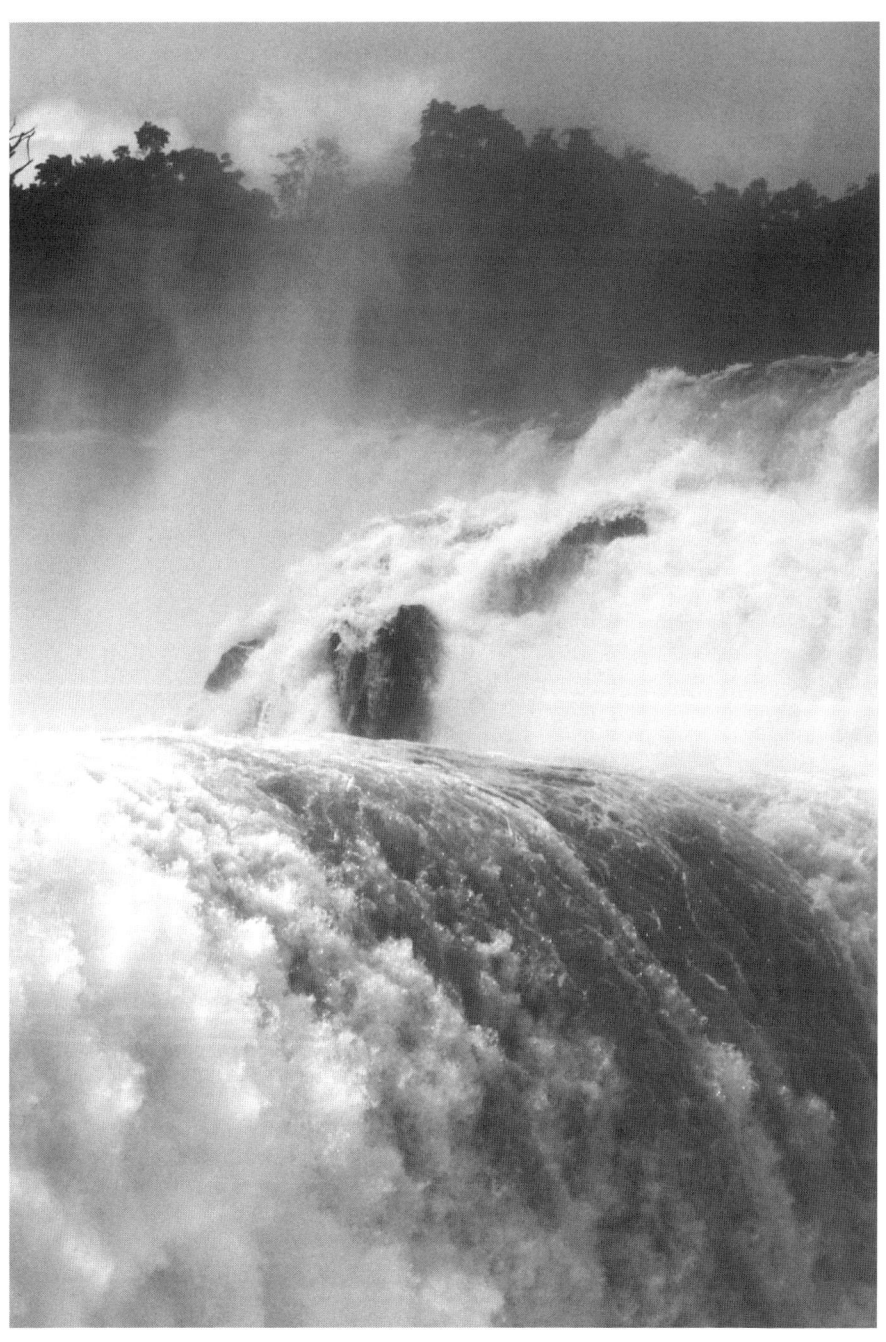

The awesome power of nature at Foz do Iguaçu, Brazil, one of the world's highest-discharge waterfalls.

Lifeboat in the Deep

The case for the precautionary principle

"It is not the critic who counts; credit belongs to the man who is actually in the arena, whose face is marred by dust and sweat and blood; who strives valiantly; who errs, who comes short again and again."

Theodore Roosevelt's words from 1910 have even more relevance today, perhaps, than they did back then. The world contains more critics now, and there are arguably fewer who are actually 'in the arena'. Leading an adventurous life is one way, I think, of living up to the noble ideal that Roosevelt identified in his Sorbonne speech. It's also a way of discovering and correcting our shortcomings, which is surely as important a dimension of the exploratory arena as the value of action itself.

Procrastination, for example, can be a feature of independent expedition planning. An unfolded map spreads across the kitchen table. The laptop blinks back at you with the finality of the 'complete booking' button; the first stage of setting out. You know the feeling – you're just about to book a trip. But you're still in two minds about it. It's a situation that can occur any time, and for all sorts of reasons. Is the trip too expensive? Can you justify being away for that long? Are you going to miss something important back home? All these questions, and many others, might raise themselves to stop you in your tracks.

The process of asking such questions at all is, I suspect, an example of a concept called the precautionary principle at work. It can trace its earliest origins in experimental British civil engineering in the 18th century, and was seriously developed in the progressive environmental legislation created in 1970s Germany (the *'Vorsorgeprinzip'*, the origin of the phrase in English). Yet there is still no single, universally accepted definition of the precautionary principle itself. It could be a qualitative, anticipatory model that might protect us against the various risks of our own actions, or it could be a more sophisticated tool for the cost-benefit analysis of a particular strategy. It's certainly an interesting means of measuring the value of action in relation to potentially high risk activities like adventure sports, where the precautionary principle might manifest itself as detailed contingency planning, and also as pragmatic decision-making in response to events or new variables during the expedition itself.

A particularly striking example of how effective these measures can be is the extraordinary story of NASA's ill-fated Apollo 13 mission in April 1970, when commander Jim Lovell, command module pilot Jack Swigert, and lunar module pilot Fred Haise pulled off what is arguably the most remarkable act of self-rescue in human history. When Apollo 13 was around 180,000 nautical miles from Earth, en route to the Moon, an oxygen tank in the service module failed and the command module had to be shut down. The scenario of using the lunar module as a 'lifeboat' for returning to Earth if there was a problem with the command module had been discussed by NASA, even though it was considered an extremely unlikely scenario. After the oxygen tank accident, lead flight director Gene Kranz at Mission Control in Houston very quickly operationalised this wildly hypothetical plan. A few days later, the three astronauts safely touched down in the South Pacific aboard Apollo 13's lunar module. It was an astonishing technical and operational achievement. The story is a clear case of the precautionary principle being used to great effect; had Mission Control not considered that unlikely scenario of using the lunar module as a 'lifeboat' and then actioned it when needed, the crew's successful self-rescue may not have happened, and Apollo 13 would have been lost in space.

For every case, of course, there's an antithesis. The story that follows shows what can happen if the precautionary principle *isn't* deployed in the planning of an adventure. In 2010, leading South African whitewater kayaker Hendrik Coetzee was intending to retire from over a decade of running the wildest rivers in sub-Saharan Africa to settle in Uganda to establish a guiding business. After being contacted by two top American kayakers, Coetzee agreed to a final 'big' expedition in one of central Africa's wildest regions: a first descent of the Ruzizi and Lukuga rivers, two obscure tributaries of the upper Congo in the politically unstable Democratic Republic of Congo (DRC).

Coetzee's expedition successfully ran the Ruzizi and took a water taxi across Lake Tanganyika, which is one of the worst places in Africa for crocodile attacks, to put in on the Lukuga, which flows out of the lake. A few days in to the Lukuga, after successful first descents of various grade V rapids, they came to a ninety-degree bend in the river that formed a wide channel where the current slowed. Here, Coetzee's American teammates looked on aghast as a truly enormous Nile crocodile launched itself from the murky water and dragged their leader and teammate into the depths. An empty kayak appeared a short time later; Coetzee's body was never found.

In their expedition planning, the team had overlooked a crucial geopolitical factor that made the Lukuga an exceptionally dangerous river to paddle. Between the mid 1990s and 2010, intermittent civil war had been the main feature of life southern DRC. This had led to an explosion in the crocodile population on rivers such as the Lukuga – both in their numbers and in the physical size of the animals – due to the fact that so many human bodies had ended up in the region's waterways. Some have speculated, gruesomely, that this also gave the local reptiles a taste for human flesh.

Generations of African explorers have traditionally considered populated areas to be safer from a wildlife perspective, since dangerous wild animals are usually shot by local people before they can proliferate. In the case of the Lukuga river, planning the expedition using this traditional assumption had deadly consequences; the Lukuga was actually *more dangerous* because it flowed through populated areas for the reasons highlighted. In Niemba, where the three kayakers stayed a few nights before Coetzee's death, 125 local people had been reported as having been taken and eaten by crocodiles in the two decades previous to the expedition.

If the precautionary principle is used in the right way, it can be a highly successful safety mechanism, as in the case of Apollo 13. If it isn't used correctly, as in the case of the Lukuga expedition, it can produce a very dangerous scenario. If there is the possibility of harm, the precautionary principle will suggest the most obvious path for the avoidance of that harm. In a simplistic form, it might suggest that manned missions into outer space should not be launched, or perhaps that sub-Saharan African rivers should never be explored in kayaks. In a political context, the precautionary principle suggests that the state should use policies to shield the public and the environment from harm; this is an extension of the idea of the social contract set out by English political philosopher Thomas Hobbes in his *Leviathan*, that the state's essential duty is to protect its citizens.

Since the spring of 2020, when multiple countries around the world began using 'lockdown' policies in an attempt to control the spread of Covid 19, a very complex conundrum arose in relation to the precautionary principle. During the course of the pandemic, it became increasingly clear to many observers that if you are implementing a policy with considerable known harms as a measure of disease control, then you need to conduct a serious analysis of the advantages of that policy in relation to its downsides. Whilst some have argued that pro-lockdown politicians and scientists were using the precautionary principle in advocating strict social restrictions, they also – universally – failed to ask the following basic question: do the benefits of this policy outweigh its enormous costs? No proper cost-benefit analysis was done by any country that introduced lockdowns, which is extraordinary given how destructive this policy can be to the young and to the poor, in particular.

The French philosopher Michel Foucault claimed that "freedom is the ontological condition of ethics". So you can only make an ethical decision, like taking a risk, if you have the freedom to act. Foucault's idea about the relationship between freedom and ethics is actually closely related to the precautionary principle, because freedom – and free access to accurate information – is essential for it to function. The survival of Apollo 13 was down to NASA actioning a radical contingency plan in response to an event they had correct information about. The three kayakers on the Lukuga didn't know the extent of the threat from the Nile crocodiles; in the absence of this important information, they were not free to arrive at the best strategy.

The factor uniting these two very different stories is the use of the precautionary principle, or its absence. Apollo 13 used it to great effect; Hendrik Coetzee wasn't so lucky, but his teammates at least immediately deployed it after the crocodile attack and got the hell out of the Lukuga. This intriguing concept can therefore be extremely useful to adventurers and explorers. Whatever you're doing out there, it's a sharp tool to keep in the box of strategic tricks.

BASE magazine, 2021

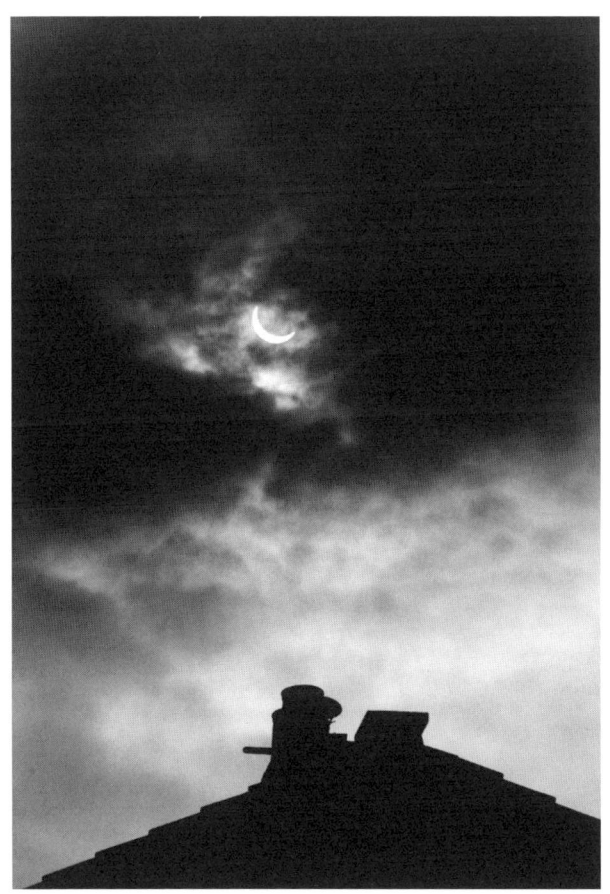

Partial eclipse of the sun, Windmill Hill, Bristol, 2015.

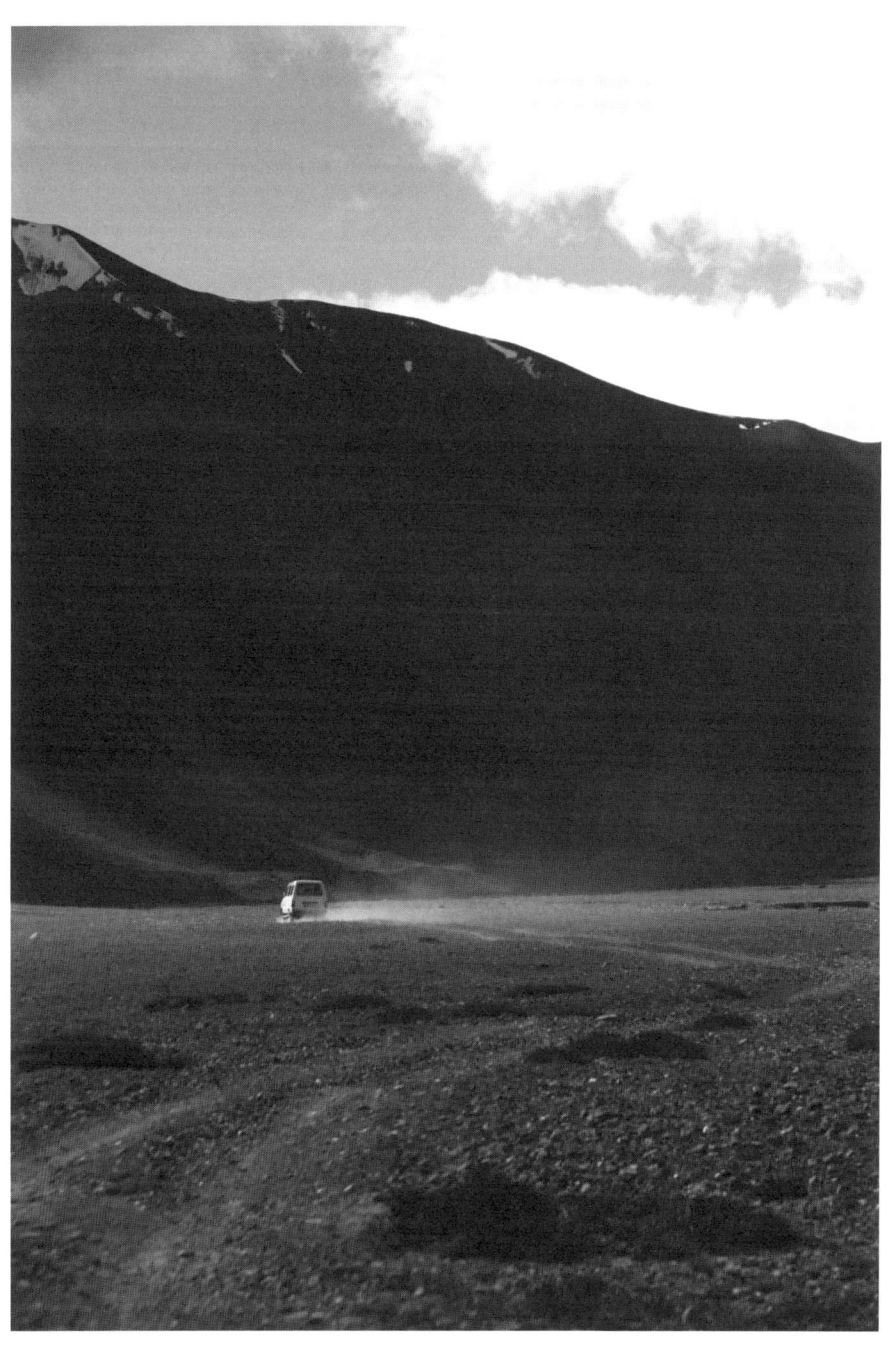

On the dusty road at circa 4500 metres in the high altitude desert of Ladakh, northern India, 2014.

Out of Town

Notes from the university of the dusty road

Many years ago, on a night flight back to the UK from Australia, I woke up suddenly. We were high over the Deccan plateau, India's great central plain. Lights glimmered like storm lanterns far below; the urban sprawl of Bangalore spread flickering lines of yellow and white into the black land, like the traces of a fire-spirit. And somewhere else down there, along the shadowed roads beyond those lights, lay something less visible: the dream of a journey. I had a vivid flashback from my mid-twenties, when I rode a motorcycle around India with my girlfriend of the time.

In *About This Life*, Barry Lopez describes being asked by a man what advice he might give his teenage daughter who wishes to become a writer. He comes back with some brilliant advice: "Tell her she will have to become someone. And tell her to get out of town, and help her do that." The point is that travelling might be the best way to find out why the world is the way it is, and possibly why you are the way you are.

All the best travel and adventure writing aspires to what Lopez calls "a literature of hope". Departing from the one-dimensional narratives of expedition reports, the best accounts of adventure portray exploration as a mystical pattern as much as a physical process. In the same way, the best travel writing somehow acknowledges that a journey is a way of understanding something as much as it's a line across a map.

Climb magazine, 2016

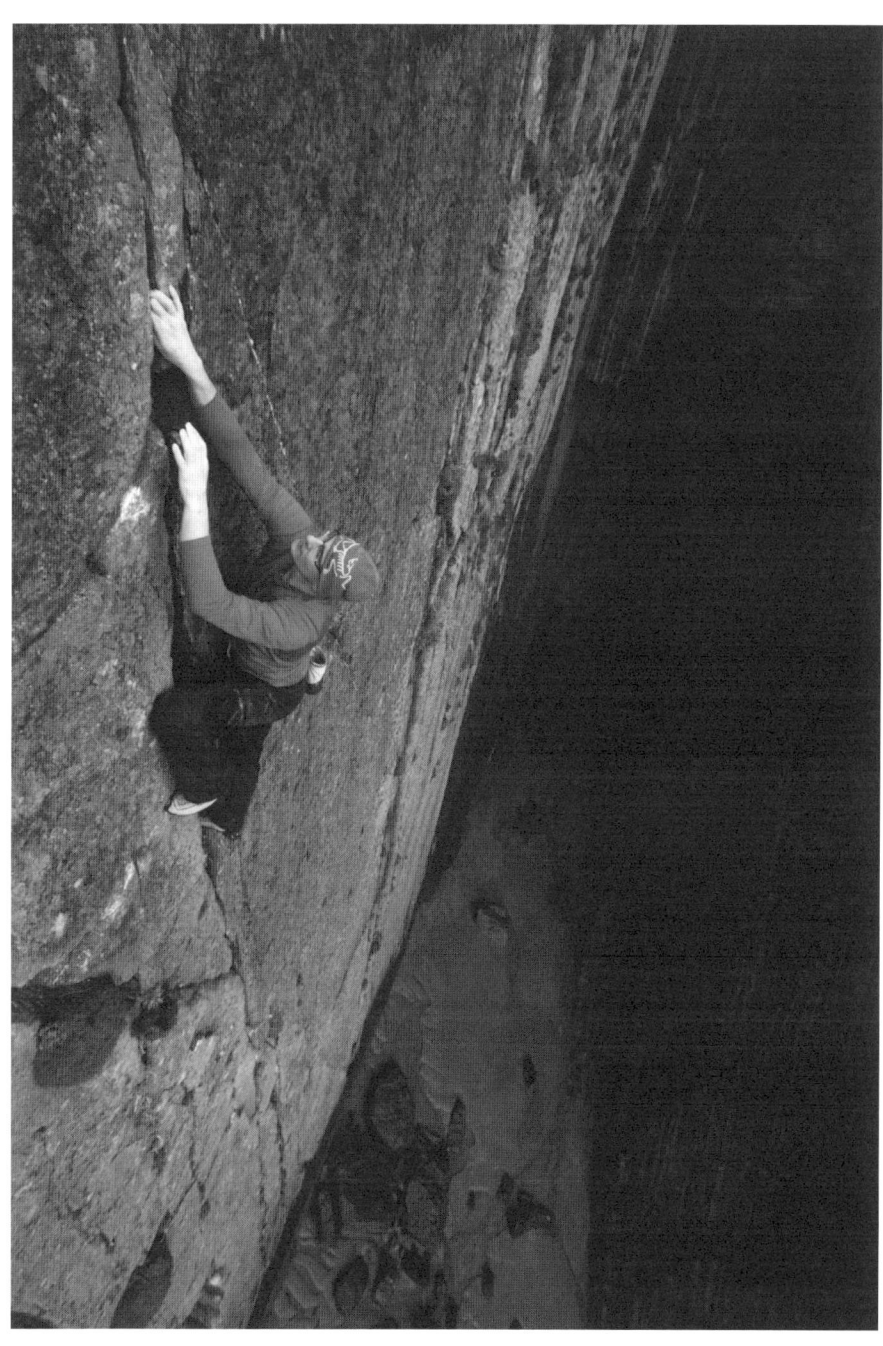

Ian Parnell soloing the British sea cliff classic Shangri-La, Baggy Point, North Devon, England.

The Weight of the Dice

On the consistent practice of the increasingly dangerous

A perfect June morning in Snowdonia. A light wind blows over Llyn Padarn as the dawn cloud breaks off the tops of the Glyders. A hunting kestrel hovers at sixty feet. There's something in the air that says it's going to be one of those special days.

I run the Snowdon Horseshoe and am back at the car before midday. The mountains are glowing gold against the diamond-blue sky. High on the south side of the Pass, the skull-shaped shadow of Cryn Las hangs over the side of Crib Goch, promising an oasis of cool in the afternoon heat. On autopilot, I stuff rock shoes and chalk bag into a lightweight pack and dash up the hill. Soon I'm standing below Joe Brown's 1953 masterpiece, *The Grooves*. It's about five hundred feet high, and one of the finest mountain rock extremes in Britain. I'd always wanted to do this route.

A small but awkward roof guards entry to a lonely dihedral. I climb nervously at first, compulsively dusting the green lichen from my rock shoes, but soon settle into the glorious rhythm of unencumbered climbing. After four hundred feet, I reach the final impasse: a delicate traverse leftwards to the shallow hanging groove that leads to the top of the wall. With every step across this narrow shelf, the exposure increases until it's like peering over the edge of a skyscraper. It is a pretty wild place to be with a rope on. Without one, I'm treading the fine line between courage and stupidity.

On the final groove, I glance down. Almost five hundred feet of mountain air separates me from the scree far below. Up here, I'm a lone dancer in a vast auditorium of silence and space. As I pull over the top of the crag, the thrill of soloing, and all the reasons why I used to solo regularly come flooding back. I'm also reminded of all of the other reasons why I've now stopped soloing almost entirely. These reasons are best summed up by a striking phrase formulated by the American sociologist Diane Vaughan: "the normalisation of deviance". She came up with it in her fascinating book about the 1986 Challenger Shuttle disaster. More generally, it could describe the process by which we do something that does not follow the accepted safety protocol (like climbing without a rope), which we get away with. Then, believing it's safe to make the same safety shortcut a second time, we do it again. Repeat this process, and something is almost certain to go wrong. The concept can be applied to climbing and other adventure sports where human error combined with the influence of probability has led to a fatal accident. One example is the story of the diver Guy Garman and his attempt to break the world depth record on open-circuit scuba with only four years' diving experience. The practice that Garman had "normalised" of doing increasingly deeper dives and just about getting away with it had led him to believe, possibly not correctly, that he was one of the world's best deep divers. This created an illusion of invulnerability, when in fact the probability that something would go seriously wrong on such an ambitious dive was extremely high according to some much more experienced divers than Garman.

Such illusions can be equally dangerous in climbing. The problem with onsight soloing is that you cannot know – or even anticipate – all of the variables that may affect your climb. You only know what they were in retrospect, after the event. As Donald Rumsfeld famously said, "it's the unknown unknowns that tend to be the difficult ones." Wingsuit flying accidents have claimed the lives of a lot of people, including many notable climbers, like Sean Leary and Dean Potter. All these accidents were probably the result of the consistent practice of the increasingly dangerous. Wingsuit proximity flying has an incredibly high fatality rate for obvious reasons. As a pilot becomes more skilled, enabling more technically complex flights, the risks increase. In many other adventure sports, experience often leads to better margins of safety due to better judgement, and so on. Not in wingsuit flying, it seems. A similar process may also take place in alpinism; a mountaineer who climbs something high and hard, then attempts something else higher and harder. Carry on this process, and something has to break, as it has for so many mountaineers over many generations.

In order to prevent situations of 'normalised deviance' in which good safety protocol is discarded and you're taking increasingly greater risks, it's a good idea to develop some clear personal ground rules. The great Canadian climber and prolific soloist Peter Croft takes a perspective on all this that remains one of most intelligent I know: "Cast the rules of the day out in your peripheral vision where you can still see them, but only as a vague reference point. This doesn't mean that the rules are gone. It might mean that you adopt a far tighter code of conduct."

If you still decide to go soloing, it's worth remembering John Bachar, one of America's most influential rock climbers. He died after a fall in 2009 from a relatively straightforward route at Mammoth Lakes, possibly after a hold broke. He had soloed thousands of routes in his life, many of them much harder than his final climb. On that day, probability was stacked against him. The final equation is a basic one: the more you solo, the greater the risks of soloing become. As Alex Honnold, the world's most accomplished soloist, points out in his book *Alone On The Wall*: "it may be that it was the sheer volume of 35 years of soloing that finally caught up with Bachar."

I spent several weeks climbing with Honnold in Rodellar, Spain, in the autumn of 2009, before he was the superstar he is today. We later climbed in the UK together. As we were driving home from the south coast one summer night in 2012, I was stretching the legs of an old E46 M3 on the deserted roads of Wiltshire. I accelerated hard out of a particularly good corner, and Alex promptly said "Wouldn't it be badass to solo El Cap?"

I knew he'd been thinking about it already for some time. This was of course half a decade before his astonishing free solo of *Freerider*. Honnold is a highly intelligent person, and he's actually a calculated, evaluative risk-taker. He's not reckless. John Bachar soloed thousands of routes over his climbing career, and eventually died soloing. Alex Honnold is the greatest soloist who's ever lived, but he's aware of the benefit of not soloing all the time: an interesting dichotomy.

Probability is an important thing to think about if you do dangerous stuff. The more you do it, the more the dice becomes weighted on the side you definitely don't want it to roll on. It's that simple, really, in the end.

Climb magazine, 2015

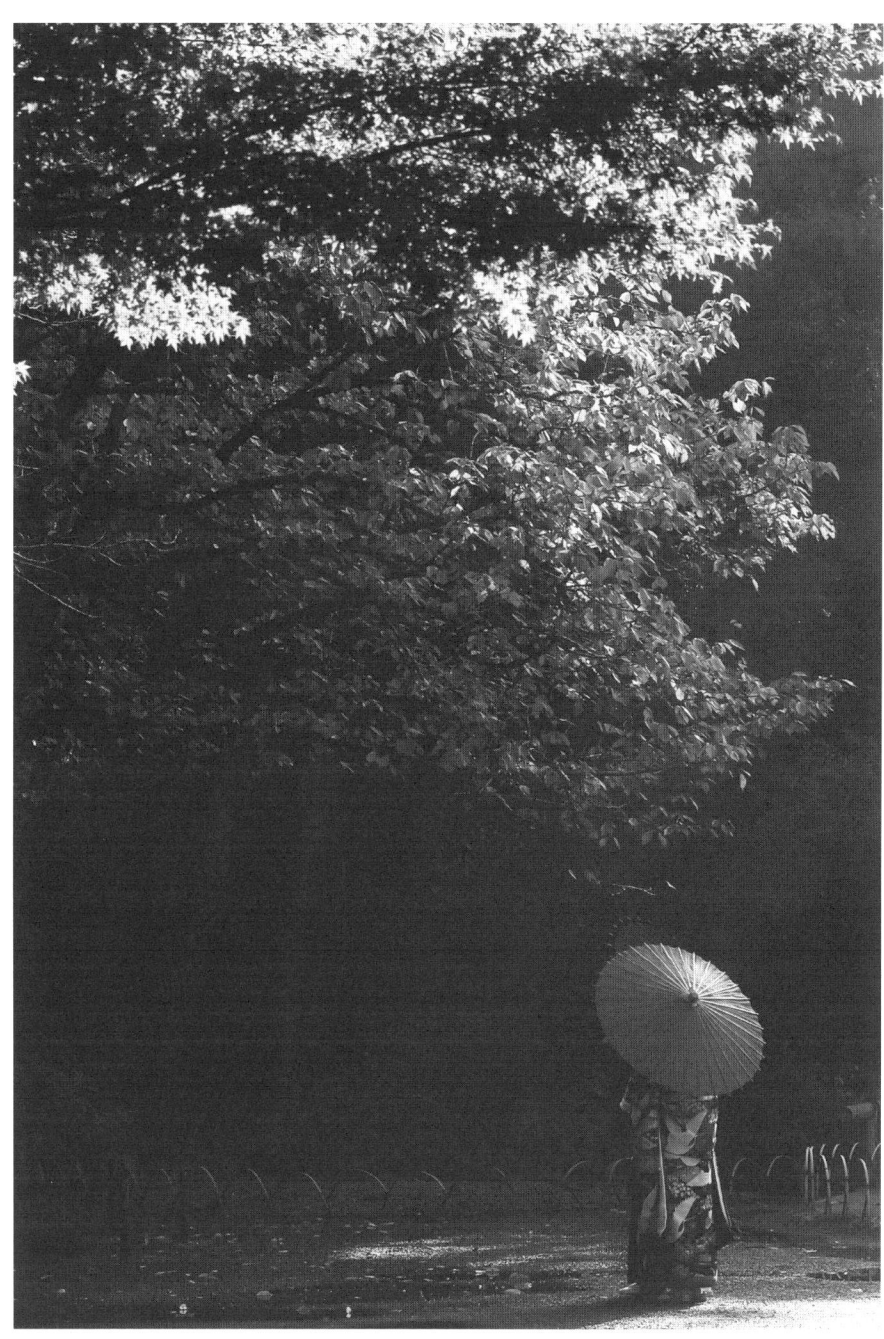

A better place to be than on social media? Offline in a garden in Kamakura, Japan.

Outside the Panopticon

The case for not using social media

When the 18th century English philosopher and social theorist Jeremy Bentham devised an innovative layout for an institutional building called the Panopticon, he could hardly have imagined that a couple of centuries later it would become a reality not as a building, but as a series of worldwide digital networks. Today, billions of people live within a similar system of control and coercion that the Panopticon enabled. It allowed a single watchman to observe all the inmates of the building at once, without the inmates knowing whether or not they were being watched. The social innovation of Bentham's design is that the inmates are compelled to modify their behaviour because they must act as if they are being watched at all times.

Jeremy Bentham's Panopticon is a powerful physical representation of the way social media works. It's not totally perfect, as in Bentham's Panopticon there is one central observer watching everyone, whereas on social media everyone can watch everyone else at once. It is, of course, to state the obvious to point out that platforms like Facebook, Twitter, Instagram and now TikTok have impacted the world profoundly. It is also worth pointing out here that no other technology, in the whole of human history, has been adopted by so many, so quickly, and on such a scale.

I hope to identify here some of the reasons why, for many people, it may be hugely beneficial to move away from social media entirely. It certainly was for me. For the purposes of full disclosure, I should make it clear that I have used social media previously, on a somewhat phlegmatic basis, but stopped using it altogether around 2018. I can honestly say that my life is better for leaving it out. My core reasons for doing so are best summarised by Chamath Palihapitiya, former vice president of user growth at Facebook: "[Social media] is eroding the core foundation of how people behave by and between each other. My solution is I just don't use these tools anymore."

Clearly many of the well-publicised problems with social media are also representative of some of the core structural failures of Meta itself as a company and its lack of real corporate governance. Mark Zuckerberg, Meta's founder and CEO, told *Time* magazine in 2010 that "The way that people think about privacy is changing a bit... What people want isn't complete privacy". It's quite a statement, isn't it? It certainly shows both the Pharaonic sense of self-entitlement and also the cynicism that seems to underpin most if not all of what this particular corporation does.

It is worth noting here that the Chinese government has already developed a full scale, comprehensively functional digital version of Bentham's Panopticon: the so-called 'social credit' system that constantly monitors the actions and behaviour of Chinese citizens. China, in this sense, is a stark warning to the rest of the world about the implications of social media technologies when they are controlled by an all-powerful agency like an authoritarian regime. Unlike the subjects of China's social credit apparatus, who cannot escape from that system, in the West we have a choice.

Facebook has been directly implicated in the facilitation of the ethnic cleansing of Rohingya Muslims in Myanmar, as well as in the promotion of extreme content that's now widely acknowledged as highly corrosive to functional democracy. One of the worst aspects of social media is the lack of interest in the well-being of young people who use social platforms from the companies that own them. There's the example of the Malaysian teenager who took her own life after asking her Instagram followers whether she should live or die. When 69% responded with the latter, she chose death. In the wake of this tragic incident, it's worth remembering that, as with many others like it, this event was a direct result of a particular corporation's digital products, and of the policies (or lack thereof) that control the way those products can be used. It is only recently that a proper public debate has begun to take place about just how corrosive social media is to young people, and to teenage girls, it seems, in particular. In his memoir *Horizon*, Barry Lopez asks "In which Western nations does a determination to address the mental, spiritual, and physical health of children override an indifference towards their fate? Or are these questions now thought of as anachronistic, no longer relevant to our situation?"

If you still believe, or would like to believe, that social platforms might be a force for good as well as evil, I cannot recommend Shoshana Zuboff's extraordinary book *The Age of Surveillance Capitalism: The Fight for a Human Future at the New Frontier of Power* more highly. In it, Zuboff forensically deconstructs the digital products and corporate strategies of Meta and Google in particular. She shows how they have succeeded in re-engineering human behaviour for profit, largely through selling private data in the so-called "behavioural futures" market. It's not very nice at all.

So is it worth checking out of social media for a while? As a longstanding sceptic, I can only advise you that doing so would be a wonderful, life-affirming thing to do. The social media information stream very quickly becomes something other than what it claims to be. In this respect and in others, social media is a profoundly deceitful kind of technology. It prioritises and channels extrinsic motivation over intrinsic motivation, and peer group recognition over the search for inner truth.

To truly succeed in life, we need to find our own vision of things, in our own time, by ourselves. This requires everything that social media is not. This process of self-discovery in isolation is exactly what the story of Christ in the wilderness is all about. Mohammed Ali captured it in another way: "the battle is won or lost far away from witnesses, out there on the road, long before I dance under those lights."

Social media oppresses free speech by cultivating echo chambers, it channels extremist ideology, it facilitates bullying and abuse, it encourages virtue-signalling, and makes people unhappy. These points are not speculation: they are facts. Perhaps worst of all, though, it takes up your precious time on Earth when you could be doing something far better – like being outside.

www.ukclimbing.com, 2019

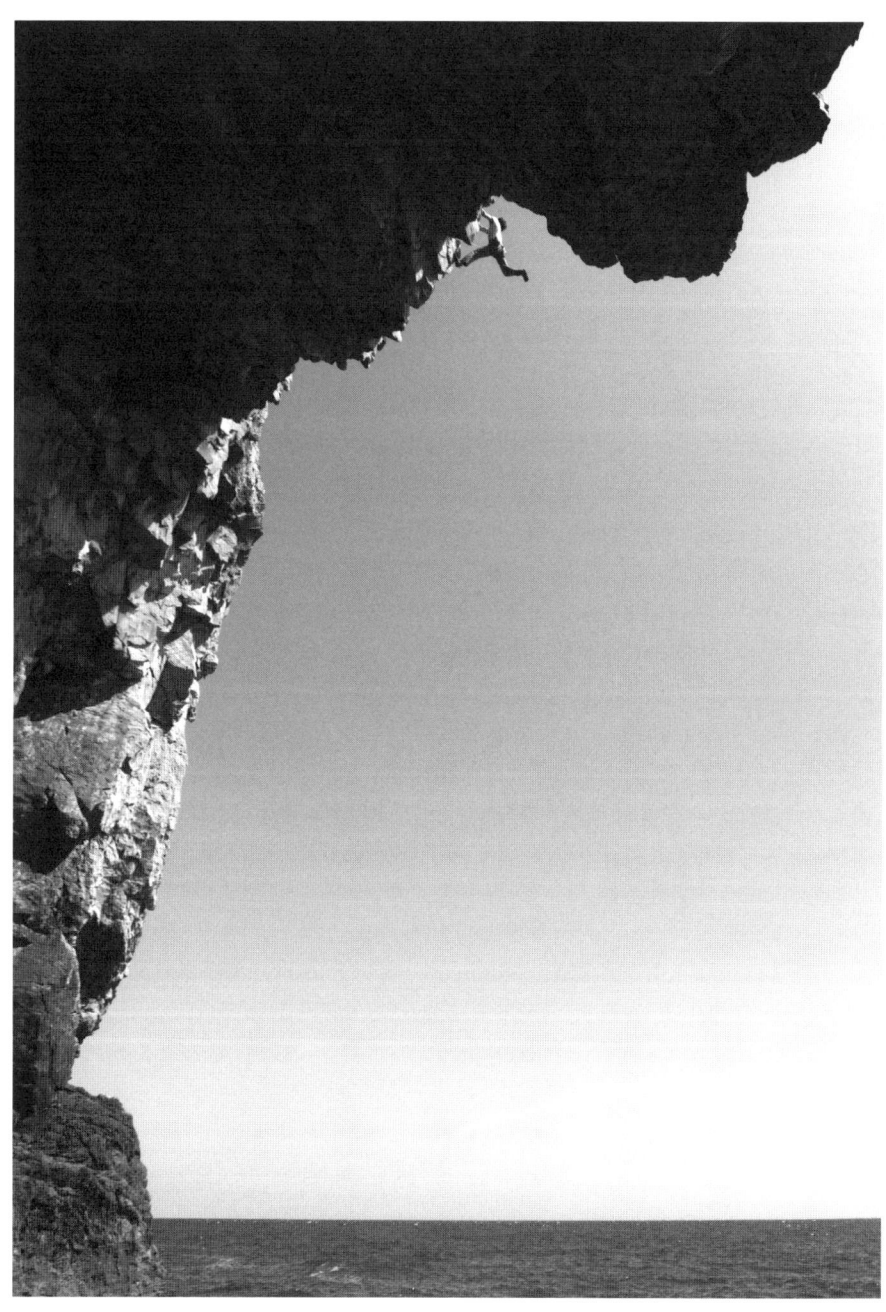

The Palace of the Brine cave at Fisherman's Ledge, Swanage, is home to some of the wildest single-pitch climbs in Britain. Accessing the cave can be complex in rough seas.

The Forbidden Terrain

The magnetic pull of the inaccessible

The green spring sea heaves as I stare into the cave, then lets out a low roar like a descending jet as the last wave recedes. Despite the rising swell, I hope the falling tide is on my side. I doubt that it is, though, as I try to approach *Palace of the Brine* at Swanage, one of the wildest single pitch climbs in Britain. Driven by an easterly wind that's amplifying the force of the ebbing tide, white horses are rising in the middle distance. Another set of waves hits the cliff, and the cave explodes in a cauldron of surging foam. I'm not going to be climbing *Palace of the Brine* today. But the route will remain. I'll just have to come back and try another day. As Hervey Voge wisely said, "the mountains will always be there, the trick is to make sure you are too".

Despite following this elegant logic, I'd still spent half the weekend in a fruitless quest to climb a route I couldn't even get to the base of. Driving home that night, I realised the fact the climb had been made temporarily inaccessible had supercharged my determination to do it. Because we want what we cannot have, I knew I would be back soon to that secret cave, like an angler to his favourite fishing-place. Stories about forbidden terrain – climbs or ventures just beyond our grasp – are the shadow country of the vertical world, and one reason that climbing is so very addictive.

Climb magazine, 2015

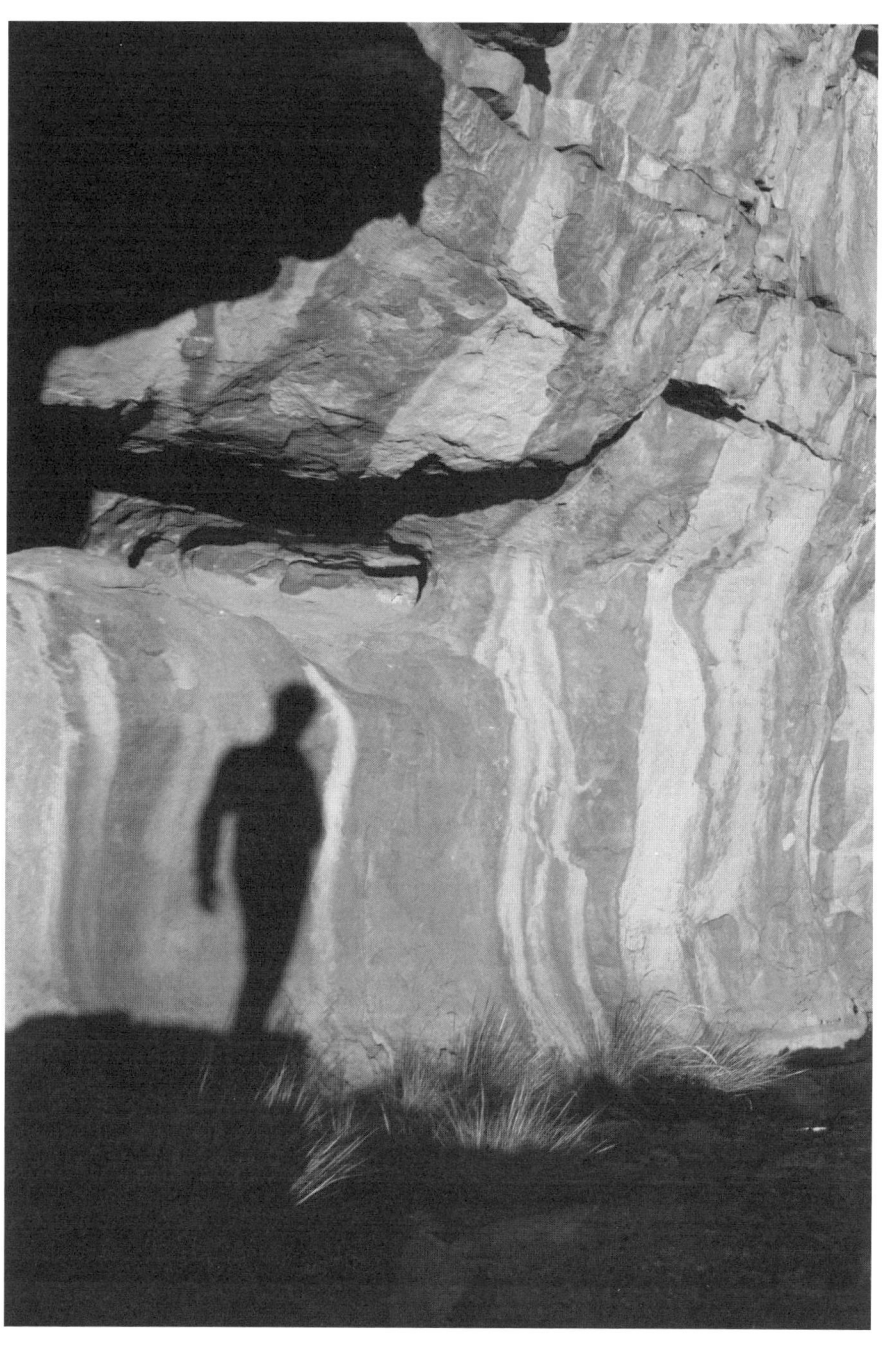

Self-portrait as a moving shadow, Cederberg mountains, South Africa.

The Lone Imagination

Solitary adventure and mystical experience

A lonely impulse of delight
Drove to this tumult in the clouds

– W. B. Yeats

Against the pale immensity of the Turner Glacier, a lone figure moves across the ice with slow, laboured steps. It is August 1975, and the short Arctic summer is almost gone. Behind him, the north face of Mount Asgard rises into the sky, the interplay of light and shadow flickering across the slender pillar that separates it from the west face to the right. Emaciated and in ragged clothes, the climber has just descended the Swiss Route in a storm after making the first ascent of that beautiful pillar in forty pitches of difficult aid and mixed climbing, completely alone and entirely self-sufficient. With no means of contacting the nearest civilisation, several days' walk away, his food has just run out. He must now make the long journey down to Pangnirtung alone, with the possibility of being hunted by polar bears an ever-present danger.

The lone figure on that glacier is Charlie Porter, one of the most remarkable climbers of the late twentieth century. A notoriously reticent character, Porter did not publicise his climb – the first modern, technical route on Mount Asgard – in any way. Nor did he give it a name. Rather appropriately, it has become known simply as *The Charlie Porter Route*. It was many months before news finally seeped out into the climbing world, and Porter's achievement was acknowledged for what it was: an astonishing feat of vision, technical innovation, endurance, and survival.

A few years after Asgard, in 1979, Charlie Porter became the first person to paddle a sea kayak around Cape Horn. And he didn't, as far as I can tell, publicise the feat in any way. This was the true measure of one of the great adventurers of American climbing. He was someone who didn't need to tell people about what he'd done; just doing it was enough for him.

The paradigm shift between the reclusive Charlie Porter and the buzzing, social-media saturated scene of today's top climbers and mountaineers reflects an intriguing process much broader and more complicated than climbing itself: the way in which the digital age is changing the way we think, the way we behave, the way we want to communicate, and what our culture considers valuable. Maybe someone like Charlie Porter doesn't exist, or at least not in the same way, in the current climbing scene.

Whilst climbers like Charlie Porter will forever remain my heroes, I don't believe that the climbing world of the 1970s when he was most active was better than that of today. In some ways, it was much worse; more closed-up, less diverse, and replete with chauvinism and factions. Climbing changes as culture changes, and climbing is becoming a more inclusive place these days. And that's a very good thing.

Does the solitude that Porter sought on that epic Asgard climb in 1975 enable a different way of seeing? Many other great adventures of modern times have been accomplished by solitary explorers like Charlie Porter. Reinhold Messner's solo Himalayan climbs, Bernard Moitessier's single handed voyages, Robyn Davidson's solo camel expedition across Australia, or Phil Harwood's hair-raising solitary descent of the Congo River from source to sea in 2008 are just a few journeys that stand out not just for being remarkable feats in themselves, but also for what they represented: a desire to explore on the most direct possible terms. Going alone to a wild place is surely a way of experiencing the world differently as much as it is a mode of travelling.

One of the greatest books ever written about personal experience in a wilderness environment is *The Snow Leopard* by the American writer Peter Matthiessen. It's an account of a journey he made to Dolpo in the Himalaya in the mid 1970s, becoming one of the very first Europeans to reach this previously forbidden region. Matthiessen's trip wasn't technically a solitary adventure, as he travelled with the field biologist George Schaller in pursuit of the snow leopard. Yet he spent a great deal of time alone in the mountains, meditating and observing the world:

"The sun is roaring, it fills to bursting each crystal of snow. I flush with feeling, moved beyond my comprehension, and once again, the warm tears freeze upon my face. These rocks and mountains, all this matter, the snow itself, the air – the earth is ringing. All is moving, full of power, full of light."

In this remarkable passage, Matthiessen seems to be suggesting that by being alone in the mountains, we become an integral part of the dynamic processes of the universe. It's strikingly similar, in a way, to the ancient Tibetan mystic Jetsun Milarepa's version of the same kind of event: "When one comes to the essence of being / The shining wisdom of reality / Illumines all like the cloudless sky."

This idea of synesthesia and heightened perception is further developed in a later passage in *The Snow Leopard*: "I grow into these mountains like a moss. Though we talk little here, I am never lonely; I am returned into myself... there is a rising of forgotten knowledge, like a spring from hidden aquifers under the earth."

This notion of a resurgence of lost knowledge is an interesting way of expressing the instructive nature of solitary experience in a wild environment. Matthiessen's book, now recognised as one of the all-time classics of wilderness writing, is actually better understood as a sort of declassified work of spiritual philosophy rather than a travel book. The fact that he makes the journey to Dolpo just after losing his wife to cancer only charges this further. Today, with the rising interest in natural navigation – using ancient skills to navigate without modern technology – and the re-wilding movement, there is most definitely a sense of the importance of the forgotten knowledge Matthiessen is talking about, and in all sorts of different guises.

There's no single motivation, of course, for a solo adventure. People go off exploring on their own for all sorts of reasons. One might argue that this paradox lies at the heart of the human story itself. There's no single explanation, for example, for the way early humans migrated and explored most of the inhabitable Earth.

All those innumerable, mysterious, astonishing journeys were made for lots of different reasons. The search for a better place has a thousand origins, and a hundred thousand meanings.

Without the company of anyone else, a solo adventure is elevated into a higher realm of experience. It is harder and more dangerous to undertake a challenging trip on your own; but that very fact, though, is part of the point of doing so. Conversation, laughter, planning, support, and reassurance all have to take place inside your own head. It's not for everyone, of course. Some people are simply not wired to do things by themselves, and there's nothing wrong with that at all. But if you're gifted with this capacity, it's worth exploiting it.

One of Britain's finest travel writers, Colin Thubron, puts together the best manifesto I have heard for going on a big trip in his book about Central Asia, *Shadow of the Silk Road:* "You go because you are still young and crave excitement, the crunch of your boots in the dust; you go because you are old and need to understand something before it's too late. You go to see what will happen."

An adventure undertaken alone becomes a vibrant dynamic with the world and the people you encounter on your journey. If you go with a companion or in a team, this dynamic still exists, but it's diluted. One thing I've taken from my own solitary adventures is that when you're alone, extraordinary things seem to happen quite often. This might be produced by heightened perception, and because you notice the world around you more, but it's true.

Despite the fact that most people in the West live without religion in their lives, we all need faith of some kind to stay alive. The ancient desire for prayer and meditation has not disappeared. Indeed, such things might be more urgent than ever as we find ourselves living in crowded urban spaces with constant digital distraction.

One of the strongest arguments of all for solitary experiences in wild places is that these encounters take us back to where we came from, reminding us that we are still hunters, fighters, and dreamers.

www.monographmedia.com, 2022

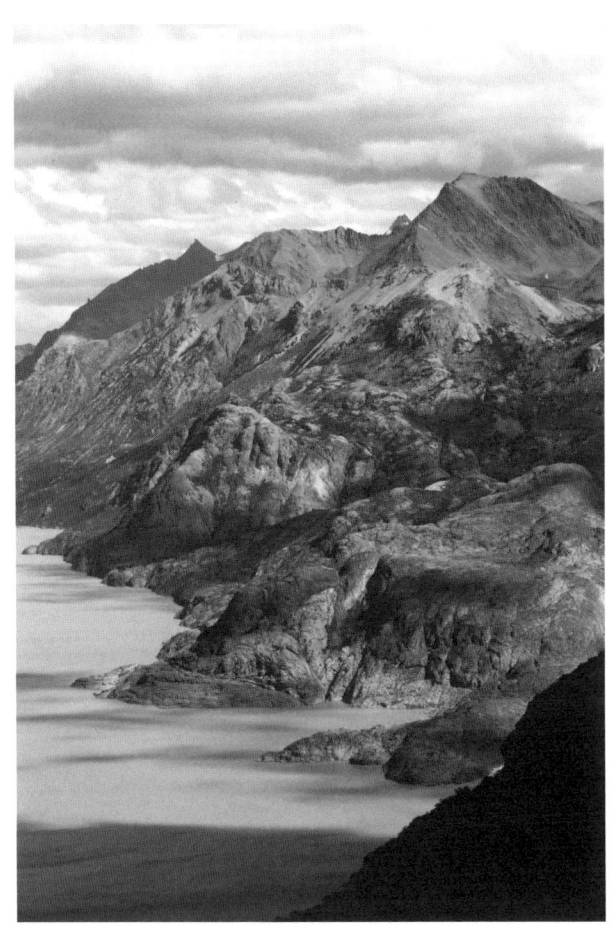

A lonely impulse of delight: looking across the wilderness towards Lago Viedma, Chalten Massif, Patagonia.

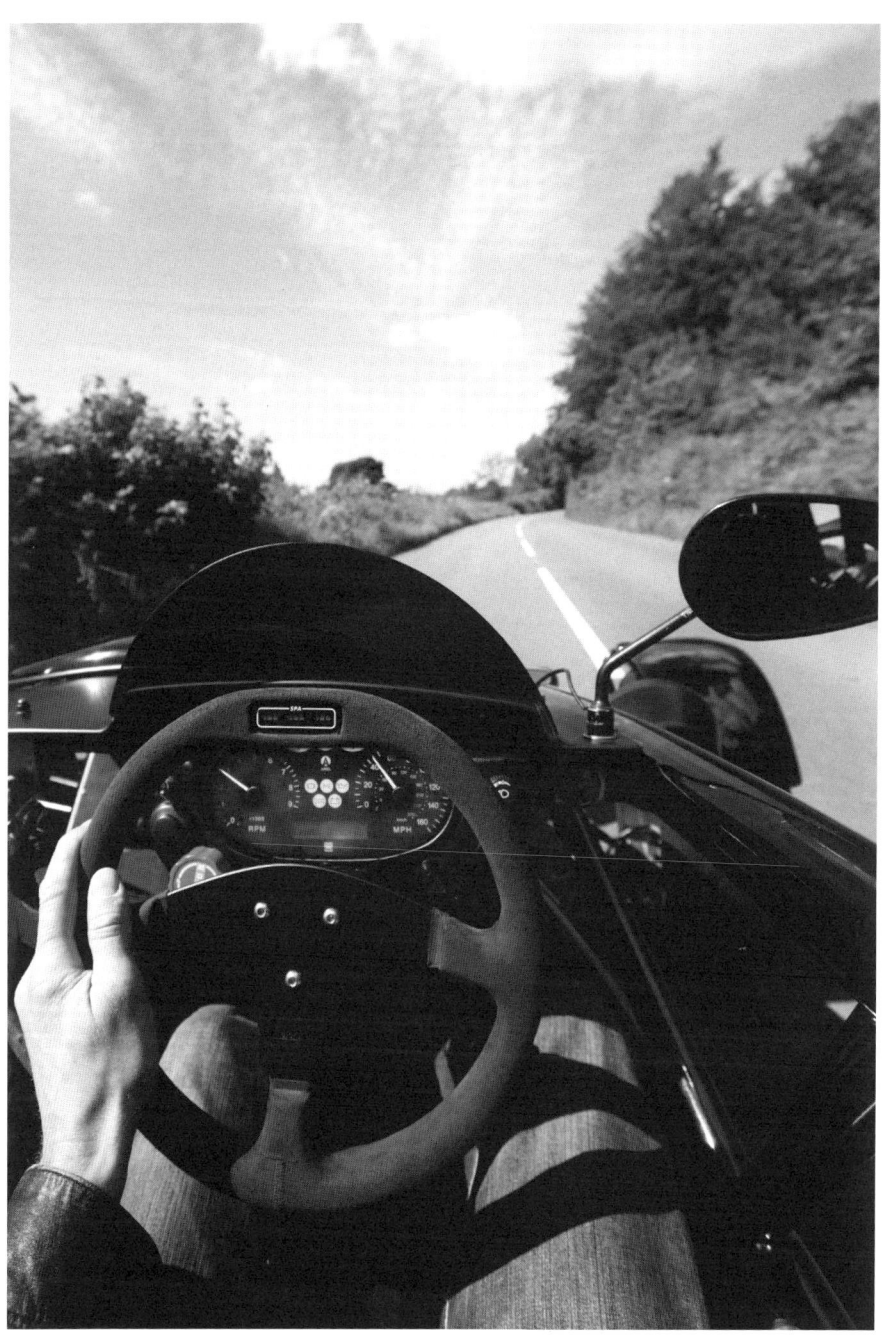

The author stretching the legs of a supercharged Ariel Atom somewhere in southeast Wales.

The Pursuit of Speed

On velocity & artistry

I slam the creaking Austin Metro into third and floor it. As the little four-pot engine howls in protest, Leo Houlding, who's leaning out of the passenger window and smoking something, glances across at the speedo. With the rev counter hovering at the red line, I direct the unfortunate piece of second-rate 1980s British engineering down the causeway about as rapidly as its seventy-two horsepower can carry it. As is the terrible way of teenage boys, I'm driving flat-out everywhere, anywhere, all the time. Leo just shakes his head and reaches for the lighter.

That was the summer of 1998. I'd just passed my driving test, and Leo and I were late for the beach party at the first of a series of legendary deep water soloing festivals that were organised on the south coast around the turn of the millennium. My boy racer antics wouldn't last long, of course. There are many parallels between climbing and driving, and young male climbers in particular tend to take similar risks behind the wheel as they do on the crags.

A few years after the Austin Metro incident, Leo had a bad climbing accident in Patagonia, and I had my own wake-up call with driving: coming around a fast corner in heavy rain on a Somerset B-road, I hit some standing water, lost control, and planted my Ford Escort RS Turbo into the back of an oncoming truck. I wrote the car off, which was a shame, but got away rather lightly with concussion and a fractured jaw. I became a much better driver afterwards. There are many hidden links between driving and climbing, and it's no coincidence that more than a few climbers, myself included, are also die-hard petrolheads. I've owned and driven a lot of fast cars, and quite a few motorbikes. I gave up motorcycling in Britain over a decade ago, because I felt the percentage of risk was too high on our busy roads. "No motorbikes" is also the advice of Nassim Taleb, author of *The Black Swan* and an expert on risk and probability, and if that's good enough for him it's good enough for me. But the fast cars will always remain.

The connection between high performance driving and technical climbing is about mastering the seemingly impossible. If you watch Aryton Senna's unbelievable first lap around Donington Park in the 1993 British Grand Prix – one of the greatest laps in F1 history – in which he overtakes four cars on a wet track to gain pole position, and then the footage of something like Adam Ondra's onsight ascent of *Il Domani* (9a), the link between motorsport and climbing becomes clearer. Both show supreme control of the near-impossible, one at high speed with marginal traction and the other upside-down in a huge limestone cave. Both have low odds of success: a crash or a fall seems more likely. The positive outcomes of both events are purely down to immense talent, fitness, and perfect timing mixed with a little bit of magic.

Johnny Dawes, one of Britain's most influential modern rock climbers, once suggested to me that a fast car in a high speed corner is "an unresolved work of art in a physical form." The dynamics of balance, strength, and precision in climbing, and the sense of speed, power, and control in driving clearly share a certain amount of common ground.

These days, knowing more about danger than I did when I was seventeen, I save really fast driving for the track. But the fascination remains. There's a strong tradition of climbers owning and driving fast cars, by the way. Scottish legend Hamish MacInnes drove a Jaguar E-Type. French rock master Patrick Edlinger, one of the world's most influential climbers in the 1980s, drove a Lancia Delta Integrale for a while, the legendary car that dominated Group B rallying at the time. Jerry Moffatt, probably the best rock climber in the world in the late 1980s, had lots of fast cars himself, and remembers Edlinger driving him around the Verdon's loop road: "I realised he was a great driver, and after warming up the engine he went flat out. However at no time did I feel scared… he had an aura of total confidence and control".

Some of the best moments in driving, as with climbing, are when you're in the right place at the right time, or more specifically on the right road, in the right car, at the right time. In driving, as in climbing, there's both risk and reward in abundance.

Certain moments in driving stand out like the best climbing memories: on the way back from North Wales late on a summer evening in a Mercedes C63, the silence of the empty road broken only by the beautiful roar of that six litre V8; braking hard on Avon Rise, the fastest part of Castle Combe circuit, in a supercharged Ariel Atom, locking up the back axle but correcting the oversteer before it becomes critical; flat-out on Silverstone's Wellington Straight in a Noble M400, chasing down a Porsche GT3 just two car lengths ahead as the November sun fires the asphalt in shades of gold; or being behind the wheel of an Ultima GTR on a twisty road, driving what is essentially a Le Mans Group C race car with a naughty grin on my face.

Confidence, skill, and control are what you need to drive a fast car at the limit, and they're also what you need to climb at the top of your game. In fact, the more you explore the links between climbing and driving the stronger they become. They both involve the application of skill in a complex, dynamic, and potentially dangerous arena. Things like motorsport and other speed sports such as downhill mountain biking, skiing, and paragliding will inevitably be attractive to risk-taking people, because along with risk they all provide the opportunity for great existential rewards. The driver, rider, climber, and pilot are all closely linked through their shared psychic space; there is much common ground between cars, bikes, planes, and climbing.

Part of the reason I started climbing when I was a teenager is because I believed in the idea of doing something that sets you free. Even back then, I knew I wanted to live my life according to this simple principle. Driving, too, can also set you free. The pursuit of speed, in one sense, is an expression of a powerful type of individuality.

Climb magazine, 2015

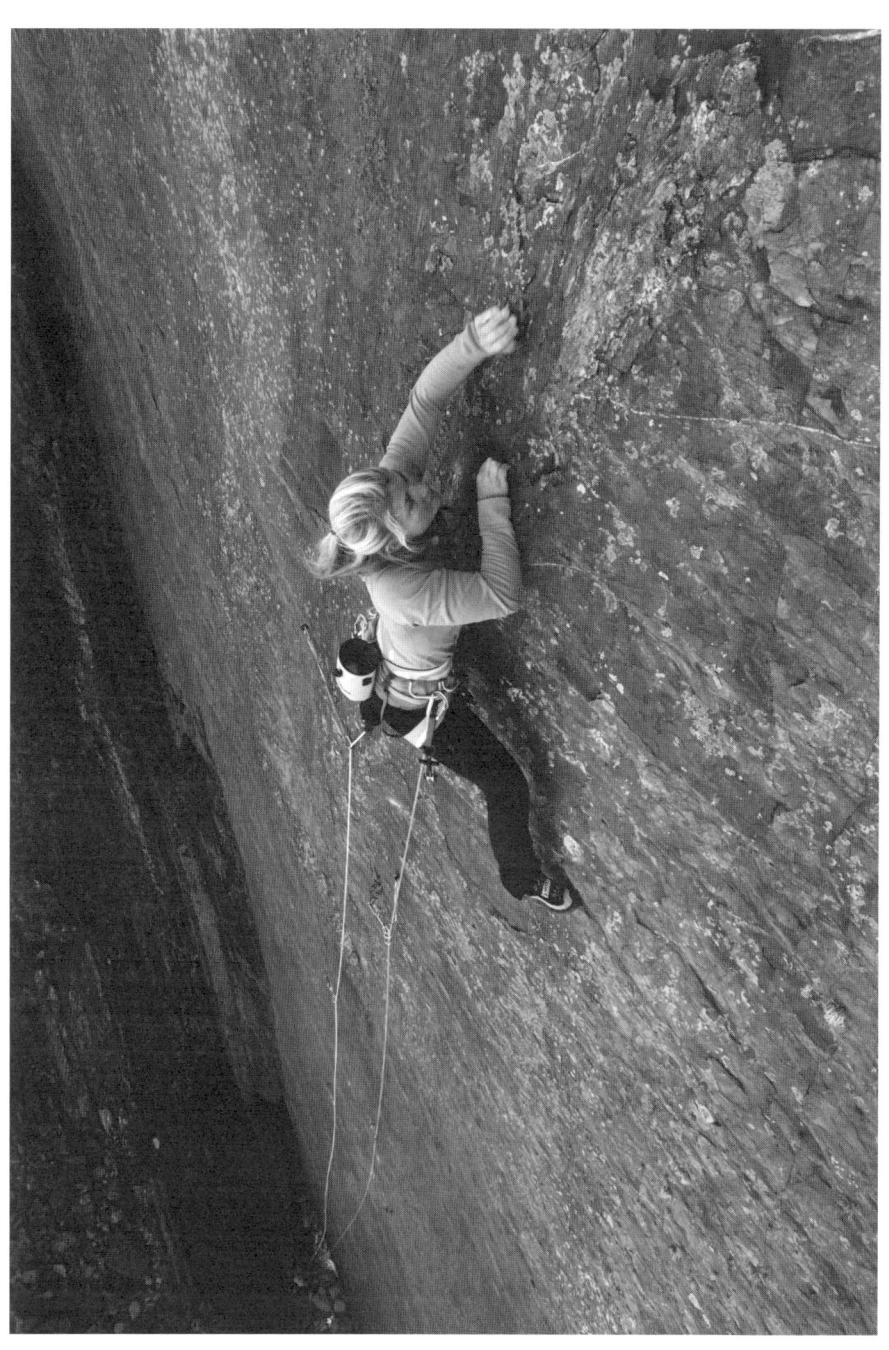

Hazel Findlay, the first British woman to climb E9, on Once Upon A Time in the Southwest (E9 6c), Devon.

The Dream & The Foil

The conundrum of sponsorship in adventure sports

In his cult book *Let My People Go Surfing*, climber and entrepreneur Yvonne Chouinard writes that "sponsorship hurts soul sports like surfing and climbing". Chouinard's point is perhaps more urgent today than it was when the book was first published in 2005. As the outdoor industry continues to grow, further monetising adventure sports, it's inevitable that sponsorship in these activities will expand. The big question, though, is whether sponsorship actually improves things for those who are sponsored. At a basic level, it provides money, which enables people to do what they want. But plenty of leading climbers and explorers have raised funds in other ways. In Chouinard's view, "sponsorship is a no-win situation for the climber in the long run. Being paid to climb forces one to compromise one's values: it encourages the alpine climber to seek routes that make good press, and it can force an otherwise wonderfully eccentric sport climber to act out a role to become more sellable in the media."

The fact that Patagonia, the apparel company that Chouinard founded, still sponsors a great many adventure athletes probably shows how valuable sponsorship is to the outdoor industry. The no-win situation for sponsored athletes arises, perhaps, when there's a lack of understanding about what sponsorship is actually for.

Leo Houlding, one of Britain's most successful professional climbers, came up with a brilliant insight into this in a discussion we once had about the subject: "Nobody actually gets paid to climb" Leo told me. His point is that companies sponsor people not for what they do as such, but for a particular image they present. This, of course, is how marketing works.

Any young, talented adventure athlete who's looking for sponsorship should bear Leo's remark in mind. Sponsorship has a great deal less to do with actual achievements than it has with a person's image – looks, personality, and the way they create publicity. All this is more true than ever before in the social media age.

Because an adventure sport like climbing is still small in relation to skiing, say, or mountain biking, some climbing brands seek to recruit talented climbers for their 'team' by offering free products to use, but no fee, in return for publicity. This arrangement of what you might call 'sponsorship-lite', whilst often useful for older and more experienced climbers with their own professional income, can be dangerous for younger climbers for one key reason. It may create the pressures of sponsorship without any of formal sponsorship's real benefits in the form of hard cash to support a certain sporting lifestyle. Some companies in the outdoor industry might encourage young people to take a view that a career as an adventure athlete is a viable option when the prospects of that happening are in fact non-existent. The former governor of the Bank of England, Mark Carney, was amusingly described as an "unreliable boyfriend" in his public statements by some parliamentarians. The same might be said of many outdoor brands in their approach to the young people they sponsor.

A deeper and more complicated conundrum underlies these issues. Sponsorship can facilitate a life of full-time climbing. But if you are lucky enough to live such a life, then what do you do afterwards? What happens when you can no longer climb as well as you once could? A lot of people who were professional climbing and mountaineering athletes in their 20s and 30s go on to reinvent themselves later in life, by becoming media personalities, business owners, and sometimes simply by starting a new career.

One key problem of sponsorship in adventure sports is that there is no clear career path for athletes once they retire from top-level performing. More organised sports have a structure in place for athletes post-retirement, but climbing does not. And it may never have. A professional tennis player knows with reasonable certainty that once they retire from the game, all kinds of opportunities will come their way. This is not the case, though, with climbing. It's still the Wild West out there.

All the most successful professional climbers, however, have been extremely talented at working with different media to tell the world about what they do. That might be the real purpose of a 'professional climber': to communicate climbing to a wider audience; not just within the climbing community, but also outside it. That's why someone like Chris Bonington was so successful in his very long career: he was brilliant at communicating the value of climbing to the general public.

Former England cricketer and sports writer Ed Smith has offered a powerful perspective on this: "What is the ultimate aim of a sportsman, or any dedicated professional? We like to think that happiness and achievement go together. In fact, far from being interchangeable, they often come into conflict… What if it's one or the other? That question casts a particular type of sporting career in a different light – the player who tastes ultimate success but does not become addicted to it; who thinks normal life is the right foil to the pressures of the job."

There are lots of examples of this in climbing. Walter Bonatti, one of the most influential alpinists in history, gave up climbing at the age of 35. Britain's Jerry Moffatt, the world's best rock climber in the late 1980s, hung up his climbing shoes when he could no longer operate at the cutting edge, and took up surfing. Both clearly understood that success in sport is a finite entity. The Formula One driver David Coulthard, who is now a successful commentator, has spoken of how moved he was when, after his first big race in 1994, the former world champion Niki Lauda came over and spoke to him. This vital process of the older generation nurturing the next is also, surely, the mark of any brilliant sporting career. Personal success on its own does not make a great sportsperson.

What, finally, is the true value of sponsorship in something like climbing? As a young climber, I covered the costs of expeditions through a couple of sponsorship contracts, and it was useful to be able to do that. But if a talented nineteen year-old asked me today if I thought being a professional climber was a good career choice, I'd tell them the truth. Just like Chouinard, I don't think it's a great idea. Ed Smith might be absolutely right that leading a more normal kind of life is the perfect foil for the dream, and also the reality, of being a talented athlete.

Climb magazine, 2015

A Brocken Spectre clearly visible at high camp at circa 4400 metres on Kilimanjaro, Tanzania.

Who's There?

On paranormal experience in extreme adventure

For more than a century, mountaineers and explorers have been aware that unusual phenomena can take place when the body is subject to conditions of exceptional hardship and extreme conditions. The combined effects of altitude, malnourishment, very low or high temperatures, and sleep deprivation can generate types of brain activity that are most often associated with illnesses such as schizophrenia, or the use of hallucinogenic drugs. Reports of extra-sensory perception, out-of-body experiences, ghostly doubles, synesthesia, disembodied voices, and spectral visions are surprisingly frequent in mountaineering literature. Adventure racers are also well aware of these visions, which most often occur at night under intense fatigue and sleep deprivation. They even give them the rather appealing name 'sleepmonsters'.

In our scientific age, most will attribute such phenomena to the extreme environmental and physiological factors outlined above. In fact, the Franciscan friar and scholastic philosopher William of Occam (1287-1347) might have been able to tell us why mountaineers and adventure athletes see weird things in the night back in the fourteenth century, when he developed a radical concept of sceptical logic now known as Occam's Razor: "entities should not be multiplied without necessity". This was a revolutionary idea and ended up being at the centre of one of the most intense intellectual debates of the late medieval period surrounding the relationship of free will and determinism.

In our case, the use of Occam's Razor would suggest that although a nocturnal apparition on a mountain *could be* a ghost, it is *highly likely not to be*. So you should look very carefully and methodically for a proper explanation for its existence before deciding that it is a supernatural event.

But the conclusion isn't always that simple. For climbers and adventurers of the past, who were often influenced by the superstitions of a pre-scientific age still dominated by Christian thought, it was often impossible to discredit supernatural activity when an inexplicable event took place in the mountains. The Romantic concept of the sublime was also highly instrumental in this process, and it strongly affected some of the key figures in the history of alpine climbing.

In 1865, at the centre of the Golden Age of alpinism in Europe, Edward Whymper saw Brocken Spectres (projections of a person's shadow on the clouds) whilst descending the Matterhorn, following the deaths of four of his companions in the infamous accident after the first ascent. He later recounted the effect of the supernatural sensibilities that took hold of him:

"…Lo! A mighty arch appeared, rising above the Lyskamm high into the sky. Pale, colourless, and noiseless… this unearthly apparition seemed like a vision from another world, and almost appalled we watched with amazement the development of two vast crosses, one on either side… The spectral forms remained motionless. It was a fearful and wonderful sight, unique in my experience, and impressive beyond description, coming at such a moment."

Whymper was likely in a heightened emotional state after the climb on which four of his comrades have perished, and found it difficult to perceive a natural optical effect in scientific terms, but saw it as laden with supernatural meaning. Whymper almost certainly didn't experience any visual hallucination as such. Rather, he perceived a visual effect common to the high mountains in an extra-sensory manner as a result of his experiences in that environment. There was no ghost on the Matterhorn that day but he may have seen the Brocken Spectre in a hallucinatory way due to the extreme circumstances.

The wider connection between wilderness experiences and spiritual understanding predates the existence of mountaineering by at least two thousand years: many Buddhist ascetics such as Jetsun Milarepa (c.1052 – 1135 AD), the great Tibetan yogi and poet, spent years living in caves in the high Himalaya in the pursuit of mediation and nirvana, a state of enlightenment free from worldly suffering and desire. Many colourful myths and legends surround Milarepa's life, including a tale in which he catapults himself to the summit of Mount Kailash, the mountain in Tibet held most sacred by Buddhists and Hindus.

A particular anecdote concerning Milarepa's meditation in the cave of Drakar Taso, near Pelgyeling Gompa in Tibet, lends force to the argument about the relationship between mountaineering and out-of-body experiences, extra-sensory perception, and related phenomena. It is well documented by various Tibetan texts that whilst meditating in Drakar Taso, Milarepa subsisted for long periods on nettle tea, and wore nothing but a thin shirt, even in winter, when the temperature in the region plummets to more than minus thirty degrees centigrade. The length of these mediations remains unknown. In practice, it seems unlikely they would have lasted more than a few days. However, it is clear from various sources that Milarepa had an extraordinary ability to regulate his own temperature by practicing a form of yoga that – allegedly – generates body heat. Did the extreme temperatures, fasting, and sleep deprivation Milarepa endured – hardships that are familiar to the modern mountaineer – lead to unusual phenomena such as out-of-body experiences and extra-sensory perception? Did his experience of such phenomena contribute to his creativity as a poet and a mystic?

The notion that Milarepa advanced of the mountain environment as a place where sensory perception is transformed is curiously similar to what's conveyed in the writings of notable mountaineers. Reinhold Messner, one of the most successful Himalayan climbers of all time, observed some of these effects on his landmark 1982 solo ascent, without oxygen, of Everest's North Ridge:

"Is that someone talking nearby? Is somebody there? Again I hear only my own heart and breathing. And yet here they are again... I jump frequently because I believe I hear voices. Perhaps it is Mallory and Irvine [two English climbers who perished hereabouts on their 1924 attempt on the mountain] With my knowledge of the circumstances surrounding their disappearance, now each noise brings a vision alive in me... I gaze at the second step and already two beings rear up in me, release phantoms; in the driving mist everything seems so near, ghostly."

Messner's sense of spectral company high on Everest has been echoed by other climbers, including Britain's Stephen Venables, who made the first British ascent of the mountain without oxygen in 1988. This experience of extra-sensory perception involving either a perceived ghost or 'other' presence seems quite common to climbers pushing the boundaries in the Greater Ranges. In October 2007, Russian climbers Valery Babanov and Sergey Kofanov made an extremely impressive alpine-style ascent of the West Pillar of Jannu in the Nepal Himalaya. In his article about the route in *Alpinist magazine*, Babanov recounts his experience of returning to the glacier in the dark after their epic climb:

"One o'clock in the morning. I feel as though I'm watching ourselves from the outside: two worn out and tortured beings, barely able to move their feet, float along the glacier like ghosts... Music has been playing in my head for several hours... I think someone is walking beside us."

Babanov's out-of-body experience and his feeling of a ghostly 'other' accompanying him is by no means unique in mountaineering circles. The legendary Polish climber Voytech Kurtyka and his Austrian partner Robert Schauer had a similar experience during their epic descent off Gasherbrum IV in 1986 after the first ascent of The *Shining Wall*, one of the most outrageous climbs ever done in the Himalaya.

The fact that Babanov and Kofanov had only eaten a few muesli bars in the previous three days, and that they'd been on the move for nineteen hours straight when they reached the glacier (they'd been on the mountain for ten days) are clearly the main factors in those supernatural experiences at the end of their Jannu climb.

Although persuasive in highlighting the causes of mountain ghost-sightings, this still doesn't provide a complete answer. Surely there are other, more subtle neurological factors? For a more specific explanation of the causes of such phenomena from a neurology perspective, we might turn to the groundbreaking research done in September 2006 by Shahar Arzy and colleagues of the University Hospital, Geneva, Switzerland:

"[We] reproduced an effect reminiscent of the doppelgänger phenomenon via the electromagnetic stimulation of a patient's brain. Focal electrical stimulation to a patient's left temporo-parietal junction was applied while she lay flat. The patient immediately felt the presence of another person... Other than epilepsy, for which the patient was being treated, she was psychologically fit."

Dr. Arzy has suggested that the left temporo-parietal junction of the brain evokes the sensation of self image; this includes body location, position, posture etc. When it is disturbed, the sensation of self-attribution is broken and may be replaced by the sensation of a foreign presence or copy of oneself nearby. The Swiss psychologist Olaf

Blanke has also suggested that the right temporo-parietal junction is important for the spatial location of the self, and that when these normal processes go awry, an out-of-body experience may arise.

Did Milarepa, mediating and fasting in his cave in the depths of the Tibetan winter, tap into this region of his brain in a manner which enabled him to regulate his body temperature and fend off hypothermia? Did the Brocken Spectres Whymper saw on the Matterhorn induce a reaction in his temporo-parietal junction, spurred on by fatigue and anxiety, which led him to perceive them as supernatural visions? What exactly happened in Messner's head, alone up there on the north ridge of Everest in 1983? And what of the brain chemistry of Kurtyka and Schauer, descending Gasherbrum IV after surviving for three days at 7500 metres? And Babanov and Kofanov after their epic climb of Jannu's West Pillar: what caused that spectral being to appear on the glacier? Exhausted brains do strange stuff, it seems.

Any detailed field research on this subject would involve highly ambitious data-collection: monitoring the brain activity of mountaineers and adventurers as they stretch the limits of endurance in the most hostile environments imaginable. It is possible but unlikely that these conditions could ever be meaningfully simulated in a laboratory.

A detailed neurological study of brain activity in extreme conditions could produce some interesting results. The psychological disturbances affecting mountaineers and adventurers at the limits of their endurance remain a deeply mysterious area. There's nothing supernatural going on as far as we know, but what's happening inside the human brain in extreme conditions is genuinely remarkable. Like the most compelling ghosts of fiction, these visions hover at the very edge of our scientific perceptions, disappearing as effortlessly as they arrived.

www.ukclimbing.com, 2008

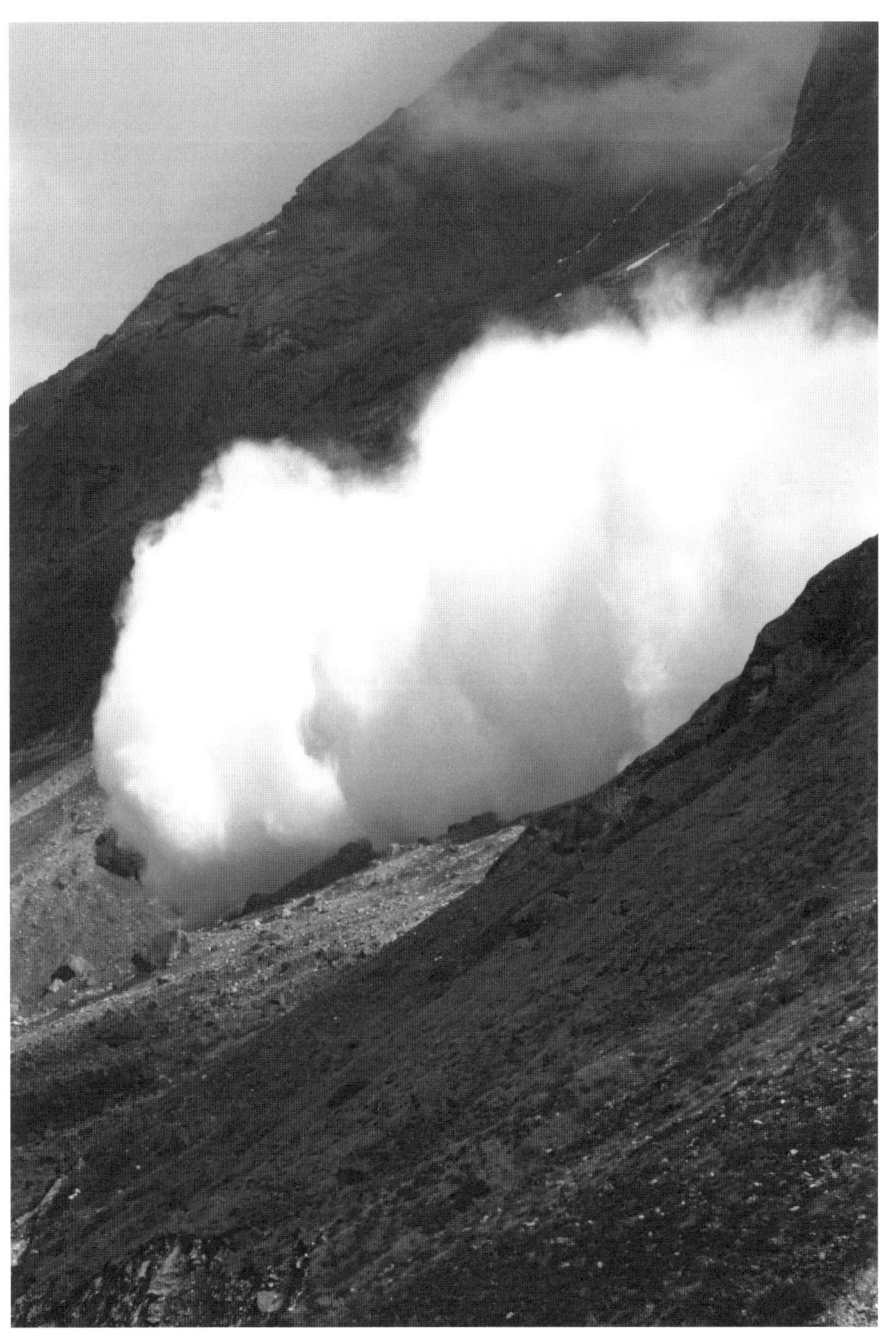

A huge avalanche thunders off the summit icefields of Kyashar (6769m), Khumbu, Nepal.

Constructive Paranoia

On good practice in dangerous places

"Stupidity", observed the Polish alpinist Wojciech Kurtyka, "means falling prey to your own illusions". During his long career of cutting-edge ascents in the world's high mountains, it's fair to say that Kurtyka probably learnt more than most about the importance of common sense in situations involving complex risks. In her superb biography of Kurtyka, *Art of Freedom*, Bernadette McDonald recalls an episode on Manaslu in 1986, when Kurtyka insisted on going down due to the extremely high avalanche risk. "[The slope above us] was loaded with ominously sparkling snow, silently waiting in the hot sun. It was like Russian roulette" Kurtyka recalled. His climbing partners continued on for a while, until they too were forced to retreat due to the conditions. "[Wojciech] couldn't agree to this approach to avalanche danger", McDonald writes. "He knew that if you played [that] game too often, the bullet would eventually find its mark."

What strategy, exactly, was Kurtyka deploying there? I'd argue it's a concept formulated by the renowned polymath and author Jared Diamond: a system for minimising exposure to objective danger which he calls *constructive paranoia*. It's an approach to risk based around the attitude of native New Guinean people, with whom he made numerous expeditions into the remote New Guinea highlands.

Diamond noticed on his expeditions how native New Guineans exercised a hyper-sensitive environmental intelligence when travelling in the jungle to mitigate any conceivable hazard. For example, they would not sleep under a dead tree, no matter how small it was. The thinking behind this is that if you camp in the forest for hundreds of nights a year, as they do, you'll know that dead trees often fall over, and you don't want to be camping under one when it does. Diamond first understood the importance of this concept of constructive paranoia, he recalls, after a boat accident in New Guinea in which the canoe he was travelling in between islands capsized, due to the boatman driving too fast and flooding the hull. He and his companions were rescued in the nick of time, shortly before sunset, clinging to the upturned canoe miles offshore in a remote part of Indonesia.

He subsequently met a man who had been scheduled to travel in the same canoe, but decided against it due to the unstable vessel and the inexperienced, gung-ho boatman. Diamond then realised he had not used the New Guinean technique of constructive paranoia when deciding whether or not to travel in the boat; he had simply boarded it, despite whatever reservations he may have had about its safety. The principle of constructive paranoia can apply, I think, to virtually any situation where there is a degree of exposure to real risk, whether on an 8000 metre peak, or choosing to board an unstable boat, or deciding to go for a swim in the sea.

For my own part, I've probably spent enough time exploring the no man's land between acceptable and unacceptable risk to understand what you need to do in order to be in a position to exercise constructive paranoia in the first place: you need to be able to identify the primary source of the risk itself. It could be a dead tree, an unstable boat – or possibly your own ambition.

This process of becoming hyper-aware of potential hazards can be applied to all adventure sports involving risk. Climbing solo was once an important discipline for me; the sense of freedom and control it gives is second to none. But I have consciously moved away from it due to the increased probability, with every climb completed this way, that something goes wrong.

Diane Vaughan's concept of "the normalisation of deviance" has been used quite widely to describe bad decision making in big organisations, and it can also be used to identify bad practice in adventure sports. For example, we do something that does not follow the accepted safety protocol such as climbing without a rope, which we then get away with. Then, believing it's safe to make the same safety shortcut a second time, we do the same thing again. Repeat this process indefinitely, and something will eventually go wrong.

Illusions can be equally dangerous in anything involving real risk: sending shuttles into space, or engaging in risk sports. It's often the new, surprising variable that catches you out. Some of the most pragmatic adventure safety guidelines I've encountered are John Willacy's tips for safe sea kayaking. Willacy, by the way, has set a round-Britain sea kayaking record, and knows his stuff. His rules define constructive paranoia: *First, if in doubt, don't go out. Second, remember what you see from the beach is 2-3 times smaller than what you get once you're out there. And third, accept the weather forecast for what it is, not what you want it to be.*

These rules reinforce how you have to be tuned in to your emotions to stay safe: apprehension can tell you a great deal. I like the way Willacy points out that you must have an awareness of your own limits, plus a pragmatic approach to conditions and the weather to succeed on your mission.

Diane Vaughan's concept of "the normalisation of deviance" is the opposite of Jared Diamond's concept of "constructive paranoia". By setting the two against each other, the real differences between recklessness and pragmatism in a wild or dangerous environment become obvious. Experimenting with the former for too long might kill you in the end, but practicing the latter could keep you alive and well for a very long time. The intriguing relationship between these two interesting concepts is a dynamic worth thinking about whenever you're heading out there, and whatever you're doing.

BASE magazine, 2019

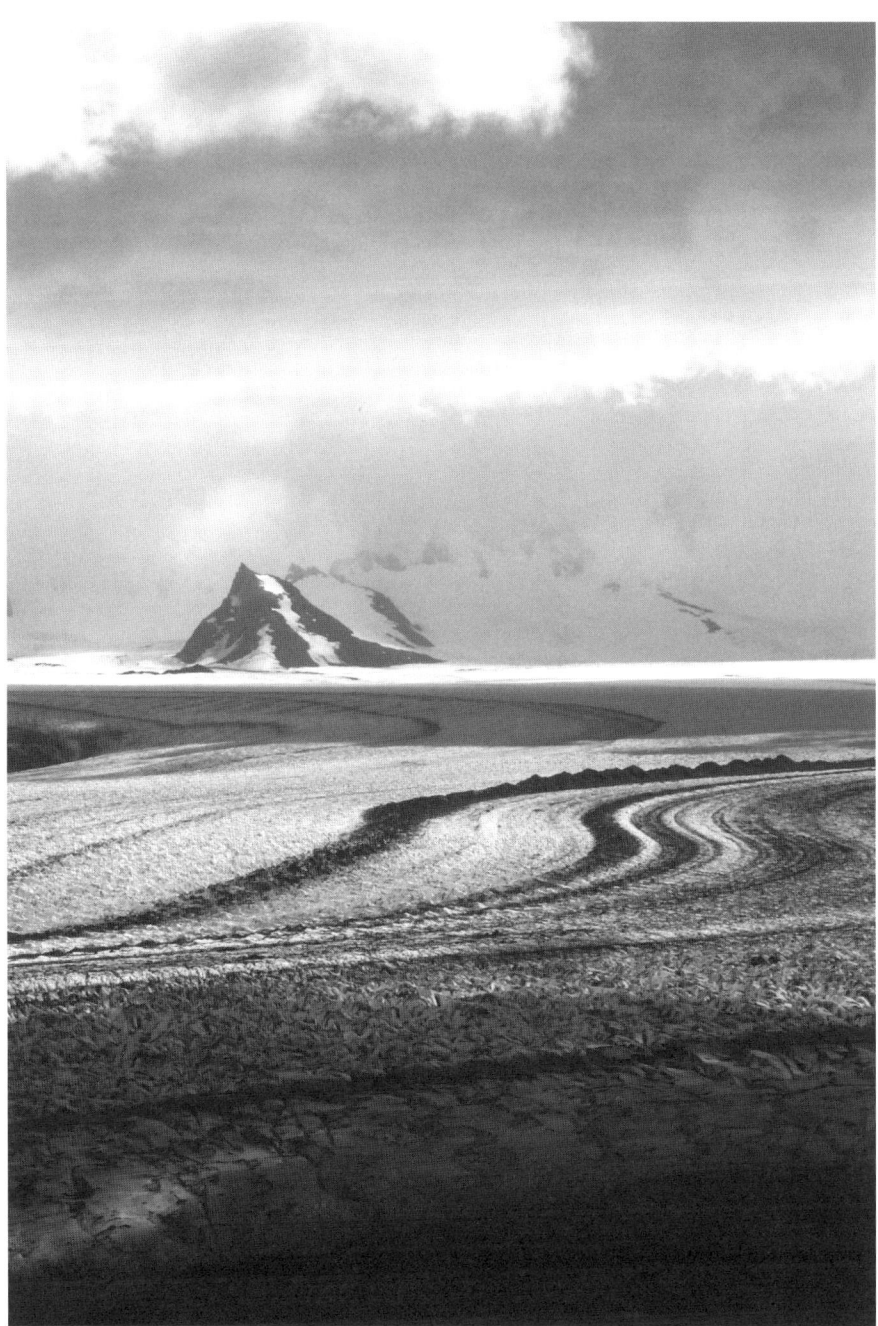

The Southern Patagonian Icesheet extends beyond Patagonia's Chalten Massif for thousands of square miles.

The Edge of Things

Notes from the way out

The land falls away into a river valley before rising up towards a line of trees on the horizon. Just before the wood, there's an open field that follows the contour of the hill. A pair of Scots pines grow tall and strong in the centre, marking out this place as distinct from the landscape that surrounds it. From the edge of the school yard, as a nine year-old kid, I would often stare at this singular piece of open ground, my eye drawn, perhaps, to its marginal quality; a place where farmland dissolves into forest, where the cultivated merges into the wild. As such, I became transfixed with this liminal field in those suspended, dreamlike moments between maths and science. It was a representation of the possibilities of the outer world, beyond the exigencies of school. It was also a representation of what I might discover one day, somewhere in the open country over those hills, as the story goes, and far away.

The extraordinary events that took place across the world during the 2020-21 Covid pandemic reminded me of that place I once contemplated from primary school, for a simple reason. The field on the edge of the woods was one thing above all else: *a way out*. If you're a naturally adventurous person, difficult and complex questions arise from a state of 'lockdown'. Enforced captivity within national or local boundaries is not a natural thing for an explorer.

How to escape? What does freedom mean? To what extent can the state control the individual? Who can you trust? What really matters in your life? Interestingly, these were also questions of enormous relevance to intellectuals and dissidents in the former USSR and 1930s Germany, and they remain of huge relevance in China today. That's a deeply revealing perspective, I think, on the policy of lockdown.

During a period when long distance travel was made, temporarily, much more inconvenient, it became more valuable than ever to explore the world closer to home; to reimagine our stomping grounds. Local exploration, in any case, often reveals more than its global counterpart. By looking into a place you already know with fresh eyes, you sometimes discover more than jumping on a plane to the other side of the world. Rather than seeking journeys to the ends of the earth, the experiences of the Covid years suggested to many of us that it's often more worthwhile to open up the local map, and to dig a bit deeper on your home turf.

Several years ago, the author and political thinker David Goodhart came up with a fascinating way of describing the human relationship with place. He defined those with a global outlook as 'anywheres'; their opposite being 'somewheres', or people with a more local outlook. For most of 2020 and into early 2021, we were all forced to become 'somewheres' for a while. This was far from an unpleasant experience, in my case at least, in the sense that it reset my relationship with the places I call home. Local captivity also reveals new places and possibilities overlooked in the rush to the next intercontinental flight. Hinterlands, after all, can be wild places.

The era of lockdowns also might have revealed the value of time spent alone or with a partner – time to reflect, to contemplate, to make plans. For many introverts, and even for socially recalcitrant extroverts, the benefits of 'social distancing' are pretty obvious. When we're alone, it's sometimes easier to get in touch with our primal instincts as human beings who were, until quite recently in evolutionary time, hunter-gatherers. Things like the need for food, shelter, warmth and so on become very apparent when on a solitary adventure.

On a solo hike along the edge of the Patagonian icesheet in 2017, I remember filling a flask from a deep, fast-flowing river that disappeared into a huge glacier. I recall thinking if I had accidentally tripped and fallen into the water, my body would have been washed, eventually, into the ice sheet itself, perhaps appearing a few thousand years from now, like the iceman Ötzi, who was discovered in September 1991 in the Ötztal Alps on the Swiss-Austrian border, perfectly preserved after over five thousand frozen years in the glacier. Alone in a wild place, you come closer to what a hunter-gatherer like Ötzi might have experienced physically and psychologically. Today, with the new interest in natural navigation, traditional bushcraft, and re-wilding, there is a resurgent sense of the importance of the environmental intelligence of our ancestors.

In a remarkable piece of writing about solitary experience in the natural world, J.A. Baker explains his fascination with observing peregrine falcons in a small area of rural Essex in the notoriously cold winter of 1962-63 in *The Peregrine*. Long overlooked, this one-off book is now regarded as a classic of British nature writing.

"I shut my eyes and tried to crystallise my will into the light-drenched prism of the hawk's mind" Baker recalls, trying to imagine what the bird he is observing is seeing, and also the way the bird is seeing what it is seeing. Baker's work, in fact, has only very recently been acknowledged as some of the best nature writing ever produced in English, perhaps partly because it was stylistically and philosophically so far advanced for its time. It's a kind of existential manifesto as well as a book about birdwatching: his pursuit of the hawk is also a pursuit of an idea about how humans might exist in the world, about the nature of perception, and about the relationship between landscape and imagination. Earlier on in the book, Baker explains his motivation for his lonely hawk-watching in a wonderful phrase: "I always longed to be part of the outward life, to be out there at the edge of things." Personally, I think this is one of the better explanations of why an adventurous, inquisitive life is something worth striving for.

The concept of adventure itself, though, is a notoriously tricky idea to define. Every easy definition seems to dissolve like silt in a glacial river when held up to real scrutiny. The best summary I know wasn't, ironically, coined by a well-known practitioner of adventure sports, but by the former Wall Street trader Naseem Taleb in his book *Skin In The Game*: "if you do not undertake a risk of real harm, reparable or potentially irreparable, from an adventure, then it is not an adventure." The notion of real harm is an interesting one. Clearly it encompasses both mortal danger and the other risks – perhaps financial and psychological – that a serious adventure might involve. I'd guess that some of the most rewarding adventures involve most, if not all, of these separate species of hazard.

Defining adventure by what it isn't might seem evasive. Yet true adventure is as much a state of mind as it is a physical experience. For real adventures to happen at all, we first need a hinterland of desire and invention. At the same time, a sense of faith in ourselves and our own capabilities is more important today than ever before. Social media has created a public sphere in which extrinsic motivation overrides intrinsic motivation; peer group recognition and response has become more important than the search for inner truth. In the context of adventure sports, as in many other areas of life, this is not an altogether positive trend.

I hope that in the future the most engaging adventure stories move away from social platforms and are delivered in new, original, and inspiring forms and through media that reward contributors first and their platforms second. There are good signs here. Substack, the user-subscription-funded platform for writers, journalists and public intellectuals, is a promising start to what will be a very long term process.

Exploration might be as natural an instinct as the need for shelter or warmth. If there's one character type that unites the disparate personalities of adventure-seekers, it's the figure of Peter Pan, the kid who never grows up. The American climber Fred Beckey, who died in 2017, was the quintessential dirtbag: he never married or had children, and was still climbing and travelling well into his nineties. Beckey's life represents the freewheeling curiosity and sense of mischief required to maintain an adventurous mindset. Whilst it might not be everyone's ideal lifestyle, Beckey proved that getting out there doesn't just keep you fit. It keeps you young, too.

At the same time, the visionary perception of children is an interesting way of understanding the creative value of exploration within the human condition. A child might look at an old oak tree in the garden and want to climb it, but an adult might look at the same tree and want to chop it down because its roots undermine the driveway. Whose priorities are the most important? Whose agenda should determine the fate of the tree? These are clearly existential as well as environmental questions.

This process of creativity through exploration takes place in many different ways, and it's one of the key reasons that self-organised, self-supported adventures appeal to free spirits and independent thinkers. Journeys at sea using small craft are a particularly good example of this: if you know what you're doing, you'll be making constant creative adjustments to take advantage of prevailing conditions, reinventing your route based on what the weather and tide are doing.

In this genre, Australian kayaker Andrew McAuley stretched the outer limits of human capability to an extraordinary degree. Already an experienced expedition kayaker and also a climber, in 2007 McAuley set out to achieve a seemingly impossible feat: to paddle a modified sea kayak alone across the Tasman Sea, the notorious thousand-mile stretch of water between Australia and New Zealand where the full force of the Southern Ocean is compressed into a perilous sea channel. The place is infamous amongst sailors for generating freak waves and mountainous seas. Incredibly, McAuley very nearly made it to New Zealand; he was lost without trace after making a distress call to the New Zealand coastguard around thirty miles off Milford Sound on South Island, after one of the most outrageous solo adventures ever undertaken. He clearly intended to have the most demanding experience imaginable, and whilst he did not physically survive his journey, in one respect he succeeded. His voyage was a triumph of solitary exploration, and it's hard not to feel more than a degree of admiration for the fact he even embarked on it. The following year, the Tasman Sea was successfully paddled by a two man team using a larger, purpose-built vessel, by the less dangerous route between New South Wales and North Island. McAuley's solo paddle between Tasmania and South Island has not, unsurprisingly, seen another attempt.

Some of the most memorable adventures are often solitary ones; they have a purity of purpose, a logistical simplicity, and a psychological intensity. Without a support structure, a solo adventure in a wild place is elevated into a different realm of human experience, and you don't need to be in the Southern Ocean in a sea kayak to find it. It could be much closer to home.

On your own, you can be truly at the edge of things; a falcon-watcher, a shape-shifter, a visionary. And on your own, too, it's easier to notice a place on the edge of the woods like the one I used to dream about a long time ago: *the way out*.

BASE magazine, 2020

Acknowledgements

"The best people possess a feeling for beauty, the courage to take risks, the discipline to tell the truth, and the capacity for sacrifice"
- Ernest Hemingway

I extend my deepest and most heartfelt thanks to all the following people, who in various ways contributed to my journey to create this book. I am a better person for knowing you all. I couldn't have done this without you.

Helen Mort, Ian Parnell, Ramon Marin, Gavin Symonds, Dorka Fekete, Tom Rowland, Kelly Vargas, Malin Holmberg, Jim Perrin, Alex Messenger, Charlotte Davies, Sarah Garnett, Amanda Symonds, Stephen Venables, Hazel Findlay, Jack Geldard, Amy Cooper, Anni Li, Naomi Wang, Charlie Woodburn, Gilly McArthur, Grant Farquhar, Nasim Eshqi, JJ Gillooly, Juliette Pitts-Crick, Mike Robertson, Camilla Peevers, Neil Gresham, Nic Sellars, Matt Perrier, Jason Porter, Katie Ives, Jon Barton, John Coefield, Amy Colson, Dave Moore, Tanya Moore, Sam Whittaker, Lena Drapella, Katja Barrueto, Mina Leslie-Wujastyk, Helen Gardiner, Grant Wright, Tim Emmett, Johnny 'Woody' Woods (RIP), Chris Savage, Joe Walczak, Bernard Newman, Neil Kennedy, Shane Ohly, Tommy Kelly, Andy Long, Adrian Baxter, Keith Bradbury, Emily Bradbury, Gill Wootton, Lisa Schulze, Tom Bodkin, Charlie Low, Dale Comley, James Marshall, Max Dutson, Doug McConnell, Andrea Ha, Dan McManus, David Wilson, Bonita Norris, Chris Hunt, Emily Graham, Jamie Hannant, Ged Desforges, Ed Chard, Giles Cornah, Robert Harding, Darren Ballinger, Tom Parsons, Paul Jeffrey, Matty Rawlinson, Neil Mawson, Kate Keltie, Matt Ward, Toby Dunn, Steve Berry, Simon Lowe, Steve Pack, Trevor Massiah, Julian Walker, Tom Briggs, Alex Honnold, Vicki Harvey, Mike Hutton, James Harrison, Paul Twomey, Rob Stanfield, Pete Robins, Mike Rolf, Bob Lamey, Steve Findlay (RIP), Steve McClure, Jerome Mowat, Joff Cook, Damian Cook (RIP), Howard Lawledge, Adrian Baxter, Madeline Cope, Jonathan Griffith, Matt Helliker, Ben O'Connor-Croft, John Bracey, Kyle Pattinson, Karl Hughes, Matt Ward, Juha Saatsi, Leonie Schaefer, Juman Al-Sayegh, Will Stanhope, Alastair Smith, Natalie Berry, Alan James, Stewart Wallis, Maryjane Wallis, Emma Millington, Tim Nicholl, Pete Oxley, Kevin Avery, Tom Richardson, Regina Davy, Chris Savage, Peter Malin, David Punter, John Lee, Mark Bessell (RIP), Martin Hurrell, Tom Hurrell, Richard White, Leo Houlding, Andy Whittaker, Richard Wheeldon, Anna Zapolska, Mark Glaister, Dave Spooner & Daniel Foley.

*I would like to extend a special thanks to my father, John,
whose intelligence and interest in the world
shaped my life from an early age.*

*I would also like to thank my partner, Dorka Fekete,
for her heroism, strength, and spirit of adventure.*

Those who inspire are travellers of eternity.

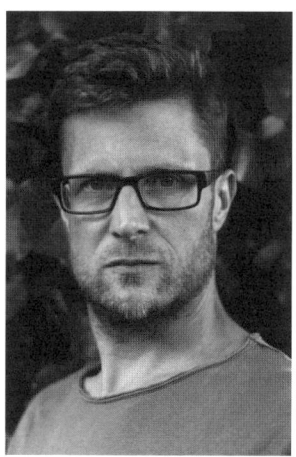

David Pickford is one of Britain's leading adventure sports writers.

A widely-published author, David has also edited two national magazines, first *Climb* and later the adventure quarterly *Base*, which he launched in 2019. David is the author of two previous books. *The Light Elsewhere: Encounters With The Elemental World* (2013), and *After the Crash and other stories* (2015).

Extreme Horizons collects some of David Pickford's best writing on climbing, adventure travel, and global exploration from the last two decades. Most of these essays were previously published in a range of magazines, journals, books, and websites. *Extreme Horizons* assembles them all together for the first time.

The power of the unfamiliar, the experience of physical risk in the environment, the nature of uncertainty, and the moral value of adventure are central themes of this far-reaching, unusual book. The author explores this fascinating terrain with unique insight and a vast frame of reference from his own experiences, and through the stories of numerous others. The essays and articles collected here document groundbreaking climbs, long distance motorcycle expeditions, exploratory journeys in faraway places, and a series of sea voyages using an ultralight craft. A final section of essays has an even wider focus, covering many other subjects relating to the adventurous life and outdoor culture. *Extreme Horizons* is a veritable treasure chest for both die-hard adventurers and casual enthusiasts alike, and essential reading for the committed armchair explorer at the same time. As a fully illustrated edition, this book is a collector's item for any outdoor enthusiast's bookshelf.